Lecture Notes in Computer Science 9761

Commenced Publication in 1973
Founding and Former Series Editors:
Gerhard Goos, Juris Hartmanis, and Jan van Leeuwen

More information about this series at http://www.springer.com/series/7407

Rachid Echahed · Mark Minas (Eds.)

Graph Transformation

9th International Conference, ICGT 2016
in Memory of Hartmut Ehrig, Held as Part of STAF 2016
Vienna, Austria, July 5–6, 2016
Proceedings

Editors
Rachid Echahed
Laboratoire LIG
Université Grenoble Alpes
Grenoble
France

Mark Minas
Universität der Bundeswehr München
Neubiberg
Germany

ISSN 0302-9743   ISSN 1611-3349   (electronic)
Lecture Notes in Computer Science
ISBN 978-3-319-40529-2   ISBN 978-3-319-40530-8   (eBook)
DOI 10.1007/978-3-319-40530-8

Library of Congress Control Number: 2016941082

LNCS Sublibrary: SL1 – Theoretical Computer Science and General Issues

Printed on acid-free paper

This Springer imprint is published by Springer Nature
The registered company is Springer International Publishing AG Switzerland

Foreword

Software Technologies: Applications and Foundations (STAF) is a federation of leading conferences on software technologies. It provides a loose umbrella organization with a Steering Committee that ensures continuity. The STAF federated event takes place annually. The participating conferences may vary from year to year, but all focus on foundational and practical advances in software technology. The conferences address all aspects of software technology, from object-oriented design, testing, mathematical approaches to modeling and verification, transformation, model-driven engineering, aspect-oriented techniques, and tools.

STAF 2016 took place at TU Wien, Austria, during July 4–8, 2016, and hosted the five conferences ECMFA 2016, ICGT 2016, ICMT 2016, SEFM 2016, and TAP 2016, the transformation tool contest TTC 2016, eight workshops, a doctoral symposium, and a projects showcase event. STAF 2016 featured eight internationally renowned keynote speakers, and welcomed participants from around the world.

The STAF 2016 Organizing Committee thanks (a) all participants for submitting to and attending the event, (b) the program chairs and Steering Committee members of the individual conferences and satellite events for their hard work, (c) the keynote speakers for their thoughtful, insightful, and inspiring talks, and (d) TU Wien, the city of Vienna, and all sponsors for their support. A special thank you goes to the members of the Business Informatics Group, coping with all the foreseen and unforeseen work (as usual :-)!

Spring 2016   Gerti Kappel

Preface

This volume contains the proceedings of ICGT 2016, the 9th International Conference on Graph Transformation. The conference was held in Vienna, Austria, during July 5–6, 2016. ICGT 2016 took place under the auspices of the European Association of Theoretical Computer Science (EATCS), the European Association of Software Science and Technology (EASST), and the IFIP Working Group 1.3, Foundations of Systems Specification. It was affiliated with STAF (Software Technologies: Applications and Foundations), a federation of leading conferences on software technologies.

The aim of the ICGT series[1] is to bring together researchers from different areas interested in all aspects of graph transformation. Graph structures are used almost everywhere when representing or modeling data and systems, not only in applied and theoretical computer science, but also in, e.g., natural and engineering sciences. Graph transformation and graph grammars are the fundamental modeling paradigms for describing, formalizing, and analyzing graphs that change over time when modeling, e.g., dynamic data structures, systems, or models. The conference series promotes the cross-fertilizing exchange of novel ideas, new results, and experiences in this context among researchers and students from different communities.

ICGT 2016 continued the series of conferences previously held in Barcelona (Spain) in 2002, Rome (Italy) in 2004, Natal (Brazil) in 2006, Leicester (UK) in 2008, Enschede (The Netherlands) in 2010, Bremen (Germany) in 2012, York (UK) in 2014, and L'Aquila (Italy) in 2015 following a series of six International Workshops on Graph Grammars and Their Application to Computer Science from 1978 to 1998 in Europe and in the USA.

This year, the conference solicited research papers that describe new unpublished contributions in the theory and applications of graph transformation, innovative case studies describing the use of graph-rewriting techniques in any application domain, and tool presentation papers that demonstrate the main features and functionalities of graph-based tools. All papers were reviewed thoroughly by at least three Program Committee members and additional reviewers. The Program Committee selected 14 papers for publication in these proceedings with an acceptance rate of 42 %. The topics of the accepted papers range over a wide spectrum, including theoretical approaches to graph transformation and their verification, analyses and compilation methods, graph queries, visual methods as well as various applications. In addition to these paper presentations, the conference program included an invited talk, given by Juergen Dingel (Queen's University, Ontario, Canada).

We would like to thank all who contributed to the success of ICGT 2016, the invited speaker Juergen Dingel, the authors of all submitted papers, as well as the members of the Program Committee and the additional reviewers for their valuable contributions

---

[1] www.graph-transformation.org.

to the selection process. We are grateful to the TU Wien and the STAF federation of conferences for hosting ICGT 2016.

Special thanks go to Barbara König, the organizer of the 7th International Workshop on Graph Computation Models (GCM 2016), a satellite workshop related to ICGT 2016 and affiliated with the STAF federation of conferences.

We also would like to acknowledge the excellent support throughout the publishing process by Alfred Hofmann and his team at Springer, in particular for the complimentary printed copies of these proceedings, which were distributed at the meeting, and the helpful use of the EasyChair conference management system.

During the preparation of the conference, we were saddened by the death of Hartmut Ehrig, one of the fathers and most productive members of the graph transformation community and co-founder of this conference series. We dedicate this volume to him. An obituary, by Hans-Jörg Kreowski, is included in these proceedings.

July 2016   Rachid Echahed
                                                               Mark Minas

Organization

Steering Committee

Michel Bauderon	LaBRI, Université de Bordeaux, France
Paolo Bottoni	Sapienza - Università di Roma, Italy
Andrea Corradini	Università di Pisa, Italy
Hartmut Ehrig †	Technische Universität Berlin, Germany
Gregor Engels	Universität Paderborn, Germany
Holger Giese	Hasso-Plattner-Institut Potsdam, Germany
Reiko Heckel (Chair)	University of Leicester, UK
Dirk Janssens	Universiteit Antwerpen, Belgium
Barbara König	Universität Duisburg-Essen, Germany
Hans-Jörg Kreowski	Universität Bremen, Germany
Ugo Montanari	Università di Pisa, Italy
Mohamed Mosbah	LaBRI, Université de Bordeaux, France
Manfred Nagl	RWTH Aachen, Germany
Fernando Orejas	Universitat Politècnica de Catalunya, Spain
Francesco Parisi-Presicce	Sapienza - Università di Roma, Italy
John Pfaltz	University of Virginia, Charlottesville, USA
Detlef Plump	The University of York, UK
Arend Rensink	University of Twente, The Netherlands
Leila Ribeiro	Universidade Federal do Rio Grande do Sul, Brazil
Grzegorz Rozenberg	Universiteit Leiden, The Netherlands
Andy Schürr	Technische Universität Darmstadt, Germany
Gabriele Taentzer	Philipps-Universität Marburg, Germany
Bernhard Westfechtel	Universität Bayreuth, Germany

Program Committee

Gábor Bergmann	Budapest University of Technology and Economics, Hungary
Paolo Bottoni	Sapienza - Università di Roma, Italy
Andrea Corradini	Università di Pisa, Italy
Juan De Lara	Universidad Autónoma de Madrid, Spain
Frank Drewes	Umeå universitet, Sweden
Rachid Echahed (Co-chair)	CNRS and Université de Grenoble Alpes, France
Claudia Ermel	Technische Universität Berlin, Germany
Holger Giese	Hasso-Plattner-Institut Potsdam, Germany
Annegret Habel	Universität Oldenburg, Germany
Reiko Heckel	University of Leicester, UK

Berthold Hoffmann	Universität Bremen, Germany
Dirk Janssens	Universiteit Antwerpen, Belgium
Barbara König	Universität Duisburg-Essen, Germany
Christian Krause	SAP SE, Potsdam, Germany
Sabine Kuske	Universität Bremen, Germany
Leen Lambers	Hasso-Plattner-Institut Potsdam, Germany
Yngve Lamo	Bergen University College, Norway
Tihamer Levendovszky	Microsoft, USA
Mark Minas (Co-chair)	Universität der Bundeswehr München, Germany
Mohamed Mosbah	LaBRI, Université de Bordeaux, France
Fernando Orejas	Universitat Politècnica de Catalunya, Spain
Francesco Parisi-Presicce	Sapienza - Università di Roma, Italy
Detlef Plump	The University of York, UK
Arend Rensink	University of Twente, The Netherlands
Leila Ribeiro	Universidade Federal do Rio Grande do Sul, Brazil
Andy Schürr	Technische Universität Darmstadt, Germany
Martin Strecker	Université de Toulouse, France
Gabriele Taentzer	Philipps-Universität Marburg, Germany
Matthias Tichy	University of Ulm, Germany
Pieter Van Gorp	Eindhoven University of Technology, The Netherlands
Hans Vangheluwe	Universiteit Antwerpen, Belgium
Bernhard Westfechtel	Universität Bayreuth, Germany
Albert Zündorf	Universität Kassel, Germany

Additional Reviewers

Becker, Jan Steffen	Kögel, Stefan
Blouin, Dominique	Leblebici, Erhan
Buchmann, Thomas	Löwe, Michael
Cabrera, Benjamin	Machado, Rodrigo
Dyck, Johannes	Maximova, Maria
Ermler, Marcus	Poskitt, Christopher M.
George, Tobias	Radke, Hendrik
Hegedüs, Ábel	Schwägerl, Felix
Heindel, Tobias	Stegmaier, Michael
Kreowski, Hans-Jörg	Vörös, András

Obituary for Hartmut Ehrig

The Graph Transformation Community Mourns for Hartmut Ehrig (1944–2016)

Hans-Jörg Kreowski

University of Bremen, Department of Computer Science
P.O.Box 33 04 40, 28334 Bremen, Germany
kreo@informatik.uni-bremen.de

Hartmut Ehrig died on March 17, 2016 at the age of 71. His death is very sad news for his colleagues and friends. The graph transformation community lost one of its pioneers, a leading, most inspiring and creative researcher, and guiding spirit to many of us.

As a student back in 1971, I attended Hartmut's seminar on categorical automata theory, soon afterwards he introduced me to the fascinating world of graph transformation and supervised my PhD thesis. This was the beginning of a long, intense and fruitful period of cooperation and friendship in which we sat together for hundreds of hours discussing and working on categorical automata theory, graph transformation and algebraic specification. I owe a lot to him.

Hartmut spent all his academic career at the Technische Universität Berlin only interrupted by longer research stays at Amherst, Yorktown Heights, Los Angeles, Leiden, Barcelona, Rome and Pisa. He studied Mathematics from 1963 to 1969, was research assistant from 1970 to 1972 at the Mathematics Department, and got his PhD in 1971. In 1972, he was appointed as assistant professor at the Computer Science Department and got his Habilitation two years later. In the same year, he was appointed as associate professor of Theoretical Computer Science and as full professor in 1985. He held this position until he retired in 2010. Besides teaching and research, he was also deeply involved in university affairs serving repeatedly as department chair and leading the Institute for Software Engineering and Theoretical Computer Science for 32 years.

He was amazingly productive. In addition to the editing and co-editing of more than 20 proceedings and handbooks, he authored and co-authored eight books and more than 400 papers in journals, proceedings, handbooks and other collective volumes while cooperating with more than 160 co-authors. Not only the sheer amount of printed outcome is striking, but also the fact that Hartmut was the driving force behind most of these publications.

If he looked into a matter, then he did not stop before he understood it in depth. In this process, he often came up with innovative formulations, views and approaches. He was a profound thinker who worked hard to disseminate his ideas. He was a great communicator attending a good many conferences, visiting numerous research groups all over the world and inviting a great number of famous and promising scientists to Berlin.

He was also a dedicated teacher who prepared many courses in Theoretical Computer Science and Mathematics for Computer Scientists including teaching materials as evidenced by his text book *Mathematisch-strukturelle Grundlagen der Informatik* and a wealth of unpublished lecture notes. At his university, he belonged to the minority of professors who experimented regularly with new principles of teaching. He invested much time and effort in the supervision of his students. In particular his over 50 PhD students always found him ready to advise and collaborate.

Hartmut largely personifies the area of graph transformation. He founded the series of International Workshops on Graph Grammars and their Application to Computer Science with Volker Claus and Grzegorz Rozenberg starting in 1978, which continued as International Conferences on Graph Transformation (ICGT) since 2002. He was the first chair of the steering committee of ICGT from 2000 to 2008. He proposed and moderated the European projects Computing by Graph Transformation I and II that played an essential role in binding the community together. And last but not least, his monograph on *Fundamentals of Algebraic Graph Transformation* and the most recent book *Graph and Model Transformation – General Framework and Applications*, as well as the co-editing of two volumes of the *Handbook of Graph Grammars and Computing by Graph Transformation* are evidence of his prominent role.

In addition to other awards, his paper on *Introduction to the Algebraic Theory of Graph Grammars (A Survey)*, which was published in the proceedings of the first international graph grammar workshop, was distinguished as the "most influential paper in 2^5 years of graph transformation" at the ICGT conference 2010, definite a fitting choice. Hartmut's contribution to the area began with his paper *Graph-Grammars:*

An Algebraic Approach (co-authored by Michael Pfender and Hans-Jürgen Schneider) presented at the SWAT conference 1973. The paper was not the first attempt to generalize rule-based string rewriting to the level of graphs as underlying structures, but it turned out to be the most influential one. This was the birth of the famous double-pushout diagram that establishes the application of a graph transformation rule. From then on, Hartmut propagated the basic ideas with enthusiasm and conviction so that his name is indissolubly associated with the approach. It became the most frequently used and the most successfully applied variant of graph transformation.

Appreciating his achievements and the services rendered to graph transformation, one should not forget that Hartmut played a similar role in algebraic specification and contributed also to the areas of automata theory, Petri net theory and formal and visual modeling in a significant way. His monographs on *Algebraic Specification Techniques and Tools for Software Development: The ACT Approach* and *Fundamentals of Algebraic Specification 1 and 2* set new standards. He started the series of international conferences on Theory and Practice of Software Develepment (TAPSOFT) in 1985, later extended to the very successful ETAPS series. Moreover, he served as vice president of the European Association of Theorectical Computer Science (EATCS) from 1997 to 2002 and also as vice president of the European Association of Software Science and Technology (EASST) since 2000.

There is a very old metaphor going back to the twelfth century that we can see further making progress in science because we can stand on the shoulders of giants. Hartmut was, is and will be such a giant for our area. What better way to honour him and his work than to follow his footsteps and aspire to achieve his level of service and dedication.

Contents

Applications

Keynote

Complexity is the Only Constant: Trends in Computing and Their Relevance to Model Driven Engineering

Juergen Dingel[(✉)]

School of Computing, Queen's University, Kingston, ON, Canada
dingel@cs.queensu.ca

Abstract. Despite ever increasing computational power, the history of computing is characterized also by a constant battle with complexity. We will briefly review these trends and argue that, due to its focus on abstraction, automation, and analysis, the modeling community is ideally positioned to facilitate the development of future computing systems. More concretely, a few, select, technological and societal trends and developments will be discussed together with the research opportunities they present to researchers interested in modeling.

1 Introduction

The development of computing is remarkable in many ways, and perhaps most of all in its progress and impact. However, due to the economic significance of computing and the pace of societal and technological change, we are constantly presented with new questions, challenges, and problems, giving us little time to reflect on how far we have come. Also, computing has become such a large and fragmented field that it is impossible to keep abreast all research developments.

This paper wants to briefly review some select past and present developments. Its main goal is to inform, stimulate, and inspire, not to convince. It will attempt to do so in a somewhat eclectic, anecdotal manner without claims of comprehensiveness, mostly driven by the author's interest, but with ample references to allow interested readers to dig deeper.

2 Complexity

> *"Complexity, I would assert, is the biggest factor involved in anything having to do with the software field."*
>
> Robert L. Glass [23]

In general, complex systems are characterized by a large number of entities, components or parts, many of which are highly interdependent and tightly coupled such that their combination creates synergistic, emergent, and non-linear behaviour [29]. One of the prime examples of a complex system is the human brain consisting, approximately, of 10^{11} neurons connected by 10^{15} synapses [11].

© Springer International Publishing Switzerland 2016
R. Echahed and M. Minas (Eds.): ICGT 2016, LNCS 9761, pp. 3–18, 2016.
DOI: 10.1007/978-3-319-40530-8_1

	Lines of code (approx.) (in million)
Operating Systems	
Windows NT 3.1 (1993)	0.5
Windows 95	11
Windows 2000	29
Windows XP (2001)	35
Windows Vista (2007)	50
Windows 7	40
Mac OS X	85
Android OS	12
Automobiles	
1981	0.05
2005	10
2015 (high end)	100
Miscellaneous	
Pacemaker	0.1
Mars Curiosity Rover	5
Firefox	10
Intuit Quickbook	10
Boeing 787	14
F-35 fighter jet	24
Large Hadron Collider	50
Facebook	60
Google (gmail, maps, etc)	2,000

Fig. 1. Approximate size of software in various products [7,48]

Figure 1 shows the size of software in different kinds of products. Noteworthy here are not only the absolute numbers, but also the rate of increase. Automotive software is a good example here. Just over 40 years ago, cars were devoid of software. In 1977, the General Motors Oldsmobile Tornado pioneered the first production automotive microcomputer ECU: a single-function controller used for electronic spark timing. By 1981, General Motors was using microprocessor-based engine controls executing about 50,000 lines of code across its entire domestic passenger car production. Since then, the size, significance, and development costs of automotive software has grown to staggering levels: Modern cars can be shipped with as much as 1 GB of software encompassing more than 100 million lines of code; experts estimate that more than 80% of automotive innovations will be driven by electronics and 90 % thereof by software, and that the cost of software and electronics can reach 40 % of the cost of a car [25].

The history of avionics software tells a similar story: Between 1965 and 1995, the amount of software in civil aircraft has doubled every two years [14]. If growth continues at this pace, experts believe that limits of affordability will soon be reached [79].

Lines of code is a doubtful measure of complexity[1]. Nonetheless, it appears fair to say the modern software is one of the most complex man-made artifacts.

2.1 Why has Complexity Increased so Much?

An enabler necessary for building and running modern software certainly is modern hardware. Today's software could not run on yesterday's hardware. The hardware industry has produced staggering advances in chip design and manufacturing which have managed to deliver exponentially increasing computing power at exponentially decreasing costs. Compared to the Apollo 11 Guidance Computer used 1969[2] a standard smart phone from 2015 (e.g., iPhone 6) has several tens of million of times the computational power (in terms of instructions per second)[3]. In 1985, an 2011 iPad2 would have rivaled a four-processor version of the Cray 2 supercomputer in performance, and in 1994, it still would have made the list of world's fastest supercomputers [45]. According to [47], the price of a megabyte of memory dropped from US$411,041,792 in 1957 to US$0.0037 in December 2015 — a factor of over 100 billion! The width of each conducting line in a circuit (approx. 15 nm) is approaching the width of an atom (approx. 0.1 to 0.5 nm).

But, it is not just technology that is getting more complex, life in general does, too. According to anthropologist and historian Josef Tainter, "the history of cultural complexity is the history of human problem solving" [73]. Societies get more complex because "complexity is a problem solving strategy that emerges under conditions of compelling need or perceived benefit". Complexity allows us to solve problems (e.g., food or energy distribution) or enjoy some benefit. Ideally, this benefit is greater than the costs of creating and sustaining the complexity introduced by the solution.

2.2 Consequences of Complexity

On the positive side, complex systems are capable of impressive feats. AlphaGo, the Go playing system that in March 2016 became the first program to beat a professional human Go player without handicaps on a full-sized board in a five-game match, was said by experts to be capable of developing its own moves: "All but the very best Go players craft their style by imitating top players. AlphaGo seems to have totally original moves it creates itself" [5], providing a great example of — seemingly or real — emergent, synergistic behaviour.

On the negative side, complexity increases risk of failure. Data on the failures of software or software development are hard to come by; according to the US

---

[1] So many alternative ones have been proposed [61] that even the study of complexity appears complex.

[2] A web-based simulator can be found at http://svtsim.com/moonjs/agc.html.

[3] https://www.quora.com/How-much-more-computing-power-does-an-iPhone-6-have-than-Apollo-11-What-is-another-modern-object-I-can-relate-the-same-computing-power-to.

National Institute of Standards and Technology, the cost of software errors in the US in 2001 was US$ 60 billion [63] and in 2012 the worldwide cost of IT failure has been estimated to be $3 trillion[4].

A recent example illustrates how subtle bugs can be and how difficult it is to build software systems correctly: Chord is a protocol and algorithm for a peer-to-peer distributed hash table first presented in 2001 [72]. The work identified relevant properties and provided informal proofs for them in a technical report. Chord has been implemented many times[5] and went on to win the SIGCOMM Test-of-Time Award in 2011. The original paper currently has over 12,000 citations on Google scholar and is listed by CiteSeer as the 9th most cited Computer Science article. In 2012, it was shown that the protocol was not correct [82].

2.3 How to Deal with Complexity

Computer science curricula teach students a combination of techniques to deal with complexity, the most prominent of which are decomposition, abstraction, reuse, automation, and analysis. Of these, abstraction, automation, and analysis lie at the heart of MDE. These principles have served us amazingly well. Examples include the development of programming languages in general, and Peter Denning's ground-breaking work on virtual memory in particular [15]. But, e.g., 'The Law of Leaky Abstractions'[6], the 'Automation Paradox' [22], and the Ariane 5 accident in 1996 [1] have also taught us that even these techniques must be used with care.

3 Developments and Opportunities

> *"I have no doubt that the auto industry will change more in the next five–10 years than it has in the last 50"*
>
> Mary Barra, GM Chairman and CEO, January 2016 [24]

> *"Only 19 % of [175] interviewed auto executives describe their organizations as prepared for challenges on the way to 2025"*
>
> B. Stanley, K. Gyimesi, IBM IBV, January 2015 [71]

Making predictions in the presence of exponential change is very difficult[7]. For instance, when asked to imagine life in the year 2000, 19th century French artists came up with robotic barbers, machines that read books to school children, and radium-based fireplaces[8]; when the concept of a personal computer

---

[4] http://www.zdnet.com/article/worldwide-cost-of-it-failure-revisited-3-trillion.
[5] At least 8 implementations are listed at https://github.com/sit/dht/wiki/faq.
[6] http://www.joelonsoftware.com/articles/LeakyAbstractions.html.
[7] http://uday.io/2015/10/15/predicting-the-future-and-exponential-growth.
[8] http://singularityhub.com/2012/10/15/19th-century-french-artists-predicted-the-world-of-the-future-in-this-series-of-postcards.

was first discussed at IBM, a senior executive famously questioned its value[9]. However, predicting further accelerating levels of change appears to be a safe bet. Increasing amounts of software are very likely to come with that, meaning there should be lots of things to do for software researchers.

The following list is highly selective and meant to complement more comprehensive treatments such as [65]. Also, we will focus most on technology; however, as pointed out in [65], more technology is not always the answer.

3.1 Semantics Engineering

Capturing the formal semantics of general purpose programming languages has been a topic of research for a long time, but the richness of these languages present challenges that limit a more immediate, practical application of the results contributing to a widespread belief that formal semantics are for theoreticians only. However, the recent interest in Domain Specific Languages (DSLs) appears to present new opportunities to leverage formal semantics. Compared to General Purpose Languages (GPLs), a DSL typically consists of a smaller number of carefully selected features. Often, semantically difficult GPL constructs such as objects, pointers, iteration, or recursion can be avoided; expressiveness is lost, but tractability is gained.

The literature contains some examples showing how this increased tractability can be leveraged to facilitate formal reasoning. For instance, automatic verifiers have been built for DSLs for hardware description [13], train signaling [18], graph-based model transformation [66], and software build systems [10].

However, the improved tractability of DSLs might also greatly facilitate the automatic generation of supporting tooling. Looking at how widely used techniques to describe the *syntax* of a language have become to generate syntax processing tools, the vision is clear: Use descriptions of the *semantics* of a language to facilitate the construction of semantics-aware tools for the execution and analysis of that language.

An Inspiring Example. This idea has already been explored in the context of programming languages [6,28,52,77]) and modeling languages [19,43,53,83] to, e.g., implement customizable interpreters, symbolic execution engines, and model checkers. However, the work in [40], in which abstract interpreters for a language are generated automatically from a description of its formal semantics, shows that more is possible. Given a description of the operational semantics of a machine-language instruction set such as x86/IA32, ARM, or SPARC in a domain-specific language called TSL, and a description of how the base types and operators in TSL are to be interpreted "abstractly" in an abstract semantic domain, the TSL tool automatically creates an implementation of an abstract interpreter for the instruction set:

$$\text{TSL : concrete semantics} \times \text{abstract domain} \longrightarrow \text{abstract semantics}$$

---

[9] http://www-03.ibm.com/ibm/history/ibm100/us/en/icons/personalcomputer.

The abstract interpreter can then be used by different analysis engines (e.g., for finding a fixed-point of a set of dataflow equations using the classical worklist algorithm, or for performing symbolic execution) to obtain an analyzer that is easily retargetable to different languages. The tool offers an impressive amount of generality by supporting different instruction sets and different analyses. It has been used to build analyzers for the IA32 instruction set that perform value/set analysis, definition/use analysis, model checking, and Botnet extraction with a precision at least as high as manually created analyzers.

Lowering Barriers, Increasing Benefit. Recent formalizations of different industrial-scale artifacts including operating system kernels [35], compilers [38], and programming languages including C [17], JavaScript [57] and Java [4] provide some evidence that large-scale formalizations are becoming increasingly feasible. Efforts are underway to make the expression, analysis, and reuse of descriptions of semantics more scalable, effective, and mainstream [21,54,62]. Paired with the increasing maturity and adoption of language workbenches such as Xtext[10], this work may allow substantial progress on the road towards the automatic generation of semantics-aware tools such as interpreters, static analyzers, and compilers. Descriptions of semantics might one day be as common and useful as descriptions of syntax are today.

3.2 Synthesis

The topic of synthesis has been receiving a lot of attention recently. For most of these efforts, 'synthesis' refers to the process of automatically generating executable code from information given in some higher level form: Examples include the generation of code that manipulates many different artifacts (e.g., bitvectors [70], concurrent data structures [69], database queries [9], data reprentations [68], or spreadsheets [26]), gives feedback to students for programming assignments [68], or implements an optimizing compiler [8]. Some of these examples use a GPL, some use a DSL. The synthesis itself is implemented either using constraint solving or machine learning. Different proposals on how to best integrate synthesis into programming languages have been made and have targeted GPLs such as Java [31,49] and DSLs [75].

Given that abstraction, automation and analysis are central to MDE, synthesis certainly also is of interest to the modeling community and the work on synthesis and its applications should be followed closely. In [74,75], the idea of "solver-aided DSLs" is introduced. The paper presents a framework in which such DSLs can be created and illustrates its use with a DSL for example-based web scraping in which the solver is used to generate an XPath expression that retrieves the desired data.

MDE features a range of activities and situations which might potentially benefit from a little help from a solver capable of finding solutions to constraints. Could the idea of synthesis and the use of solvers facilitate, e.g.,

---

[10] https://eclipse.org/Xtext.

- the development of models via extraction or autocompletion,
- the support for partial models with incomplete or uncertain information,
- the analysis of models,
- the refinement of models via, e.g., the generation of substate machines from interface specifications,
- the generation of correct, efficient code from models,
- the generation of different views from a model?

How could synthesis be leveraged in language workbenches that generate supporting tools such as analyzers and code generators, or in model transformation languages and engines that support different transformation intents [44]?

Some attempts to leverage synthesis for, e.g., model creation [36], transformation authoring [2], design space exploration [27] already exist, but the topic hardly seems exhausted. Indeed, some of the technical issues Selic mentions in [65] might be mitigated using synthesis including dealing with abstract, incomplete models, model transformation, and model validation.

3.3 Reconciling Formal Analysis and Evolution

There is a fundamental conflict between analysis and evolution: As soon as the model evolves (changes), any analysis results obtained on the original version may be invalidated and the analysis may have to be rerun. Unfortunately, both seem unavoidable not just in the context of MDE, but software engineering in general.

Most analyses require the creation of supporting artifacts that represent analysis-relevant information about the model. For instance, software reverse engineering tools collect relevant information about the code in a so-called fact repository typically containing a collection of tuples encoding graphs [34]; most static analysis tools require some kind of dependence graph, and test case generation tools often rely on symbolic execution trees.

When the cost of the analysis rises, the motivation to avoid a complete re-analysis after a change and to leverage information about the nature of the change to optimize the analysis increases as well. In general, aspects of this topic are handled in the literature on impact analysis [39]; however, the analyses considered typically are either manual (comprehension, debugging) or rather narrow (regression testing, software measurement via metrics), and do not consider, e.g., static analyses or analyses based on formal methods.

Two Approaches. Assuming the analysis requires supporting artifacts, there are, in principle, at least two ways of reconciling analysis and evolution [33]:

1. Artifact-oriented (Fig. 2): The goal here is to update the supporting artifact A_1 as efficiently as possible, but in such a way that it becomes fully reflective of the information in the changed program. To this end, the impact of the change Δ on the artifact original artifact A_1 is determined, and the parts of the artifact possibly affected are recomputed, while leaving parts known to be unaffected unchanged. Then, the updated artifact A_2 can be used as before to perform all

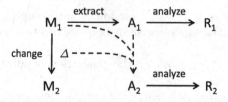

Fig. 2. Artifact-oriented approach to reconcile analysis and evolution. M_i, A_i, and R_i denote, respectively, a model, the artifact extracted from the model to support the analysis, and the analysis result

analyses it is meant to support. For instance, for analyses based on dependence graphs such as slicing or impact analysis, the parts of the graph affected by the change are updated and the result is used to recompute the result. Similarly, for a dead code analysis (or test case generation) using a symbolic execution tree (SET), affected parts of the tree would be updated to produce a tree corresponding to the changed program. In this approach, the savings come from avoiding the reconstruction of parts of the supporting artifact A_2.

2. Analysis-oriented (Fig. 3): Here, the focus is on updating the result of the analysis as efficiently as possible, rather than the supporting artifact. To this end, the impact of the change Δ on the analysis result is determined, and the parts of the analysis that may lead to a different result due to the change are redone, ignoring any parts known to produce the same result. For instance, when impact analysis is used during regression testing, only tests for executions that were introduced by the change are run; tests covering unaffected executions are ignored [60]. In this approach, the focus is on reestablishing the analysis result R_2 as some combination $R_2 = op(\Delta, R_1, R'_2)$ of the previous result R_1 and the partial result R'_2. E.g., an analysis-oriented optimization of the dead code analysis mentioned above (or test case generation) would use the most efficient means to determine dead code in (or test cases for) the affected parts and the construction of the full SET for the changed program may not be necessary for that; in this case, R_1 would be the dead code in (or test cases for) M_1 and the partial result R'_2 would be the dead code in (test cases for) the parts of the model introduced by the change; the operation $op(\Delta, R_1, R'_2)$ would return the union of R'_2 and the dead code (test cases) in R_1 not impacted by the change. In this case, the savings come from avoiding unnecessary parts of the analysis.

Comparing the two approaches, we see an interesting tradeoff: The first approach does not speed up the analysis itself (only the update of the supporting artifact). However, it results in a complete supporting artifact (e.g., dependence graphs, SET) that can then be used for whatever analyses it supports (e.g., different static analyses for dependence graphs, and, test case generation, dead code analysis for SETs). Moreover, the result of the analysis of the changed model does not rely on the result of the analysis of the original program at all. The second approach speeds up the analysis itself, but since it focusses on the changed parts, it is partial only. E.g., the updated program can only be concluded to be free of dead code, if the second *and* the first analysis say so.

In sum, the second approach is more restricted compared to the first, but might well hold additional optimization potential. Recent research on program analysis using formal techniques has begun to explore these possibilities, and analysis-oriented approaches to optimize model checking [81] and symbolic execution [58] have been developed. Inspired by these proposals, we have developed prototypes that use both approaches to optimize the symbolic execution of Rhapsody statemachines [33]. Results indicate that both approaches are complementary and effective in different situations.

3.4 Open Science

In 2010, two Harvard economists published a paper entitled "Growth in a Time of Debt" in a non-peer reviewed journal which provided support for the argument that excessive debt is bad for growth. The paper was used by many policy makers to back up their calls for fiscal austerity. However, in 2013, the paper was shown to have used flawed methodology and to not support the authors' conclusions[11].

Reproducibility. Examples of research producing doubtful results due to unintended or even intended flaws in the data or methodology have been going through the media recently and many disciplines have begun to investigate the reproducibility of their research results. For instance, a study in economics showed that 78 % of the 162 replication studies conducted "disconfirm a major finding from the original study" [16]. A study focusing on research in Computer Systems [12], examined 601 papers from eight ACM conferences and five journals: of the papers with results backed by code, the study authors were able to build the system in less than 30 min only 32 % of the time; in 54 % of cases the study authors failed to build the code, but the paper authors said that the code does build with reasonable effort.

The U.S. President Steps In. However, it has been pointed out in prominent places that in many disciplines these days reproducibility means the availability of programs and data [30,50,64]. In other words, since software, programming,

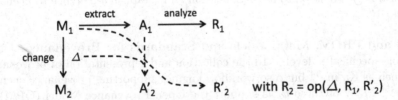

Fig. 3. Analysis-oriented approach to reconcile analysis and evolution. The analysis result R_2 for M_2 is obtained by combining the result for M_1 with the partial result R'_2

---

[11] A discussion of the paper and the controversy it caused can be found at https://en.wikipedia.org/wiki/Growth_in_a_Time_of_Debt.

and the use and manipulation of data plays such a central role in so many disciplines, some of the problems with reproducibility in other disciplines are due to limitations in programming, software, and the use and manipulation of data, that is, they are due to problems that the computing community is at least partially responsible for and should put on its research agenda[12]. About a year ago, the world's most powerful man has done exactly that with an executive order to create a "National Strategic Computing Initiative" which includes accessibility and workflow capture as central objectives [56].

A Good Start: Encouraging Artifact Submission. The research community has begun to adjust with, e.g., no less than four events devoted to reproducibility at the 2015 Conference for High Performance Computing, Networking, Storage and Analysis (SC'15)[13], and Eclipse's Science Working Group announcing specific initiatives (Eclipse Integrated Computational Environment and Data Analysis Workbench). However, more should be done and promoting the value of artifact submission at workshops, conferences, and journal appears to be a good place to start. According to [12], 19 Computer Science conferences have participated in an artifact submission and evaluation process between 2011 and 2016, including PLDI'15, OOPSLA'15, ECOOP'15, and POPL'16, but more need to join. The availability of the artifacts that research is based and their integration into the scientific evaluation process should be the norm, not the exception.

3.5 Provenance

A topic closely related to open science and reproducibility is provenance. In general, data provenance refers to the description of the origins of a piece of data and the process by which it was created or obtained with the goal to allow assessments of quality, reliability, or trustworthiness. It has traditionally been studied in the context of databases, but has also been used for data found on the web or data used in scientific experiments. Domains of application include

- *science*, to make data and experimental results more trustworthy and experiments more reproducible,
- *business*, to demonstrate ownership, responsibility, or regulatory compliance and facilitate auditing processes, and
- *software development*, to aid certification and establish adherence to licensing rules.

OPM and PROV: Metamodels and Standards for Provenance. There are tools specifically devoted to the collection and representation of provenance data such as Karma[14] but also workflow engines supporting provenance such as Kepler[15]. Many of these tools support the Open Provenance Model (OPM), a

---

[12] Computers are even said to have "broken science", https://www.eclipsecon.org/na2016/session/how-computers-have-broken-science-and-how-we-can-fix-it.

[13] http://sc15.sueprcomputing.org.

[14] http://d2i.indiana.edu/provenance_karma.

[15] https://kepler-project.org.

data model (i.e., metamodel) for provenance information [51] based on directed, edge-labeled, hierarchical graphs with three kinds of nodes representing things (Artifact, Agent, and Process) and five kinds of edges representing causal relationships (used, wasGeneratedBy, wasControlledBy, wasTriggeredBy, and wasDerivedFrom). OPM graphs are subject to well-formedness constraints, can contain time information, and have inference rules (allowing, e.g., the inclusion of derived information via transitive edges) and operations (for, e.g., union, intersection, merge, renaming, refinement and completion) associated with them. A formal semantics of OPM graphs published recently views them as temporal theories on the temporal events represented in the graph [37], but does not account for Agents. OPM has been a major influence in the design of the PROV family of documents by the World Wide Web Consortium (W3C) [78] which not only defines a data model, but also corresponding serializations and other supporting definitions to enable the interoperable interchange of provenance information in heterogeneous environments such as the Web.

Open-ended Opportunities. There appears to be a lot of opportunity for researchers with background in graph transformation, formal methods, or modeling to advance the state-of-the-art in provenance. Many established topics (e.g., formal semantics, constraint solving, traceability, querying, language engineering for graphical DSMLs, and model management), but also emerging topics (e.g., the use of models and modeling to support inspection, certification and compliance checking [20,46,55] and data aggregation and visualization [41,42,48,76]) appear potentially relevant. Moreover, no approaches have been found to build models that allow the quantification of the quality or trustworthiness of data. In case of producer/consumer relationships, service level agreements guaranteeing data with a certain level of quality might also be of interest.

3.6   Open Source Modeling Tools

The need to improve MDE tooling has been expressed before [32,65,80]. At the same time, significant efforts to develop industrial-strength open source modeling tools and communities that support and sustain them are currently being made. Sample tools include AutoFocus[16], xtUML[17], Papyrus[18] [3], and PapyrusRT[19] [59].

The development and availability of complete, industrial-strength open source MDE tools is a radical shift from past practices and presents both exciting opportunities and substantial challenges for everybody interested in MDE, regardless of whether they use the tools for industrial development, research, or education. Due to the importance of tooling to the success of MDE, this shift has the potential to provide a much-needed stimulus for major advances in its adoption, development, and dissemination.

---

[16] http://www.fortiss.org/en/about-us/alle-news/autofocus-3.

[17] https://xtuml.org.

[18] https://eclipse.org/papyrus.

[19] https://www.eclipse.org/papyrus-rt.

4 Conclusion

> *"We can only see a short distance ahead, but we can see plenty there that needs to be done."*
>
> Alan Turing

As we continue to entrust more and more complex functions and capabilities to software, our ability to build this software reliably and effectively should increase as well. Much more work is needed to make this happen and this paper has suggested some starting points.

The fragmentation that plagues many research areas is harmful. Any scientific community should keep an open mind and remain willing to learn from others about existing and new problems and potentially new ways to solve them [67].

Acknowledgment. This work was supported by the Natural Sciences and Engineering Research Council of Canada (NSERC), and by the Ontario Ministry of Research and Innovation (MRI).

References

1. Ariane 5 flight 501 failure, report by the inquiry board (1996). http://esamultimedia.esa.int/docs/esa-x-1819eng.pdf
2. Baki, I., Sahraoui, H.: Multi-step learning and adaptive search for learning complex model transformations from examples. ACM Trans. Softw. Eng. Methodol. (2016) (in print)
3. Barrett, R., Bordeleau, F.: 5 years of 'Papyrusing' – migrating industrial development from a proprietary commercial tool to Papyrus (invited presentation). In: Workshop on Open Source Software for Model Driven Engineering (OSS4MDE 2015), pp. 3–12 (2015)
4. Bogdănaş, D., Roşu, G.: K-Java: a complete semantics of Java. In: ACM SIGPLAN/SIGACT Symposium on Principles of Programming Languages (POPL 2015), pp. 445–456. ACM, January 2015
5. Borowiec, S., Lien, T.: AlphaGo beats human Go champ in milestone for artificial intelligence. Los Angeles Times, 12 March 2016
6. Borras, P., Clement, D., Despeyroux, T., Incerpi, J., Kahn, G., Lang, B., Pascual, V.: Centaur: the system. In: ACM SIGSoft/SIGPlan Software Engineering Symposium on Practical Software Development Environments (SDE 1987) (1987)
7. Charette, R.N.: Why software fails. IEEE Spectr. **42**(9), 42–49 (2005)
8. Cheung, A., Kamil, S., Solar-Lezama, A.: Bridging the gap between general-purpose and domain-specific compilers with synthesis. In: Summit oN Advances in Programming Languages (SNAPL 2015) (2015)
9. Cheung, A., Solar-Lezama, A., Madden, S.: Optimizing database-backed applications with query synthesis. ACM SIGPLAN Not. **48**(6), 3–14 (2013)
10. Christakis, M., Leino, K.R.M., Schulte, W.: Formalizing and verifying a modern build language. In: Jones, C., Pihlajasaari, P., Sun, J. (eds.) FM 2014. LNCS, vol. 8442, pp. 643–657. Springer, Heidelberg (2014)
11. Chudler, E.H.: Neuroscience for kids. https://faculty.washington.edu/chudler/what.html

12. Collberg, C., Proebsting, T.A.: Repeatability in computer systems research. Commun. ACM **59**(3), 62–69 (2016)
13. Cook, B., Launchbury, J., Matthews, J.: Specifying superscalar microprocessors in Hawk. In: Workshop on Formal Techniques for Hardware and Hardware-like Systems (1998)
14. Potocki de Montalk, J.P.: Computer software in civil aircraft. Cockpit/Avionics Eng. **17**(1), 17–23 (1993)
15. Denning, P.J.: Virtual memory. ACM Comput. Surv. **2**(3), 153–189 (1970)
16. Duvendack, M., Palmer-Jones, R.W., Reed, W.R.: Replications in economics: a progress report. Econ. Pract. **12**(2), 164–191 (2015)
17. Ellison, C., Roşu, G.: An executable formal semantics of C with applications. In: ACM SIGPLAN/SIGACT Symposium on Principles of Programming Languages (POPL 2012), pp. 533–544 (2012)
18. Endresen, J., Carlson, E., Moen, T., Alme, K.-J., Haugen, Ø., Olsen, G.K., Svendsen, A.: Train control language - teaching computers interlocking. In: Computers in Railways XI. WIT Press (2008)
19. Engels, G., Hausmann, J.H., Heckel, R., Sauer, S.: Dynamic meta modeling: a graphical approach to the operational semantics of behavioral diagrams in UML. In: Evans, A., Caskurlu, B., Selic, B. (eds.) UML 2000. LNCS, vol. 1939, pp. 323–337. Springer, Heidelberg (2000)
20. Falessi, D., Sabetzadeh, M., Briand, L., Turella, E., Coq, T., Panesar-Walawege, R.K.: Planning for safety standards compliance: a model-based tool-supported approach. IEEE Softw. **29**(3), 64–70 (2012)
21. Felleisen, M., Findler, R.B., Flatt, M.: Semantics Engineering with PLT Redex. MIT Press, Cambridge (2009)
22. Geer, D.E.: Children of the magenta. IEEE Secur. Priv. **13**(5) (2015)
23. Glass, R.L.: Sorting out software complexity. Commun. ACM **45**(11), 19–21 (2002)
24. GM: GM chairman and CEO addresses CES. https://www.gm.com/mol/m-2016-Jan-boltev-0106-barra-ces.html, 6 Jan 2016
25. Grimm, K.: Software technology in an automotive company – major challenges. In: International Conference on Software Engineering (ICSE 2003) (2003)
26. Gulwani, S., Harris, W., Singh, R.: Spreadsheet data manipulation using examples. Commun. ACM **55**, 97–105 (2012)
27. Hendriks, M., Basten, T., Verriet, J., Brassé, M., Somers, L.: A blueprint for system-level performance modeling of software-intensive embedded systems. Softw. Tools Technol. Transf. **18**, 21–40 (2016)
28. Henriques, P.R., Pereira, M.J.V., Mernik, M., Lenic, M., Gray, J., Wu, H.: Automatic generation of language-based tools using the LISA system. IEE Proc. Softw. **152**, 54–69 (2005)
29. Homer-Dixon, T.: The Ingenuity Gap. Vintage Canada (2001)
30. Ince, D.C., Hatton, L., Graham-Cumming, J.: The case for open computer programs. Nature **482**, 485–488 (2012)
31. Jeon, J., Qiu, X., Foster, J.S., Solar-Lezama, A.: Jsketch: sketching for Java. In: Joint Meeting of the European Software Engineering Conference and the ACM SIGSOFT Symposium on the Foundations of Software Engineering (ESEC/FSE 2015) (2015)
32. Kahani, N., Bagherzadeh, M., Dingel, J., Cordy, J.R.: The problems with Eclipse modeling tools: a topic analysis of Eclipse forums, April 2016 (submitted)
33. Khalil, A., Dingel, J.: Incremental symbolic execution of evolving state machines. In: ACM/IEEE International Conference on Model Driven Engineering Languages and Systems (MODELS 2015) (2015)

34. Kienle, H.M., Mueller, H.A.: Rigi – an environment for software reverse engineering, exploration, visualization, and redocumentation. Sci. Comput. Prog. **75**, 247–263 (2010)
35. Klein, G., Elphinstone, K., Heiser, G., Andronick, J., Cock, D., Derrin, P., Elkaduwe, D., Engelhardt, K., Kolanski, R., Norrish, M., Sewell, T., Tuch, H., Winwood, S.: Formal verification of an OS kernel. In: ACM SIGOPS Symposium on Operating Systems Principles (SOSP 2009), pp. 207–220. ACM (2009)
36. Koksal, A.S., Pu, Y., Srivastava, S., Bodik, R., Piterman, N., Fisher, J.: Synthesis of biological models from mutation experiments. In: ACM SIGPLAN/SIGACT Symposium on Principles of Programming Languages (POPL 2013) (2013)
37. Kwasnikowska, N., Moreau, L., Van den Bussche, J.: A formal account of the Open Provenance Model. ACM Trans. Web **9**, 10:1–10:44 (2015)
38. Leroy, X.: Formal verification of a realistic compiler. Commun. ACM **52**(7), 107–115 (2009)
39. Li, B., Sun, X., Leung, H., Zhang, S.: A survey of code-based change impact analysis techniques. Softw. Test. Verification Reliab. **23**, 613–646 (2012)
40. Lim, J., Reps, Th.: TSL: a system for generating abstract interpreters and its application to machine-code analysis. ACM Trans. Program. Lang. Syst. **35**(1), 4:1–4:59 (2013)
41. Lima, M.: Visual complexity website. http://www.visualcomplexity.com/vc
42. Lima, M.: The Book of Trees: Visualizing Branches of Knowledge Hardcover. Princeton Architectural Press (2014)
43. Lu, Y., Atlee, J.M., Day, N.A., Niu, J.: Mapping template semantics to SMV. In: IEEE/ACM International Conference on Automated Software Engineering (ASE 2004) (2004)
44. Lúcio, L., Amrani, M., Dingel, J., Lambers, L., Salay, R., Selim, G.M.K., Syriani, E., Wimmer, M.: Model transformation intents and their properties. Softw. Syst. Model., 1–38 (2014)
45. Markoff, J.: The iPad in your hand: as fast as a supercomputer of yore. New York Times article based on interview with Dr. Jack Dongarra, 9 May 2011. http://bits.blogs.nytimes.com/2011/05/09/the-ipad-in-your-hand-as-fast-as-a-supercomputer-of-yore
46. Mayr, A., Plösch, R., Saft, M.: Objective safety compliance checks for source code. In: Companion Proceedings of the 36th International Conference on Software Engineering, ICSE Companion 2014 (2014)
47. McCallum, J.C.: Memory prices (1957–2015). http://www.jcmit.com/memoryprice.htm. Accessed Mar 2016
48. McCandless, D.: Information is beautiful: Million lines of code. http://www.informationisbeautiful.net/visualizations/million-lines-of-code
49. Milicevic, A., Rayside, D., Yessenov, K., Jackson, D.: Unifying execution of imperative and declarative code. In: International Conference on Software Engineering (ICSE 2011) (2011)
50. Monroe, D.: When data is not enough. Commun. ACM **58**(12), 12–14 (2015)
51. Moreau, L., Clifford, B., Freire, J., Futrelle, J., Gil, Y., Groth, P., Kwasnikowska, N., Miles, S., Missier, P., Myers, J., Plale, B., Simmhan, Y., Stephan, E., van den Bussche, J.: The open provenance model core specification (v1.1). Future Gener. Comput. Syst. **27**(6), 743–756 (2011)
52. Mosses, P.: Sis: A compiler-generator system using denotational semantics. Technical report 78-4-3, Department of Computer Science, University of Aarhus (1978)

53. Muller, P.-A., Fleurey, F., Jézéquel, J.-M.: Weaving executability into object-oriented meta-languages. In: Briand, L.C., Williams, C. (eds.) MoDELS 2005. LNCS, vol. 3713, pp. 264–278. Springer, Heidelberg (2005)

54. Mulligan, D.P., Owens, S., Gray, K.E., Ridge, T., Sewell, P.: Lem: Reusable engineering of real-world semantics. SIGPLAN Not. **49**(9), 175–188 (2014)

55. Nair, S., de la Vara, J.L., Melzi, A., Tagliaferri, G., de-la-Beaujardiere, L., Belmonte, F.: Safety evidence traceability: problem analysis and model. In: Salinesi, C., Weerd, I. (eds.) REFSQ 2014. LNCS, vol. 8396, pp. 309–324. Springer, Heidelberg (2014)

56. The President of the United States: Executive order: creating a national strategic computing initiative, 29 July 2015. https://www.whitehouse.gov/the-press-office/2015/07/29/executive-order-creating-national-strategic-computing-initiative

57. Park, D., Ştefănescu, A., Roşu, G.: KJS: a complete formal semantics of JavaScript. In: SIGPLAN Conference on Programming Language Design and Implementation (PLDI 2015), pp. 346–356. ACM, June 2015

58. Person, S., Yang, G., Rungta, N., Khurshid, S.: Directed incremental symbolic execution. In: ACM SIGPLAN Conference on Programming Language Design and Implementation (PLDI 2011) (2011)

59. Posse, E.: PapyrusRT: modelling and code generation (invited presentation). In: Workshop on Open Source Software for Model Driven Engineering (OSS4MDE 2015) (2015)

60. Ren, X., Shah, F., Tip, F., Ryder, B.G., Chesley, O.: Chianti: A tool for change impact analysis of Java programs. In: ACM Conference on Object-Oriented Programming, Systems, Languages, and Applications (OOPSLA 2004) (2004)

61. Riguzzi, F.: A survey of software metrics. Technical report DEIS-LIA-96-010, Università degli Studi di Bologna (1996)

62. Roşu, G., Şerbănuţă, T.F.: An overview of the K semantic framework. J. Logic Algebraic Prog. **79**(6), 397–434 (2010)

63. RTI: The economic impacts of inadequate infrastructure for software testing. Technical report Planning Report 02-3, National Institute of Standards & Technology (NIST), May 2002

64. Schuwer, R., van Genuchten, M., Hatton, L.: On the impact of being open. IEEE Softw. **32**, 81–83 (2015)

65. Selic, B.: What will it take? A view on adoption of model-based methods. Softw. Syst. Model. **11**, 513–526 (2012)

66. Selim, G.M.K., Lúcio, L., Cordy, J.R., Dingel, J., Oakes, B.J.: Specification and verification of graph-based model transformation properties. In: Giese, H., König, B. (eds.) ICGT 2014. LNCS, vol. 8571, pp. 113–129. Springer, Heidelberg (2014)

67. Shapiro, S.: Splitting the difference: the historical necessity of synthesis in software engineering. IEEE Ann. Hist. Comput. **19**(1), 20–54 (1997)

68. Singh, R., Gulwani, S., Solar-Lezama, A.: Automated feedback generation for introductory programming assignments. ACM SIGPLAN Not. **48**, 15–26 (2013). ACM

69. Solar-Lezama, A., Jones, C., Bodik, R.: Sketching concurrent data structures. ACM SIGPLAN Not. **43**, 136–148 (2008). ACM

70. Solar-Lezama, A., Rabbah, R., Bodík, R., Ebcioğlu, K.: Programming by sketching for bit-streaming programs. ACM SIGPLAN Not. **40**, 281–294 (2005). ACM

71. Stanley, B., Gyimesi, K.: Automotive 2025 – industry without borders. Technical report, IBM Institute for Business Value, January 2015. http://www-935.ibm.com/services/us/gbs/thoughtleadership/auto2025

72. Stoica, I., Morris, R., Karger, D., Kaashoek, F.M., Balakrishnan, H.: Chord: a scalable peer-to-peer lookup service for internet applications. In: ACM Conference on Applications, Technologies, Architectures, and Protocols for Computer Communications (SIGCOMM 2001), pp. 149–160 (2001)
73. Tainter, J.A.: Complexity, problem solving, and sustainable societies. In: Costanza, R., Segura, O., Martinez-Alier, J. (eds.) Getting Down to Earth: Practical Applications of Ecological Economics. Island Press (1996)
74. Torlak, E., Bodik, R.: Growing solver-aided languages with Rosette. In: ACM International Symposium on New Ideas, New Paradigms, and Reflections on Programming & Software, Onward! 2013, pp. 135–152 (2013)
75. Torlak, E., Bodik, R.: A lightweight symbolic virtual machine for solver-aided host languages. In: ACM SIGPLAN Conference on Programming Language Design and Implementation (PLDI 2014) (2014)
76. Tufte, E.: Beautiful Evidence. Graphics Press, Cheshire (2006)
77. den van Brand, M.G.J., van Deursen, A., Heering, J., de Jong, H.A., de Jonge, M., Kuipers, T., Klint, P., Moonen, L., Olivier, P.A., Scheerder, J., Vinju, J.J., Visser, E., Visser, J.: The ASF+SDF meta-environment: a component-based language development environment. In: Wilhelm, R. (ed.) CC 2001. LNCS, vol. 2027, p. 365. Springer, Heidelberg (2001)
78. W3C Working Group. PROV-Overview: An overview of the PROV family of documents. In: Groth, P., Moreau, L. (eds.) W3C Working Group Note. W3C (2013)
79. Ward, D.: Avsis system architecture virtual integration program: proof of concept demonstrations. In: INCOSE MBSE Workshop, 27 January 2013
80. Whittle, J., Hutchinson, J., Rouncefield, M., Heldal, R.: Industrial adoption of model-driven engineering: are the tools reallythe problem? In: ACM/IEEE International Conference on Model-Driven Engineering Languages and Systems (MODELS 2013) (2013)
81. Yang, G., Dwyer, M., Rothermel, G.: Regression model checking. In: International Conference on Software Maintenance (ICSM 2009), pp. 115–124. IEEE (2009)
82. Zave, P.: Using lightweight modeling to understand Chord. ACM SIGCOMM Comput. Commun. Rev. **42**(2), 50–57 (2012)
83. Zurowska, K., Dingel, J.: A customizable execution engine for models of embedded systems. In: Roubtsova, E., McNeile, A., Kindler, E., Gerth, C. (eds.) BM-FA 2009-2014. LNCS, vol. 6368, pp. 82–110. Springer, Heidelberg (2015)

Foundations

Sesqui-Pushout Rewriting with Type Refinements

Michael Löwe[✉]

FHDW Hannover, Freundallee 15, 30173 Hannover, Germany
michael.loewe@fhdw.de

Abstract. Sesqui-pushout rewriting is an algebraic graph transformation approach that provides mechanisms for vertex cloning. If a vertex gets cloned, the original and the copy obtain the same context, i.e. all incoming and outgoing edges of the original are copied as well. This behaviour is not satisfactory in practical examples which require more control over the context cloning process. In this paper, we provide such a control mechanism by allowing each transformation rule to refine the underlying type graph. We discuss the relation to the existing approaches to controlled sesqui-pushout vertex cloning, elaborate a basic theoretical framework, and demonstrate its applicability by a practical example.

1 Introduction

Sesqui-pushout graph transformation (SqPO) [2] is a relatively new variant in the family of algebraic graph rewriting frameworks. It extends the double-pushout (DPO) [4,5] and the single-pushout approach (SPO) [6,8] by mechanisms for object cloning including the complete context of the object, which are all incoming and outgoing edges in the case of graphs. Many practical examples, however, require more control over the cloning process. In many cases, only edges of specific types shall be cloned. In this paper, we propose a new mechanism for controlled object cloning by allowing each rule to refine the underlying type graph in a way that is suitable for the cloning performed by the rule.

The paper is organised as follows: Sect. 2 recapitulates sesqui-pushout rewriting in a categorical set-up and presents major results that shall be valid in any extension. The example in Sect. 3 motivates the mechanisms that are introduced in Sect. 4. Section 5 formulates a categorical framework for the new approach and shows that many results known from the standard approach carry over. Finally, Sect. 6 discusses related work and future research.

2 Standard SqPO-Rewriting

In this section, we present sesqui-pushout rewriting in a categorical set-up. We require that the underlying category $\mathcal{C}$ satisfies the following conditions:

C1 $\mathcal{C}$ has all finite limits and co-limits.

© Springer International Publishing Switzerland 2016
R. Echahed and M. Minas (Eds.): ICGT 2016, LNCS 9761, pp. 21–36, 2016.
DOI: 10.1007/978-3-319-40530-8_2

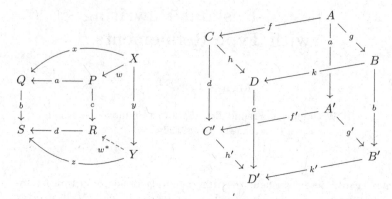

Fig. 1. Final pullback complement and commutative cube property

C2 Every pair $(a : P \to Q, b : Q \rightarrowtail S)$ of morphisms with monic b has a final pullback complement.

C3 Pushouts along monomorphisms are van-Kampen: In a commutative cube as in Fig. 1, where g' is monic, (h', k') is pushout of (f', g'), and (g, a) and (f, a) are pullbacks of (g', b) and (f', d) resp., we have: (h, d) and (k, b) are pullbacks of (h', c) and (k', c) resp., if and only if (h, k) is pushout of (f, g).

A pair $(c : P \to R, d : R \to S)$ as in the left part of Fig. 1 is *final pullback complement* (*FPC*) of the pair $(a : P \to Q, b : Q \to S)$, if (a, c) is pullback of (b, d) and for each collection of morphisms (x, y, z, w), where (x, y) is pullback of (b, z) and $a \circ w = x$, there is a unique w^* with $d \circ w^* = z$ and $c \circ w = w^* \circ y$. Note the following special cases of final pullback complement situations:

F1 For every morphism $f : P \to Q$, (id_P, f) is FPC of (f, id_Q) and vice versa.

F2 In a commutative cube as in the right part Fig. 1, where (b, k') is FPC of (k, c), (f, g) is pullback of (k, h), and (d, h) is pullback of (c, h') the following compatibility condition between pullbacks and final pullback complements holds: (a, f') is FPC of (f, d), if and only if (f', g') is pullback of (k', h').[1]

Final pullback complements possess the following composition and decomposition properties (for proofs compare [11]):

F3 Horizontal composition and decomposition: Let $c \circ k = k' \circ b$ and $b \circ g = g' \circ a$ in the right part of Fig. 1 and let (b, k') be FPC of (k, c): (a, g') is FPC of (g, b), if and only if $(a, k' \circ g')$ is FPC of $(k \circ g, c)$.

F4 Vertical composition: If (g, b) and (k, c) are FPCs of (a, g') and (b, k') respectively in the right part of Fig. 1, then $(k \circ g, c)$ is FPC of $(a, k' \circ g')$.

Condition C3 guarantees:[2]

F5 Pushouts in $\mathcal{C}$ preserve monomorphisms.

---

[1] For a proof of the if-part see [9].
[2] Compare [4,7].

F6 Pushouts along monomorphisms are pullbacks.

For a compact notion of sesqui-pushout rewriting, we pass over from $\mathcal{C}$ to the span category $\mathcal{C}^{\leftarrow\rightarrow}$ of $\mathcal{C}$. A *concrete span* is a pair of $\mathcal{C}$-morphisms $(p : K \rightarrow P, q : K \rightarrow Q)$. Two spans (p_1, q_1), (p_2, q_2) are equivalent, if there is isomorphism i with $p_1 \circ i = p_2$ and $q_1 \circ i = q_2$; $[(p, q)]_{\equiv}$ denotes the class of spans equivalent to (p, q). The *category of abstract spans* $\mathcal{C}^{\leftarrow\rightarrow}$ has the same objects as $\mathcal{C}$ and equivalence classes of spans as morphisms. The identity for an object $A \in \mathcal{C}^{\leftarrow\rightarrow}$ is defined by $\mathrm{id}_A^{\mathcal{C}^{\leftarrow\rightarrow}} = [(\mathrm{id}_A, \mathrm{id}_A)]_{\equiv}$. And composition of $[(p, q)]_{\equiv}$ and $[(r, s)]_{\equiv}$ such that $\mathrm{codomain}(q) = \mathrm{codomain}(r)$ is given by $[(r, s)]_{\equiv} \circ_{\mathcal{C}^{\leftarrow\rightarrow}} [(p, q)]_{\equiv} = [(p \circ_{\mathcal{C}} r', s \circ_{\mathcal{C}} q')]_{\equiv}$ where (r', q') is a pullback of (q, r). A span composition is *strong*, written $[(r, s)]_{\equiv} \bullet [(p, q)]_{\equiv}$, if (q', r) is final pullback complement of (r', q).

Note that there is the natural and faithful embedding functor $\iota : \mathcal{C} \rightarrow \mathcal{C}^{\leftarrow\rightarrow}$ defined by identity on objects and $(f : A \rightarrow B) \mapsto [\mathrm{id}_A : A \rightarrow A, f : A \rightarrow B]$ on morphisms. In the following, the composition of a span $(p, q) \in \mathcal{C}^{\leftarrow\rightarrow}$ with a morphism $m \in mathcalC$, i.e. $(p, q) \circ m$ (or $m \circ (p, q)$), is the span defined by $(p, q) \circ \iota(m)$ (resp. $\iota(m) \circ (p, q)$). By a slight abuse of notation, we write $[d : A' \rightarrow A, f : A' \rightarrow B] \in \mathcal{C}$ if d is an isomorphism. Direct derivations in sesqui-pushout rewriting are special strong compositions of spans.

Definition 1 (Standard Rule and Derivation). *A rewrite rule p is a morphism in $\mathcal{C}^{\leftarrow\rightarrow}$, i.e. $p = (l : K \rightarrow L, r : K \rightarrow R)$. A match for p is a monic $\mathcal{C}$-morphism $m : L \rightarrowtail G$. The* direct derivation *with p at match m is constructed in two steps, compare Fig. 2:*

1. $(m \langle l \rangle, l \langle m \rangle)$ *is final pullback complement of (l, m).*
2. $(m \langle p \rangle, r \langle m \rangle)$ *is pushout of $(m \langle l \rangle, r)$.*

In a direct derivation, G is the source, *$p@m$ is the* target, *the span $p \langle m \rangle = (l \langle m \rangle, r \langle m \rangle)$ is called the* trace, *and $m \langle p \rangle$ is also referred to as co-match.*

Remarks. A derivation is determined up to isomorphism by the match. This is due to the fact that FPCs and pushouts are unique up to isomorphism. Since pullbacks preserve monomorphisms, $m \langle l \rangle$ is a monomorphism and Fact F5 provides monic $m \langle p \rangle$. Due to Condition C2, rules are applicable at every match. The direct derivation in Fig. 2 constitutes a special commutative diagram in $\mathcal{C}^{\leftarrow\rightarrow}$, i.e. $m \langle p \rangle \bullet p = p \langle m \rangle \bullet m$, with pullback $(l, m \langle l \rangle)$ of $(l \langle m \rangle, m)$.

Since traces are morphisms in $\mathcal{C}^{\leftarrow\rightarrow}$, they can be used as rules and, by Fact F4 and the composition property of pushouts, we immediately obtain:

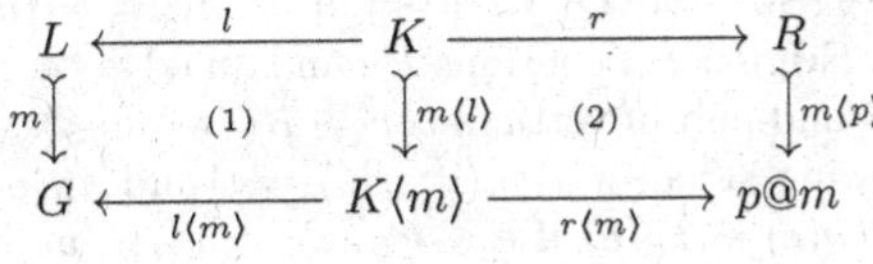

$$
\begin{array}{ccccc}
L & \xleftarrow{\quad l \quad} & K & \xrightarrow{\quad r \quad} & R \\
\downarrow{\scriptstyle m} & (1) & \downarrow{\scriptstyle m\langle l\rangle} & (2) & \downarrow{\scriptstyle m\langle p\rangle} \\
G & \xleftarrow{\quad l\langle m\rangle \quad} & K\langle m\rangle & \xrightarrow{\quad r\langle m\rangle \quad} & p@m
\end{array}
$$

Fig. 2. Direct transformation

Proposition 2 (Standard Derived Rule). *If $p\langle m\rangle$ is the trace of a deriva-tion with rule p at match m and n is match for $p\langle m\rangle$, then $(p\langle m\rangle)\langle n\rangle = p\langle n\circ m\rangle$ and $(n\circ m)\langle p\rangle = n\langle p\langle m\rangle\rangle\circ m\langle p\rangle$.*

Since rules are spans, they can be composed and decomposed. Condition C3 guarantees that rule composition and decomposition carries over to derivations.

Proposition 3 (Composition of Standard Derivations). *If m is a match for $p'\circ p$, $(p'\circ p)\langle m\rangle = p'\langle m\langle p\rangle\rangle\circ p\langle m\rangle$ and $m\langle p'\circ p\rangle = m\langle p\rangle\langle p'\rangle$.*

The proposition is a direct consequence of Theorem 7 in [11]. Together with Proposition 2, it provides the fundaments for a rich theory.[3]

3   Example: Version Management

As an example that demonstrates the power of SqPO-rewriting, we present a model for version management of decomposed components. It uses the category $\mathcal{G}$ of graphs and graph morphisms or, more precisely, the slice category $\mathcal{G}\downarrow T$ of all graphs wrt. a type graph $T\in\mathcal{G}$. A *graph* $G = (V; E; s, t : E\to V)$ has a set V of vertices, a set E of edges, and source and target mappings s and t. A *graph morphism* $h : G\to H$ is a pair $(h_V : G_V\to H_V, h_E : G_E\to H_E)$ of mappings with $s_H\circ h_E = h_V\circ s_G$ and $t_H\circ h_E = h_V\circ t_G$. $\mathcal{G}$ satisfies Conditions C1–C3.[4]

The left part of Fig. 3 depicts the type graph for the version management system. The right part shows a sample instance graph.[5] The sample contains

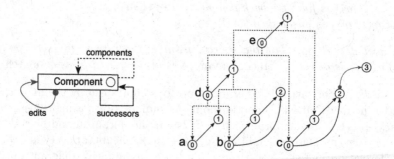

Fig. 3. Version management: model and instance

---

[3] Compare for example [11].

[4] Given arbitrary morphism $a : P\to Q$ and monomorphism $b : Q\rightarrowtail S$, the final pullback complement $(c : P\rightarrowtail R, d : R\to S)$ is constructed as follows, compare [2]:
Vertices: $R_V = P_V\uplus(S_V - b_V(Q_V))$, $c_V = \mathrm{id}_{P_V}$, $d_V = b_V(a_V(v))$ if $v\in P_V$ and $d_V = \mathrm{id}_{S_V}$ otherwise. **Edges:** R_E contains P_E and an edge "copy" (v, e, v') for every edge $e\in S_E - b_E(Q_E)$ and pair of vertices $v, v'\in R_V$ with $s_S(e) = d_V(v)$ and $t_S(e) = d_V(v')$ with the following structure: $s_R(v, e, v') = v$ and $s_R(e) = s_P(e)$ if $e\in P_E$, $t_R(v, e, v') = v'$ and $t_R(e) = t_P(e)$ if $e\in P_E$, $c_E = \mathrm{id}_{P_E}$, and $d_E(v, e, v') = e$ and $d_E(e) = b_E(a_E(e))$ if $e\in P_E$.

[5] The typing is indicated by the graphical symbols.

integrate: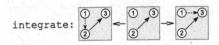

Fig. 4. Evolution of decomposed components

5 components, the components a, b and c are elementary and the components d and e are decomposed into a and b respectively d and c. In the example, only component c has an editable version, namely c3. Decomposed components evolve by integrating successor versions of their components. The corresponding rule is depicted in Fig. 4.

If a new editable version x' of a component x is created, it shall integrate the same components as x. Here the copy mechanism of sesqui-pushout rewriting shall be applicable. But it is not, since x' shall only have copies of the outgoing components-edges of x but neither of the incoming components-edges nor of outgoing or incoming successor-edges. A similar problem arises, when an editable version x' gets ready to be published. Then it shall become successor of the version x it has been spun off and of all predecessor versions of that x. This means that the successor-relation shall be transitive, for example to skip some versions in component integration, compare Fig. 4. Here, we do not need a complete copy of x, we only need a copy of all successor-edges pointing to x.

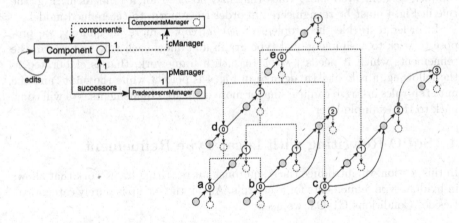

Fig. 5. Refined version management: model and instance

Problems of this type can be tackled by using a refined type graph. The left part of Fig. 5 shows a suitable refinement for the problems described above. The right part of the figure shows the instance of Fig. 3 in the refined version. The trick is that we provide internal structure to each component by two one-to-one relations. Now, we have a port-object for every Component that handles incoming successors-edges, namely a node the type of which is PredecessorsManager, and a port that handles outgoing components-edges, namely a node of type

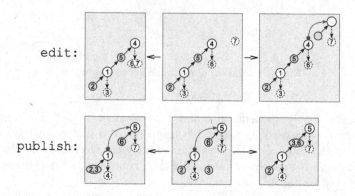

Fig. 6. Rules for editing and publishing

ComponentsManager. Having these structures at hand, we can formulate the rules for the creation of a new and for the publication of an existing editable version, compare Fig. 6.[6]

Although this approach works quite satisfactory, there are a lot of drawbacks of such a *global* type graph refinement. First of all, the model gets more complex and many additional consistence conditions come into play, like the one-to-one relations in Fig. 5, that have to be preserved by all rules. Thus, rules get more complex as well, even those rules that had no need for a refinement. E.g. the rule in Fig. 4 must be reengineered in order to conform to the refined model.

In order to tackle the problem of *partial copies* more adequately, we propose to stick to a simple global type graph and allow each rule to perform the refinements which it needs *locally*. In such a framework, that is elaborated in the following, a rule like integrate in Fig. 4 is perfect while the edit- and the publish-rules in Fig. 6 can use smaller more dedicated refinements. We will come back to the example later.

4   SqPO-Rewriting with Local Type Refinement

In this section, we introduce the sesqui-pushout rewriting framework that allows individual type refinements for every rule. Again, the set-up is purely categorical. Besides Conditions C1–C3, we need:

C4 The underlying category $\mathcal{C}$ has epi-mono-factorisations.
C5 Pullbacks in $\mathcal{C}$ preserve epimorphisms.

By $\mathcal{C}^{\rightarrow}$, we denote the arrow category over $\mathcal{C}$.[7] And, for any given (type) object $T \in \mathcal{C}$, $\mathcal{C} \downarrow T$ denotes the slice category of all objects under T.[8]

---

[6] Note the edit-rule copies the node labelled "6, 7" from the left- to the right-hand side, and the publish-rule copies the node labelled "2, 3" from left to right.

[7] The objects of $\mathcal{C}^{\rightarrow}$ are all morphisms of $\mathcal{C}$ and a morphism i from $f : A \rightarrow B$ to $g : C \rightarrow D$ is a pair $(i_A : A \rightarrow C, i_B : B \rightarrow D)$ of $\mathcal{C}$-morphism such that $i_B \circ f = g \circ i_A$.

[8] $\mathcal{C} \downarrow T$ is the restriction of $\mathcal{C}^{\rightarrow}$ to morphisms with co-domain T. Note that every category $\mathcal{C}$ in our set-up is equivalent to $\mathcal{C} \downarrow F$, where F is the final object in $\mathcal{C}$.

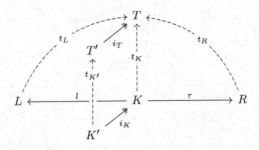

Fig. 7. Rule

Definition 4 (Rule). *Given an object T in $\mathcal{C}$, a T-typed rewrite rule $p = ((l : t_K \to t_L, r : t_K \to t_R), i : t_{K'} \twoheadrightarrow t_K)$ consists of a $\mathcal{C} \downarrow T$-span (l, r) and an epic[9] type refinement $i \in \mathcal{C}^{\to}$, compare Fig. 7.*

A rule is applicable at a monic match morphism $m : t_L \to t_G$. The derivation shall comprise 3 phases, namely (1) the refinement of the rule and the match to the type T', (2) the SqPO-rewriting with the refined rule at the refined match, and (3) the abstraction of the derivation back to type T.

Definition 5 (Derivation). *The* direct derivation *with a rewrite rule $p = (l : t_K \to t_L, r : t_K \to t_R, i : t_{K'} \twoheadrightarrow t_K)$ at a monic match morphism $m : t_L \rightarrowtail t_G$ consists of the trace $p(m) = (l \langle m \rangle : t_{K\langle m \rangle} \to t_G, r \langle m \rangle : t_{K\langle m \rangle} \longrightarrow t_{p@m})$ and the co-match $m \langle p \rangle : t_R \to t_{p@m}$ in $\mathcal{C} \downarrow T$ together with an epic type refinement $i \langle m \rangle : t'_{K'\langle m' \rangle} \twoheadrightarrow t_{K\langle m \rangle}$ which are constructed as follows, compare Fig. 8:*

1. Refinement: *Let $(i_L, t'_{L'})$ and $(i_R, t'_{R'})$ be the pullbacks of (i_T, t_L) and (i_T, t_R) resp. and l' and r' the morphisms making the diagram commute. Call $p' = (l', r')$ the i-refined rule of $p = (l, r)$. Let $(i_G, t'_{G'})$ be the pullback (i_T, t_G) and m' the unique morphisms such that (i_L, m') is pullback of (i_G, m).*
2. Derivation: *Let $(l \langle m \rangle_A, r \langle m \rangle_A) : G \to p@m_A$ be the trace and $m \langle p \rangle_A : R \to p@m_A$ the co-match of the sesqui-pushout derivation with p at m. Let $(l' \langle m' \rangle, r' \langle m' \rangle) : G' \to p'@m'$ be the trace and $m' \langle p' \rangle : R' \to p'@m'$ the co-match of the sesqui-pushout derivation with p' at m'. And let $i \langle m \rangle_A$ and $i_{p@m_A}$ be the morphisms into the final pullback complement of (l, m) and from the pushout of $(r', m' \langle l' \rangle)$ making the resulting diagram commute.*
3. Abstraction: *Construct $(i \langle m \rangle, d)$ and $(i_{p@m}, d_A)$ as epi-mono-factorisations of $i \langle m \rangle_A$ and $i_{p@m_A}$. Set $l \langle m \rangle = l \langle m \rangle_A \circ d$. And, finally, let $m \langle l \rangle$, $r \langle m \rangle$, and $m \langle p \rangle$ be the diagonal morphisms making the diagram commute.*

Although the construction for a direct derivation is rather complex, it possesses good properties that can lead to a rich theory and are investigated in the following. The first result is obvious, namely that the new rewriting mechanism subsumes simple sesqui-pushout rewriting.

---

[9] I.e. both components are epimorphisms.

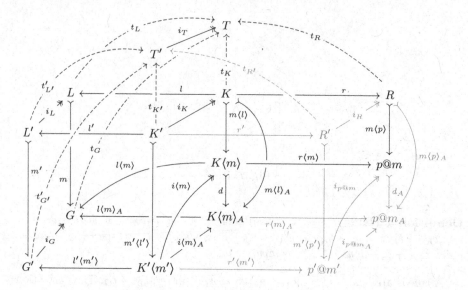

Fig. 8. Direct derivation

Proposition 6. *For a rule $p = (l : t_K \to t_L, r : t_K \to t_R, (i_T, i_K) : t_{K'} \twoheadrightarrow t_K)$ such that $(i_K, t_{K'})$ is pullback of (i_T, t_K), the traces and co-matches of the direct derivations according to Definitions 1 and 5 coincide.*

Proof. If $(i_K, t_{K'})$ is pullback of (i_T, t_K), (i_K, l') becomes pullback of (i_L, l) in Definition 5, compare Fig. 8. By Fact F2, $(i\langle m\rangle_A, l'\langle m'\rangle)$ is pullback of $(l\langle m\rangle_A, i_G)$. Finally, Condition C5 provides that $i\langle m\rangle_A$ is epi- and d is isomorphism.   □

The following lemma shows that the type refinement produces an effect on the derivation's left-hand side only. The right-hand side is a simple pushout.

Proposition 7. *In Definition 5, $(r\langle m\rangle, m\langle p\rangle)$ is pushout of $(m\langle l\rangle, r)$.*

Proof. Since d_A is monic, $m\langle p\rangle \circ r = r\langle m\rangle \circ m\langle l\rangle$.

If $(r^* : K\langle m\rangle \to X, m\langle l\rangle^* : R \to X)$ is pushout of $(m\langle l\rangle, r)$, there is monic[10] $d_X : X \to p@m_A$ with $d_X \circ r^* = r\langle m\rangle_A \circ d$ and $d_X \circ m\langle l\rangle^* = m\langle p\rangle_A$ and there is $e_X : p'@m' \to X$ from the pushout object $p'@m'$ with $e_X \circ r'\langle m'\rangle = r^* \circ i\langle m\rangle$ and $e_X \circ m'\langle p'\rangle = m\langle l\rangle^* \circ i_R$.

The morphism e_X turns out to be epic: $f \circ e_X = g \circ e_X$ implies $f \circ e_X \circ r'\langle m'\rangle = g \circ e_X \circ r'\langle m'\rangle$ and $f \circ e_X \circ m'\langle p'\rangle = g \circ e_X \circ m'\langle p'\rangle$. The first leads to $f \circ r^* \circ i\langle m\rangle = g \circ r^* \circ i\langle m\rangle$ and (i) $f \circ r^* = g \circ r^*$, since $i\langle m\rangle$ is epic. The second leads to $f \circ m\langle l\rangle^* \circ i_R = g \circ m\langle l\rangle^* \circ i_R$ and (ii) $f \circ m\langle l\rangle^* = g \circ m\langle l\rangle^*$, since i_R is epic. Properties (i) and (ii) provide $f = g$, since $(r^*, m\langle l\rangle^*)$ is pushout.

---

[10] Morphism d_X is monic, since decomposition of pushouts provides that $(d_X, r\langle m\rangle_A)$ is pushout of (r^*, d), and pushouts preserve monomorphisms due to Fact F5.

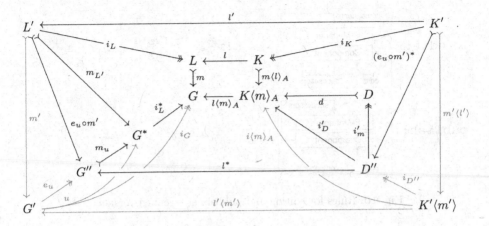

Fig. 9. Local derivation

Since $d_X \circ e_X \circ m' \langle p' \rangle = d_X \circ m \langle l \rangle^* \circ i_R = m \langle p \rangle_A \circ i_R = i_{p@m_A} \circ m' \langle p' \rangle$ and $d_X \circ e_X \circ r' \langle m' \rangle = d_X \circ r^* \circ i \langle m \rangle = r \langle m \rangle_A \circ d \circ i \langle m \rangle = r \langle m \rangle_A \circ i \langle m \rangle_A = i_{p@m_A} \circ r' \langle m' \rangle$, we can conclude that $d_X \circ e_X = i_{p@m_A}$.

Therefore, (e_X, d_X) is an epi-mono-factorisation of $i_{p@m_A}$ which provides $X \cong p@m$. Uniqueness of diagonals leads to $r^* = r \langle m \rangle$ and $m \langle l \rangle^* = m \langle p \rangle$. □

We conclude this section by an important observation, namely that a derivation in sesqui-pushout rewriting with local type refinements has local effects only. For a rewrite, we do not have to refine the complete source object, it is sufficient to refine the part that is in the image of the match.

Consider Fig. 9. It depicts the left-hand side of a rule, namely $l : K \to L$, its refinement $l' : K' \to L'$, the match $m : L \to G$, and the refined match $m' : L' \to G'$. While Definition 5 constructed the final pullback complement $(m' \langle l' \rangle, l' \langle m' \rangle)$ of l' and m' (grey in Fig. 9), we now construct the "local" FPC $(m_{L'} : L' \to G^*, i_L^* : G^* \to G)$ of the match m and the refinement of the rule's left-hand side, namely i_L. Since (m', i_L) is pullback of (i_G, m), we obtain morphism u which makes the diagram commute. By factorisation of u into epic $e_u : G' \to G''$ and monic $m_u : G'' \to G^*$, we construct a local refinement of G, i.e. $i_L^* \circ m_u : G'' \to G$, and a locally refined match $e_u \circ m'$.[11] Note that $(\mathrm{id}_{L'}, e_u \circ m')$ is pullback of $(m_u, m_{L'})$, since m_u is monic, and $(\mathrm{id}_{L'}, m')$ is pullback of $(e_u, e_u \circ m')$ by decomposition of pullbacks. Now, construct the FPC $((e_u \circ m')^*, l^*)$ of $(l', e_u \circ m')$ which provides the morphism i'_D making the diagram commute.

We show that the epi-mono-factorisation $(i'_m : D'' \twoheadrightarrow D, d : D \rightarrowtail K \langle m \rangle_A)$ of this morphism provides the same sub-object of $K \langle m \rangle_A$ as the epi-mono-factorisation of $i \langle m \rangle_A$. The argument is straightforward, since there is the morphism $i_{D''} : K' \langle m' \rangle \to D''$ mediating between the "global" and the "local" FPC and making the diagram commute. Since $(\mathrm{id}_{L'}, m')$ is pullback of $(e_u, e_u \circ m')$, $(\mathrm{id}_{K'}, l')$ is

---

[11] Note that the object G'' cannot be typed in the refined type of the rule, since it contains unrefined parts, namely the parts outside $m(L)$.

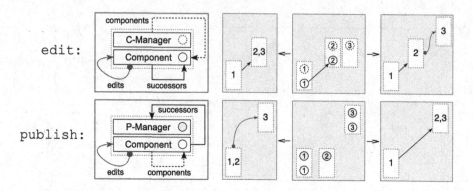

Fig. 10. Rules for editing and publishing – refined version

pullback of $(l', \mathrm{id}_{L'})$, and $(m' \langle l' \rangle, l' \langle m' \rangle)$ as well as $((e_u \circ m')^*, l^*)$ are FPCs, Fact F2 makes sure that $(i_{D''}, l' \langle m' \rangle)$ is pullback of (e_u, l^*) and Condition C5 guarantees that i_D'' is epimorphism. Therefore the pair $(i'_m \circ i_{D''}, d)$ is epi-mono-factorisation of $i \langle m \rangle_A$, q.e.d.

Example 8 (Version Management – Refined Version). Using the mechanisms introduced in this section, we can specify the rules for the version management system described in Sect. 3: We use the standard type graph given in Fig. 3. The rule for component integration is given in Fig. 4. It does not specify any type-refinement and works as it is. The rules for editing and publishing are depicted in Fig. 10. The `edit`-rule refines the type graph such that outgoing `components`-relations of a component can be handled separately, compare middle object of the rule span for `edit` in Fig. 10. The `publish`-rule uses a different type refinement such that incoming `successors`-relations can be handled and copied separately.

5   The Category of Type-Refined Spans

In this section, we define a composition operator on rewrite rules which leads to the category of type-refined spans. We show that rule composition and decomposition carries over to composition and decomposition of direct derivations, i.e. that theorems like Propositions 2 and 3 are also valid in the new framework.

Let $T \in \mathcal{C}$ be a fixed type object. Two rules

$$p_1 = \left((l_1 : t_{K_1} \to t_L, r_1 : t_{K_1} \to t_R), (i_{T'}, i_1) : (t_{K'_1} : K'_1 \to T') \to t_{K_1} \right) \text{ and}$$
$$p_2 = \left((l_2 : t_{K_2} \to t_L, r_2 : t_{K_2} \to t_R), (i_{T''}, i_2) : (t_{K'_2} : K'_2 \to T'') \to t_{K_2} \right)$$

with common domain and co-domain are equivalent if there is a triple

$$(j_K : K_2 \to K_1, j_{K'} : K'_2 \to K'_1, j_T : T'' \to T')$$

of isomorphisms, such that the resulting diagram commutes, i.e. $l_1 \circ j_K = l_2$, $r_1 \circ j_K = r_2$, $t_{K_1} \circ j_K = t_{K_2}$, $i_{T'} \circ j_T = i_{T''}$, $j_T \circ t_{K'_2} = t_{K'_1} \circ j_{K'}$, and $j_K \circ i_2 =$

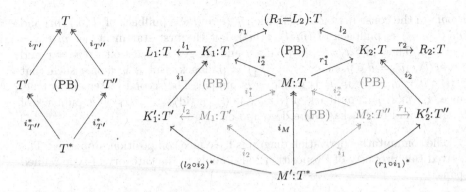

Fig. 11. Rule composition

$i_1 \circ j_{K'}$. In this case, we write $p_1 \equiv p_2$ and denote the class of rules that are equivalent to p by $[p]_\equiv$. Such a class is called *abstract type-refined span*. These spans can be composed.

Definition 9 (Composition). *Given two abstract typed-refined spans $p_1 = [((l_1, r_1), (i_{T'}, i_1))]$ and $p_2 = [((l_2, r_2), (i_{T''}, i_2))]$ such that the co-domain of r_1 coincides with the domain of l_2, the composition $p_2 \circ p_1$ of p_1 and p_2 is defined by $[((l_1 \circ l_2^*, r_2 \circ r_1^*) (i_{T'} \circ i_{T''}^*, i_M))]$, where $(i_{T''}^* : T^* \to T', i_{T'}^* : T^* \to T'')$, $(r_1^* : M \to K_2, l_2^* : M \to K_1)$, and $((r_1 \circ i_1)^* : M' \to K_2', (l_2 \circ i_2)^* : M' \to K_1')$ are the pullbacks of the pairs $(i_{T'}, i_{T''})$, (l_2, r_1), and $(l_2 \circ i_2, r_1 \circ i_1)$ resp. and $t_M : M \to T$, $t_{M'} : M' \to T^*$, and $i_M : M' \to M$ are the unique morphisms making the diagram commute, compare Fig. 11.*

Note that the composition operator is defined independent of the choice of representatives, since pullbacks of isomorphic diagrams are isomorphic. The composition is associative due to composition/decomposition properties of pullbacks.

The composition operator of Definition 9 gives rise to the category $\mathcal{C}^{\leftarrow \overset{T}{\uparrow} \rightarrow}$ of abstract type-refined spans under T, which has the same objects as $\mathcal{C} \downarrow T$ and abstract type-refined spans of type T as morphisms. The identity on $A \in \mathcal{C} \downarrow T$ is given by $\text{id}_A = [((\text{id}_A, \text{id}_A), (\text{id}_T, \text{id}_A))]$. We call a type-refined span $p = [((l, r), i)]$ *total*, if l and i are isomorphisms and *co-total* if r is isomorphism. Note that any morphism in $\mathcal{C} \downarrow T$ one to one corresponds to a total morphism in the category of abstract type-refined spans. With this categorical background, a rewrite rule p is just a morphism, a match is a monic and total morphism, and a direct derivation is a special commutative diagram.

Definition 10 (Refined Trace). *Let $p(m)$ be the trace, $m \langle p \rangle$ the co-match, and $i \langle m \rangle$ the type refinement in a derivation with rule p at match m as given in Definition 5. Then the $\mathcal{C}^{\leftarrow \overset{T}{\uparrow} \rightarrow}$-morphism $p \langle m \rangle = (p(m), i \langle m \rangle)$ is called refined trace.*

Proposition 11 (Direct Derivation). *If $p \langle m \rangle$ and $m \langle p \rangle$ are trace and co-match in a derivation with rule p at match m, then $p \langle m \rangle \circ m = m \langle p \rangle \circ p$.*

Proof. In the construction in Definition 5, $(l, m \langle l \rangle)$ is pullback of $(l \langle m \rangle, m)$ and $(m' \langle l' \rangle, i_K)$ is pullback of $(m \langle l \rangle, i \langle m \rangle)$. For the first statement, let $m \circ x = l \langle m \rangle \circ y$. Then $m \circ x = l \langle m \rangle_A \circ d \circ y$ and we get u such that $l \circ u = x$ and $d \circ m \langle l \rangle \circ u = d \circ y$, since $(l, m \langle l \rangle_A)$ is pullback. But d is monic such that $m \langle l \rangle \circ u = y$. Since $m \langle l \rangle$ is monic, u is unique. The second statement is true, since $(i_K, m' \langle l' \rangle)$ is pullback of $(i \langle m \rangle_A, m \langle l \rangle_A)$, $(\mathrm{id}_{K' \langle m' \rangle}, i \langle m \rangle_A)$ is pullback of $(d, i \langle m \rangle)$, and pullbacks can be decomposed.   □

The commutative derivation diagrams have nice composition properties. The derived rule property of Proposition 2 also holds in the category of type-refined spans.

Theorem 12 (Derived Rule). *If $p \langle m \rangle$ and $m \langle p \rangle$ are trace and co-match in a derivation and n is match for $p \langle m \rangle$, then $(p \langle m \rangle) \langle n \rangle = p \langle n \circ m \rangle$ and $(n \circ m) \langle p \rangle = n \langle p \langle m \rangle \rangle \circ m \langle p \rangle$.*

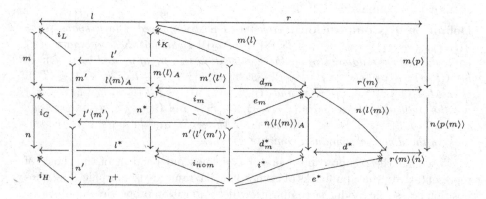

Fig. 12. Derivation with derived rule

Proof. The situation is depicted in Fig. 12: The rule $p = (l, r, i_K)$ has been applied at match m resulting in trace $(l \langle m \rangle_A \circ d_m, r \langle m \rangle, e_m)$ and the co-match $m \langle p \rangle$, where (e_m, d_m) is the factorisation of the morphism i_m the mediator between the two FPCs of the derivation $p@m$. The application of $p \langle m \rangle$ at match n results in the trace $(l^* \circ d_m^* \circ d^*, r \langle m \rangle \langle n \rangle, e^*)$ and co-match $n \langle p \langle m \rangle \rangle$, where $(n \langle l \langle m \rangle \rangle_A, l^* \circ d_m^*)$ is FPC of $(l \langle m \rangle = l \langle m \rangle_A \circ d_m, n)$, $(n' \langle l' \langle m' \rangle \rangle, l^+)$ is FPC of $(l' \langle m' \rangle, n')$, and (e^*, d^*) is the factorisation of i^* which is the mediator between these two pullback complements. Due to Condition C2 and Fact F3, we can decompose the pullback complement $(n \langle l \langle m \rangle \rangle_A, l^* \circ d_m^*)$ into two FPCs, i.e. (n^*, l^*) and $(n \langle l \langle m \rangle \rangle_A, d_m^*)$. Since FPCs preserve monomorphisms[12], d_m^* is monic. By Fact F4, $(n^* \circ m \langle l \rangle_A, l^*)$ and $(n' \langle l' \langle m' \rangle \rangle \circ m' \langle l' \rangle, l^+)$ are FPCs. Thus, we obtain the mediator $i_{n \circ m}$ which is subject to an epi-mono-factorisation

---

[12] Compare [10].

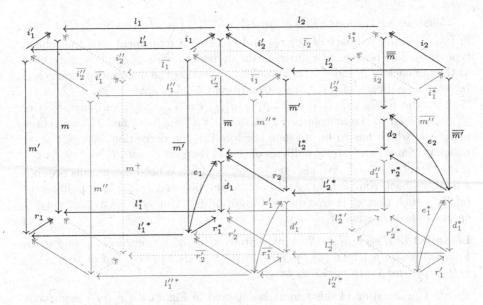

Fig. 13. Derivation composition for co-total morphisms

in the derivation with p at $n \circ m$. Since $i_{n \circ m}$ and $d_m^* \circ i^*$ are two morphisms into the FPC $(n^* \circ m \langle l \rangle_A, l^*)$ which coincide under postfix composition with l^* and prefix composition with $n' \langle l' \langle m' \rangle \rangle \circ m' \langle l' \rangle$, we conclude $i_{n \circ m} = d_m^* \circ i^*$. But then $(e^*, d_m^* \circ d^*)$ is the epi-mono-factorisation of $i_{n \circ m}$ which results in $l^* \circ d_m^* \circ d^* = l \langle n \circ m \rangle$ and $n \langle l \langle m \rangle \rangle \circ m \langle l \rangle = (n \circ m) \langle l \rangle$. Finally, the composition property of pushouts concludes the proof.   □

For the proof of the horizontal composition property, compare Proposition 3, we need to investigate some special cases first.

Lemma 13. *If $p_2 \circ p_1$ is composition of co-total morphisms, then $(p_2 \circ p_1) \langle m \rangle = p_2 \langle m \langle p_1 \rangle \rangle \circ p_1 \langle m \rangle$ and $m \langle p_2 \circ p_1 \rangle = m \langle p_1 \rangle \langle p_2 \rangle$, for every match m for p_1.*

Proof. Figure 13 depicts the situation. The black part shows the derivation with p_1 and p_2, i.e. $p_1 = (l_1, i_1)$, $p_2 = (l_2, i_2)$, $p_1 \langle m \rangle = (l_1^* \circ d_1, e_1)$, $m \langle p_1 \rangle = \overline{m}$, $p_2 \langle \overline{m} \rangle = (l_2^* \circ d_2, e_2)$, and $\overline{\overline{m}} = \overline{m} \langle p_2 \rangle = m \langle p_1 \rangle \langle p_2 \rangle$. By constructing $(\overline{l_2}, i_1^*)$ as pullback of (i_1, l_2) and $(\overline{i_2}, \overline{i_1^*})$ as pullback of (i_1^*, i_2), we obtain $p_2 \circ p_1 = (l_1 \circ l_2, i_2 \circ \overline{i_1^*})$.

The four morphisms $i_2'', i_1', \overline{i_2''}$, and $\overline{i_1'}$ with $(\overline{i_2''}, \overline{i_1'})$ as pullback of (i_2'', i_1') are the refinements of the left-hand side of p_1 and m', m'', and m^+ are the refinements of the match m, such that the cube under $i_2'', i_1', \overline{i_2''}$, and $\overline{i_1'}$ is a cube of pullbacks. If $(\overline{i_2'}, \overline{i_1})$ is constructed as pullback of (i_2', i_1) and l_1'' and l_2'' as well as $\overline{l_1}$ and $\overline{l_2}$ are the morphisms making the diagram commute, then $l_1'' \circ l_2''$ is the refinement of $l_1 \circ l_2$ and $(i_2', \overline{l_1})$ and $(l_1'', \overline{i_2''})$ are pullbacks of (l_1, i_2'') and $(l_1', \overline{i_2''})$ resp. and $(l_2'', \overline{i_1^*})$ is pullback of $(l_2', \overline{i_1})$.

Constructing the four FPCs of (m, l_1), (m', l_1'), $(m^+, \overline{l_1})$, and (m'', l_1'') results in $(\overline{r_2'}, \overline{r_1^*})$ being pullback of (r_2', r_1^*) and r_2' being a refinement wrt. the refined type of p_2. The refinement r_2 used in the derivation with p_2 can be constructed as the pullback of d_1 and r_2'. This provides e_1' making the diagram commute and $(e_1', \overline{r_2'})$ the pullback of (r_2, e_1). By Condition C5, e_1' is epic.

Now the FPCs $(d_1'', l_2^{*'})$ of (l_2^*, d_1), (d_1^*, l_2^+) of $(l_2''^*, d_1')$, and $(\overline{m''}, l_2''^*)$ of (l_2'', m''^*) lead to the morphism $r_2'^*$ with $r_2'^* \circ d_1^* = d_1'' \circ r_2^*$ and $r_2'^* \circ r_1'$ is the morphism which has to be epi-mono-factored in the derivation with $p_2 \circ p_1$ at match m. Since $(m''^*, \overline{i_1})$ is pullback of $(\overline{m'}, e_1')$, we get e_1^* with $l_2'^* \circ e_1^* = e_1' \circ l_2''^*$ and $e_1^* \circ \overline{m''} = \overline{m'} \circ \overline{i_1^*}$. We also get $d_1^* \circ e_1^* = r_1'$ since both morphisms are into a final pullback complement. And Fact F2 implies that $(e_1^*, l_2''^*)$ is pullback of $(e_1', l_2'^*)$ such that e_1^* is epic due to Condition C5. But now $(e_2 \circ e_1^*, d_1'' \circ d_2)$ is epi-mono-factorisation of $r_2'^* \circ r_1'$.   □

Lemma 14. *If $p_2 \circ p_1$ is the composition of a total morphism p_1 and a co-total morphism p_2, then $(p_2 \circ p_1)\langle m \rangle = p_2 \langle m \langle p_1 \rangle \rangle \circ p_1 \langle m \rangle$ and $m \langle p_2 \circ p_1 \rangle = m \langle p_1 \rangle \langle p_2 \rangle$, for every match m for p_1.*

Proof. The situation of this lemma is depicted in Fig. 14: $(n, r_1\langle m \rangle)$ is pushout of (r_1, m), i.e. constitutes the direct derivation with rule p_1 at match m. The co-total rule p_2 is represented by (l_2, i_2). The pullbacks (l_2^*, r_1^*) of (r_1, l_2) and $(\overline{r}_1, \overline{i}_2)$ of (i_2, r_1^*) define the rule $p_2 \circ p_1 = (l_2^*, r_1^*, \overline{i}_2)$. The application of this composition at match m is defined by the trace $(l_2^* \langle m \rangle \circ d^*, r_1^* \langle m \rangle, i \langle m \rangle)$ and the co-match $n \langle p_2 \circ p_1 \rangle$, i.e. $\overline{i}_2$ is the FPC-mediator of the derivation, $(i \langle m \rangle, d^*)$ its epi-mono-factorisation, and $(r_1^* \langle m \rangle, n \langle p_2 \circ p_1 \rangle)$ is pushout of $(m \langle l_2^* \rangle, r_1^*)$. Let, finally, i_2^* be the FPC-mediator of the derivation with p_2 at $n = m \langle p_1 \rangle$.

Consider the cube defined by refinements i_L, i_G, i_G', and i_L'. Its top, back, front, and bottom faces are pullbacks and the right face is pushout and pullback by Fact F6. Condition C3 implies that the left face is pushout and pullback. Now consider the inner and outer cube in Fig. 14 defined by $(l_2, l_2^*, l_2 \langle n \rangle_A, l_2^* \langle m \rangle)$ and $(l_2', l_2^{*'}, l_2' \langle n' \rangle, l_2^{*'} \langle m' \rangle)$. In both cubes, the left face is pushout and pullback, the top face is pullback, and the front and back faces are FPCs. By Condition C3 and Fact F2, their bottom faces are pullbacks and their right faces are pushouts and pullbacks. Therefore, we obtain monic d, pushout (d, g') of $(d^*, r_1^* \langle m \rangle)$, and $i \langle n \rangle$ with $i \langle n \rangle \circ \overline{r}_1' = r_1^* \langle m \rangle \circ i \langle m \rangle$ and $i \langle n \rangle \circ n' \langle l_2' \rangle = n \langle p_2 \circ p_1 \rangle \circ i_2$. We know $d \circ i \langle n \rangle = i_2^*$, since both morphisms coincide under prefix composition with $\overline{r}_1'$ and $n' \langle l_2' \rangle$. A similar argument as in Proposition 7 shows that $i \langle n \rangle$ is epimorphism and $(i \langle n \rangle, d)$ is epi-mono-factorisation of i_2^*.   □

Theorem 15 (Composition). *If m is a match for $p' \circ p$, then $(p' \circ p)\langle m \rangle = p' \langle m \langle p \rangle \rangle \circ p \langle m \rangle$ and $m \langle p' \circ p \rangle = m \langle p \rangle \langle p' \rangle$.*

Proof. Consequence of Lemmata 13 and 14 and the fact that pushouts compose.

Theorems 12 and 15 demonstrate that the extension of sesqui-pushout rewriting presented in this paper is as well-behaved as the standard approach as far as rule composition and decomposition is concerned. This provides a good fundament for future research wrt. subrules, remainders and amalgamation.

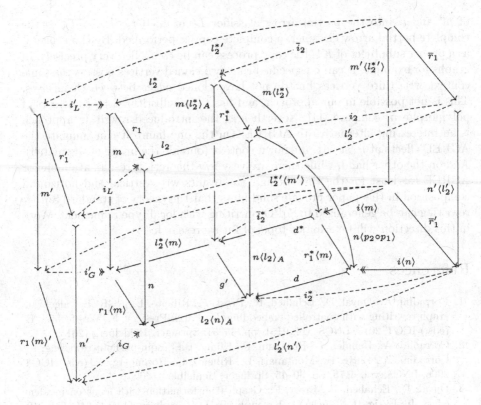

Fig. 14. Derivation composition of total and co-total morphism

6   Related Work and Future Research

There are two major other approaches to controlled sesqui-pushout cloning, namely rewriting on polarised graphs [3] and the AGREE framework [1]. Sesqui-pushout rewriting of polarised graphs allows to specify for every vertex in the middle graph K of a rule $(l : K \to L, r : K \to R)$ if it allows incoming edges only, outgoing edges only, or incoming and outgoing edges. This specifies, whether a copy of a vertex obtains the full context, the incoming context only, or the outgoing context only. Thus, SqPO-rewriting of polarised graphs is a special case of the mechanism presented here. Polarisation of graphs can be represented by passing from $\mathcal{G} \downarrow F$, i.e. the comma category of all graphs under the final graph (node with a singleton loop) to $\mathcal{G} \downarrow E$ where E is the graph with two vertices connected by a singleton edge. E is a refinement of F in our sense.

AGREE controls the cloning by a monomorphism $t : K \rightarrowtail T_K$ from middle object K of a rule $(l : K \to L, r : K \to R)$ typically into a subobject of the partial arrow classifier K^* of K. By interpreting a monic match $m : L \rightarrowtail G$ as a partial arrow $m' = (m, \mathrm{id}_L)$ from the source object G and $l' = (t, l)$ as a partial arrow from T_K into the rules left-hand side, the cloning in AGREE is performed by the pullback of $\overline{m} : G \to L^*$ and $\overline{l} : T_K \to L^*$ which are the totalisation of m' and l' into the partial arrow classifier L^* of L. If $t : K \rightarrowtail T_K$ is the complete partial arrow classifier, a complete copy is performed. By choosing T_K as a proper subobject of K^*, the copy process can be controlled very precisely. In graphs for example, it can be specified that two cloned vertices possess the same context wrt. third vertices but do not have cloned edges between themselves. This is not possible in our approach, and has to be substituted by some sort of polymorphism, compare [11]. Nevertheless, the introduced rewriting approach is an interesting alternative to AGREE. On the one hand, it can simulate the AGREE-effects at least wrt. unknown context (outside the image of the match). And on the other hand, while there are very few theoretical results available for AGREE (at least for the time being), the results wrt. vertical and horizontal composition in this paper provide solid hints that the theory of standard SqPO-rewriting can be generalised to SqPO-rewriting with local type refinement. Work in this direction will be a major topic of future research.

References

1. Corradini, A., Duval, D., Echahed, R., Prost, F., Ribeiro, L.: AGREE – algebraic graph rewriting with controlled embedding. In: Parisi-Presicce, F., Westfechtel, B. (eds.) ICGT 2015. LNCS, vol. 9151, pp. 35–51. Springer, Heidelberg (2015)
2. Corradini, A., Heindel, T., Hermann, F., König, B.: Sesqui-pushout rewriting. In: Corradini, A., Ehrig, H., Montanari, U., Ribeiro, L., Rozenberg, G. (eds.) ICGT 2006. LNCS, vol. 4178, pp. 30–45. Springer, Heidelberg (2006)
3. Duval, D., Echahed, R., Prost, F.: Graph transformation with focus on incident edges. In: Ehrig, H., Engels, G., Kreowski, H.-J., Rozenberg, G. (eds.) ICGT 2012. LNCS, vol. 7562, pp. 156–171. Springer, Heidelberg (2012)
4. Ehrig, H., Ehrig, K., Prange, U., Taentzer, G.: Fundamentals of Algebraic Graph Transformation. Springer, Heidelberg (2006)
5. Ehrig, H., Pfender, M., Schneider, H.J., Graph-grammars: an algebraic approach. In: FOCS, pp. 167–180. IEEE (1973)
6. Kennaway, R.: Graph rewriting in some categories of partial morphisms. In: Ehrig, H., Kreowski, H.-J., Rozenberg, G. (eds.) Graph-Grammars and Their Application to Computer Science. LNCS, vol. 532, pp. 490–504. Springer, Heidelberg (1990)
7. Lack, S., Sobocinski, P.: Adhesive and quasiadhesive categories. ITA **39**(3), 511–545 (2005)
8. Löwe, M.: Algebraic approach to single-pushout graph transformation. Theor. Comput. Sci. **109**(1&2), 181–224 (1993)
9. Löwe, M.: Graph rewriting in span-categories. Technical report 2010/02, FHDW-Hannover (2010)
10. Löwe, M.: A unifying framework for algebraic graph transformation. Technical report 2012/03, FHDW-Hannover (2012)
11. Löwe, M.: Polymorphic sesqui-pushout graph rewriting. Technical report 2014/02, FHDW-Hannover (2014)

Parallelism in AGREE Transformations

Andrea Corradini[1]($\boxtimes$), Dominique Duval[2]($\boxtimes$), Frederic Prost[3]($\boxtimes$),
and Leila Ribeiro[4]($\boxtimes$)

[1] Dipartimento di Informatica, Università di Pisa, Pisa, Italy
andrea@di.unipi.it
[2] LJK - Université Grenoble Alpes and CNRS, Grenoble, France
Dominique.Duval@imag.fr
[3] LIG - Université Grenoble Alpes and CNRS, Grenoble, France
Frederic.Prost@imag.fr
[4] INF - Universidade Federal Do Rio Grande Do Sul, Porto Alegre, Brazil
leila@inf.ufrgs.br

Abstract. The AGREE approach to graph transformation allows to specify rules that clone items of the host graph, controlling in a fine-grained way how to deal with the edges that are incident, but not matched, to the rewritten part of the graph. Here, we investigate in which ways cloning (with controlled embedding) may affect the dependencies between two rules applied to the same graph. We extend to AGREE the classical notion of parallel independence between the matches of two rules to the same graph, identifying sufficient conditions that guarantee that two rules can be applied in any order leading to the same result.

1 Introduction

Graph Transformations (GT) are very much used to specify systems where concurrency and non-determinism are present. For instance GT has been used to model the evolution of biological systems [6], chemical reactions [15] and also concurrent models of computations [8]. From this perspective a major concern is to investigate how the application of different rules may affect each other. There are two classical questions:

1. **(parallel independence)** Given two rules with matches in the same graph G, are they independent? That is, can they be applied in any order (or even in parallel) with the same result?
2. **(sequential independence)** Given a sequence of two rewrite steps, is the second step independent of the first? That is, could the second rule be applied first, followed by the application of the first rule, leading to the same result?

In this paper we shall consider parallel independence only. In the classical setting, where typically rules are injective, two rewrite steps are parallel independent if their matches overlap only on items that are preserved by both. In other words, they *are not* parallel independent if there is a conflict of the following types:

This work has been partly funded by projects CLIMT (ANR/(ANR-11-BS02-016), VeriTeS (CNPq 485048/2012-4 and 309981/2014-0), PEPS égalité (CNRS).

R. Echahed and M. Minas (Eds.): ICGT 2016, LNCS 9761, pp. 37–53, 2016.
DOI: 10.1007/978-3-319-40530-8_3

delete-delete: two rules try to delete the same item. In this case the conflict is symmetric and it means that the rules are mutually exclusive.

preserve-delete: a rule deletes an item that is preserved by the other. In this situation the conflict is asymmetric because the application of the rule that deletes the item prevents the other to occur, but not the other way around.

Parallel independence is usually formalized, in the algebraic approaches to GT, making reference to the following diagram. The rewrite step using rule 2 at match m_2 is said to be (parallel) independent from the rewrite step using rule 1 at match m_1 if there exists a morphism m_{2d} such that $m_2 = g_1 \circ m_{2d}$ (and symmetrically for rule 1). That is, it is still possible to apply rule 2 after rule 1 has been applied, using the "same" match, and in this case the *Local Church-Rosser Theorem* shows that the resulting graph is the same.

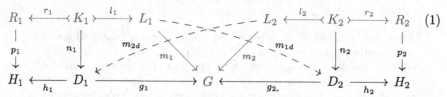

$$\tag{1}$$

Those problems have been studied in many GT approaches: double pushout (DPO) [5], single pushout [12], sesqui-pushout (SqPO) [4], reversible sesqui-pushout [7], with negative application conditions [13], borrowed contexts [1] and nested application conditions [11]. To our knowledge, in all these approaches (but for [7]) rules are required to be *linear*, i.e. both the left- and the right-hand side have to be monomorphisms. In this paper we address the problem of parallel independence for the AGREE approach [3]. The main feature of AGREE rewriting is the ability to clone matched items, like in the SqPO approach that it extends, but with the possibility of specifying how edges incident to the image of the match can be handled. Because of this feature the analysis of parallel independence becomes quite more complex than in the other approaches, since new kinds of conflicts between rewrite steps arise.

The paper is organized as follows. We start with an informal introduction to AGREE in Sect. 2, showing how, from a programmer point of view, AGREE rewrite rules can be specified by exploiting the ability both to clone items, and to control the embedding of the preserved or cloned items in the context. In Sect. 3 we recall from [3] the formal definition of AGREE rewriting. Then we present in Sect. 4, through several canonical counter-examples, how new types of conflicts may arise due to cloning. Those counter-examples will motivate the assumptions needed for the main result that is stated and proved in Sect. 5. Finally we conclude and sketch future developments in Sect. 6.

2   Controlling the Embedding in AGREE

AGREE is a GT approach: states are represented by graphs and transitions are specified by rules. When specifying a transition between states using an AGREE

rule (see the left diagram of (2)), the designer describes with the *left-hand side*
L the items that must be present to trigger the application of the rule. The
morphism l from the *gluing graph* K to L describes which items of L will be
preserved, cloned or deleted. More precisely an item of L is *deleted* if it is not
in the image of l, it is *preserved* if it is the image of exactly one item of K
along l, and it is *cloned* if it is the image of more than one item along l. The
morphism r to the *right-hand side* R, that we assume to be mono in this paper,
defines the items that will be created, i.e. those not in the image of r. Finally
the embedding component T_K, which is typical of AGREE, is used to describe
how the preserved or cloned items are embedded in the rest of the state graph.

To apply a rule to a graph G (see the right diagram of (2)), first an image of L
in G has to be found (a *match*).[1] Then, basically, all items from G are removed,
preserved or cloned according to the rule (using L, K and T_K), and new items
are added according to r. In the following, we explain intuitively how to specify
the embedding component of a rule, and how rule application is performed in
the case of typed graphs. The formal definitions will be given in Sect. 3.

$$
\begin{array}{ccccc}
L \xleftarrow{\ l\ } K \xrightarrow{\ r\ } R & \qquad & L \xleftarrow{\quad l \quad} K \xrightarrowtail{\quad r \quad} R & \qquad (2) \\[2pt]
\ \ \downarrow{\scriptstyle t} & & {\scriptstyle m}\downarrow \quad\text{\scriptsize deletion+cloning}\quad \downarrow{\scriptstyle n} \quad \text{\scriptsize creation} \quad \downarrow{\scriptstyle p} & \\[2pt]
\ \ T_K & & G \xleftarrow{\quad g \quad} D \xrightarrowtail{\quad h \quad} H &
\end{array}
$$

The embedding component describes how to handle the context, i.e. the part
of G that is not in the image of the match. To specify this component, we first
build a graph containing the gluing graph K and all possible ways in which it is
connected to the rest. This is done by a construction called *partial map classifier*
for K, denoted by $T(K)$ [2]. This graph contains the following classes of items:

(i) **preserved items:** K (the gluing graph);
(ii) **independent context items:** a copy of the type graph (to describe the
     part of the state graph that is not touched by the rule application);
(iii) **gluing context edges:** one instance of each type of edge (from the type
     graph) for each pair of nodes of K; and
(iv) **embedding context edges:** one instance of each type of edge (from the type
     graph) for each pair made of a node in class (i) and a node in class (ii).

We call the items in classes (ii) to (iv) $\star$-items. They are used to represent the
context: given any graph X with a map to $T(K)$, we can *classify* X's items
into items that represent K's items (whose images are in (i)) and items that are
context (whose images are in (ii)–(iv)). Now, to obtain the embedding component
of a rule, one can specify how the preserved or cloned items are embedded in the
rest of the graph by removing from $T(K)$ all items that should not be maintained
when the rule is applied, or adding some items to obtain more copies of specific
elements of the state graph.

For example, consider the graphs in Fig. 1. TG is a type graph having two
nodes and two different edges. The items to be preserved/cloned by a rule are

---

[1] In the AGREE approach matches have to be monic.

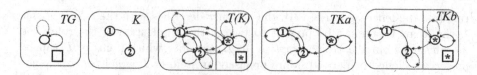

Fig. 1. Embedding Component Examples

shown in graph K (here the left- and right-hand sides are not important since we only want to illustrate the embedding component). To obtain $T(K)$ we follow the steps previously described. As a graphical notation, we use arrows with tips at both sides to depict two edges, one in each direction, all $\star$-items are marked with $\star$ and there is a vertical bar dividing $T(K)$ as follows: we put to the right of the bar the items of (ii) (the copy of the type graph), to the left the items of (i) (graph K) and (iii) (all possible kinds of edges among K's nodes), and connect the left and right sides with items of (iv) (all possible types of edges between items of K and of the copy of the type graph). Choosing for a rule any embedding component TK that does not include all the items of (ii) (or adds some elements to (ii)) would result in a rule with *non-local effects*. For example, if TKa is the embedding component of a rule with gluing graph K, the application of this rule would remove from a graph all square nodes and dashed edges, even if not in the image of the match, because TKa specifies that the context should not have those items. Instead the embedding component TKb has a *local effect* since the part to the right side of the bar is a copy of TG. The application of this rule would remove dashed edges between node ① and the rest of the graph, and solid edges between node ② and the rest of the graph (only one edge between nodes ① and ② would remain, since this edge is in K).

In the rest of this section we consider *local rules* only, that is, the embedding component must include a copy of the type graph.[2] For simplicity we also assume that the embedding component is included in $T(K)$ (even if the formal development does not require it), thus it is obtained from $T(K)$ by deleting only items belonging to (iii) and (iv). For a simpler graphical representation of the embedding component, to the right of the vertical bar we draw only the nodes of the type graph (considering the edges implicitly there), since only the nodes of (ii) are needed to specify how the gluing graph is connected to the context.

To illustrate the AGREE approach and the effect of the embedding component, we will model the generation of Sierpinski triangles. A Sierpinski triangle is a well-known fractal in which an equilateral triangle is divided into smaller equilateral triangles in a controlled way, given by a rule like the one depicted in Fig. 2(a). Applying this rule repeatedly and fairly a convenient number of times leads to shapes like the ones shown in Fig. 2(b).

In [16] the generation of Sierpinski triangles was used as a case study to compare different graph transformation tools. There, a triangle was modeled as a graph with three nodes and three edges, and each step of the generation deleted

---

[2] We refer the interested reader to [3] for a formal definition of locality.

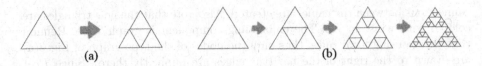

Fig. 2. (a) Sierpinski generation rule (b) Sierpinski triangles generation

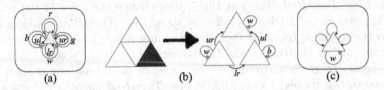

Fig. 3. (a) Type Graph (b) Graph Representation (c) Start Graph

edges and created new nodes and edges. Here, instead, we consider a triangle as a single node, and each step will split (or clone) the triangle into three other ones and create suitable connections (edges) among them. To control how many times the splitting process should occur, we use a special kind of edge: the number of dashed loops on a node indicates how many times the splitting process can be applied. To make the example more interesting, we will color the triangles: a gray loop on a triangle indicates its color (*b* for black, *g* for green, *r* for red and *w* for white). Moreover, there will be three different edges that are used to connect triangles, called *ur* (up-right), *ul* (up-left) and *lr* (left-right). The corresponding type graph is shown in Fig. 3(a). Figure 3(b) depicts a Sierpinski triangle and its corresponding graph representation, and Fig. 3(c) presents a possible start state for the generation of Sierpinki triangles of order 3.

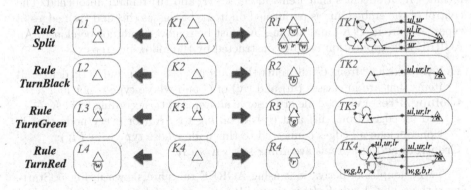

Fig. 4. Rules for generation and coloring of Sierpinski triangles.

To model the splitting of the triangle we use rule **Split** of Fig. 4: whenever there is a triangle that may be split (has a dashed loop) *L1*, it is split and new connections between the copies are created. Also note that the new triangles are colored with white in $R1$. The embedding component is graph $TK1$. Remind that we are using local rules, thus only the nodes of the type graph of Fig. 3(a) are drawn to the right of the bar (the edges are implicitly there). Since $TK1$ does not contain gray loops on the left node, any possible color of the matched node is deleted by the application of the rule. The same happens in all other rules, but for **TurnRed**. The last three rules change the color of the triangle to black, green or red in slightly different ways. Rule **TurnBlack** adds a black color to the matched triangle after removing any colored or dashed loop, thus preventing further splitting. Rule **TurnGreen** changes the color of a triangle (of any color) to green and requires the presence of at least one dashed loop, and preserves all dashed loops. Finally rule **TurnRed** changes the color of a triangle to red, if the triangle was white, and keeps all existing connections (the embedding component of this rule is the partial maps classifier of $K4$).

When a match $m : L \to G$ from an AGREE rule to a graph G is found, this match induces a partition of G's items into the following classes:

(i) **preserved/cloned items:** items in the image of K,
(ii) **independent context items:** the items that are neither in the image of L nor are connections to items in the image of L,
(iii) **gluing context edges:** edges not in the image of L that connect nodes in the image of K,
(iv) **embedding context edges:** edges not in the image of L that connect nodes in the image of K to other nodes in G (not in the image of L),
(v) **deleted items:** items in the image of L and not in the image of K,
(vi) **dangling edges:** edges that connect nodes marked for deletion (in class (v)) to other nodes (not in (v)).

The embedding component TK of a local rule is a subgraph of $T(K)$ that includes K, it specifies that items in classes (i) and (ii) remain untouched. The control of the embedding is performed on items of classes (iii) and (iv): edges of types that are in $T(K)$ and not in TK must be removed/can not be cloned. An AGREE rule application can be constructed by the following steps:

Deletion: delete from G all items that are in classes (v) and (vi); delete all items that are in classes (iii) and (iv) of G and whose type is not in TK;
Cloning/Preserving: clone or preserve all items of (i) according to $l : K \to L$, and all edges from (iii) and (iv) according to TK (for every node that is cloned, clone all edges connected to this node whose type is in TK);
Creation: add new items according to $r : K \rightarrowtail R$.

Two examples of derivations using AGREE rules are shown in Fig. 5. Starting with graph $G1$ rule **Split** is applied (to the right), deleting one of the dashed loops of the triangle and splitting the triangle in three (cloning also the rest of the dashed loops and creating a white-loop in each of the resulting triangles). To the left, rule **TurnBlack** is applied, removing all the dashed arrows from the triangle. These extra deletion effects are specified in the corresponding embedding components (see $TK2$ in Fig. 4).

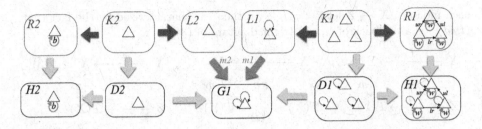

Fig. 5. AGREE rewrite steps using rules *TurnBlack* (left) and *Split* (right)

3   The AGREE approach to graph transformation

In this section we recall basic definitions of the AGREE approach to rewriting [3]. We assume the reader to be familiar with categorical notions used in the algebraic approaches to GT (including pushouts, pullbacks and their properties). The following definition will be useful in the technical development in Sect. 5.

Definition 1 (reflection). *Given arrows $A \rightarrowtail m \rightarrow C \leftarrow f - B$, we say that* (the image of) *A is reflected identically by f* (to B) *if the square below to the right is a pullback for some mono $A \rightarrowtail B$ or equivalently if the pullback object of f and m is isomorphic to A.*

Intuitively, this means that f is an iso when restricted to the image of A. If objects are concrete structures like graphs, then every item of the image of A in C has exactly one inverse image along f in B.

$$\begin{array}{ccc} A & \xrightarrow{\ id\ } & A \\ \downarrow & \lrcorner & \downarrow m \\ B & \xrightarrow{\ f\ } & C \end{array}$$

We assume that the category in which GT is performed has *partial maps classifiers* [2] (needed for the definition of AGREE rewriting [3]) and is *adhesive*, the latter assumption being standard for the results about parallelism [10,14].

Definition 2 (partial map classifier). *Let $\mathbf{C}$ be a category with pullbacks along monos. A partial map over $\mathbf{C}$, denoted $(i, f) : Z \to Y$, is a span $(i : X \rightarrowtail Z, f : X \to Y)$ in $\mathbf{C}$ with i mono, up to the equivalence relation $(i', f') \sim (i, f)$ when there is an isomorphism h with $i' \circ h = i$ and $f' \circ h = f$. Category $\mathbf{C}$ has a partial map classifier (T, η) if T is a functor $T : \mathbf{C} \to \mathbf{C}$ and η is a natural transformation $\eta : Id_{\mathbf{C}} \xrightarrow{\cdot} T$, such that for each object Y and each partial map $(i, f) : Z \to Y$ there is a unique arrow $\varphi(i, f) : Z \to T(Y)$ such that (i, f) is a pullback of $(\varphi(i, f), \eta_Y)$ (see the left diagram of (3)).*[3]

Then η_Y is mono for each Y, T preserves pullbacks, and η is *cartesian*, which means that for each $f : X \to Y$ the span (η_X, f) is a pullback of $(T(f), \eta_Y)$. For

---

[3] Intuitively, a partial map classifier provides a bijective correspondence between partial maps over $\mathbf{C}$ from object Z to Y and arrows of $\mathbf{C}$ from Z to $T(Y)$, given by $[(i, f)] \iff \phi(i, f)$, as described by the left diagram of (3).

each mono $i : X \rightarrowtail Z$ let $\bar{\imath} = \varphi(i, id_X)$; then $\bar{\imath}$ is characterized by the fact that (i, id_X) is a pullback of $(\bar{\imath}, \eta_X)$ (right diagram of (3)).

$$
\begin{array}{ccc}
\begin{array}{ccc}
X & \xrightarrow{f} & Y \\
i\downarrow & PB & \downarrow \eta_Y \\
Z & \xrightarrow{\varphi(i,f)} & T(Y)
\end{array}
&
\begin{array}{ccc}
X & \xrightarrow{f} & Y \\
\eta_X\downarrow & PB & \downarrow \eta_Y \\
T(X) & \xrightarrow{T(f)} & T(Y)
\end{array}
&
\begin{array}{ccc}
X & \xrightarrow{id_X} & X \\
i\downarrow & PB & \downarrow \eta_X \\
Z & \xrightarrow{\bar{\imath}} & T(X)
\end{array}
\end{array}
\qquad (3)
$$

By composing the right and middle squares we get the left one, which proves that for each partial map $(i, f) : Z \rightharpoonup Y$:

$$
\varphi(i, f) = T(f) \circ \bar{\imath} \qquad (4)
$$

For the definition of adhesivity, we stick to the seminal work [14]. Since then adhesivity has been generalized in several variants and sometimes in subtly different ways: for a recollection of such notions the reader is referred to [10].

Definition 3 (adhesive category). *A category* **C** *is adhesive if it has all pullbacks, pushouts along monos, and if each pushout along a mono, like the square to the left below, is a Van Kampen square, i.e. if for any commutative cube as below to the right, where the pushout is the bottom face and the back faces are pullbacks, we have: the top face is a pushout if and only if the front faces are pullbacks.*

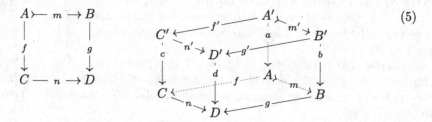

(5)

We recall that in an adhesive category pushouts preserve monos, and pushouts along monos are also pullbacks; pullbacks preserve monos in any category.

Definition 4 (AGREE rewriting). *Let* **C** *be an adhesive category with a partial map classifier* (T, η), *An AGREE rule is a triple of arrows with the same source* $\rho = (K \xrightarrow{l} L, K \rightarrowtail^{r} R, K \rightarrowtail^{t} T_K)$, *with* r *and* t *mono. Arrows* l *and* r *are the* left- *and* right-hand side, *respectively, and* t *is the embedding. A match of rule* ρ *is a mono* $L \rightarrowtail^{m} G$. *An AGREE rewrite step* $G \Rightarrow_{\rho,m} H$ *is constructed as follows (see diagram (6)). First* $G \xleftarrow{g} D \xrightarrow{n'} T_K$ *is the pullback of* $G \xrightarrow{\bar{m}} T(L) \xleftarrow{t'} T_K$. *It follows that there is a unique* $n : K \to D$ *such that* $n' \circ n = t$, $g \circ n = m \circ l$ *and* (l, n) *is a pullback of* (m, g), *and that* n *is mono. Then* $R \rightarrowtail^{p} H \xleftarrow{h} D$ *is the pushout of* $D \xleftarrow{n} K \xrightarrow{r} R$.

$$
\begin{array}{ccccc}
T(L) & \xleftarrow{\ \ l'=\varphi(t,l)\ \ } & T_K & & (6) \\
\eta_L \Big\updownarrow & PB & \Big\updownarrow t & & \\
\overline{m}\ L & \xleftarrow{\ \ l\ \ } & K & \xrightarrow{\ \ r\ \ } & R \\
\Big\downarrow m & PB & n \Big\downarrow \ \ n' & PO & \Big\downarrow p \\
G & \xleftarrow{\ \ g\ \ } & D & \xrightarrow{\ \ h\ \ } & H
\end{array}
$$

The assumptions of Definition 4 are satisfied by the categories of graphs, of *typed graphs* (defined as a slice category), and by toposes in general.

Notice that, differently from [3], we stick to rules with monic right-hand side, thus rules which possibly model the cloning of items, but not their merging. This choice is supported by the observation that matches must be monic in AGREE, and thus even if a monic morphism, say, $m_{1d} : L_1 \rightarrowtail D_2$ can be found (see diagram (1)), its composition with a non-monic $h_2 : D_2 \to H_2$ would not necessarily result in a legal (i.e., monic) match of L_1 in H_2. The analysis of such more complex situations is left as future work.

Finally it is worth recalling that as proved in [3], AGREE rewriting coincides with SqPO rewriting [4] for rules where $T_K = T(K)$.

4   Analysis of Independence of Rewrite Steps

As stated in the Introduction, *parallel independence* is a condition on two rewrite steps from the same graph that ensures that they can be applied sequentially in both orders, leading to the same result. We formalize this last property with the following notion of *commutativity*, also known as *diamond property*.

Definition 5 (Commutativity of rewrite steps). *Let ρ_1 and ρ_2 be two rules and for $i \in \{1,2\}$ let m_i be a match for ρ_i in G. We say that the rewrite steps $G \Rightarrow_{\rho_1,m_1} H_1$ and $G \Rightarrow_{\rho_2,m_2} H_2$ commute if there exist an object H and matches m_{12} of ρ_1 in H_2 and m_{21} of ρ_2 in H_1 such that $H_1 \Rightarrow_{\rho_2,m_{21}} H$ and $H_2 \Rightarrow_{\rho_1,m_{12}} H$.*

We discussed two possible kinds of conflicts that could prevent commutativity in classical approaches to GT: **preserve-delete** (one of the rules deletes some item that is preserved by the other) and **delete-delete** (two rules delete the same item). In AGREE we still have these kinds of conflicts. But we have to investigate what is the impact of using the embedding component TK of the rules, and of allowing the cloning of items.

Let us now consider some examples illustrating different kinds of situations that may occur in AGREE derivations. These examples are meant to show that, although cloning is a kind of preservation, the application of a rule that clones may hinder the application of a rule that uses the cloned items, since the

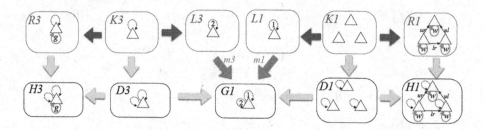

Fig. 6. Rewrite steps with rules *TurnGreen* and *Split*: Clone-Use Confict

match may become non-deterministic (leading to different results) and items that belong to the context of one rule application may be changed in a way that prevents the other rule from being applied. Moreover, even if no cloning is used, rules may get into conflict due to the treatment of context items specified in the TK component of one of the rules.

Cloning vs Use. Consider the derivations shown in Fig. 6, where rules *Turn-Green* (left) and *Split* (right) are applied to graph $G1$ (indices indicate the match). The application of rule *TurnGreen* just changes the color of the triangle and, after applying this rule, it would still be possible to apply *Split* to the *same* match (that is, it is possible to extend $m1$ to $H3$) and the result would be a graph with 3 white triangles (and corresponding edges). However, if *Split* is applied first, we would have three possibilities to match rule *TurnGreen* that would be extensions of $m1$. By choosing any of them, the result would be a graph with 3 triangles, two white and one green, i.e. the results would not be the same.

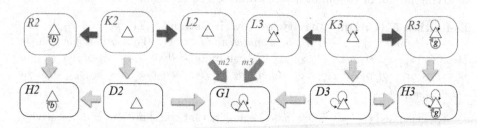

Fig. 7. Applying *TurnBlack* and *TurnGreen*: Context Deletion–Preservation Confict

Context Deletion vs Preservation. Now consider the derivations shown in Fig. 7, where rules *TurnBlack* (left) and *TurnGreen* (right) are applied to graph $G1$. The application of rule *TurnGreen* just changes the color of the triangle and, after applying this rule, it would still be possible to apply *TurnBlack* to the *same* match (that is, it is possible to extend $m2$ to $H3$) and the result would be a graph with only one black triangle. But if *TurnBlack* is applied first, all dashed loops are removed (as specified by $TK2$) preventing *TurnGreen* from being applied. None of these rules clones items, thus this kind of conflict depends only on the embedding component of the rules: if the embedding components were $TK2 = T(K2)$ and $TK3 = T(K3)$, no conflict would arise because all context items would be preserved by both rules.

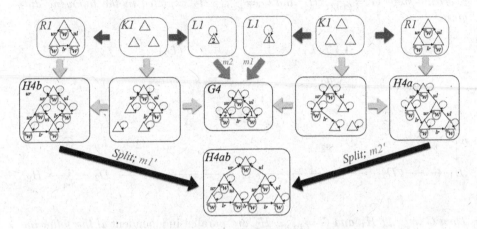

Fig. 8. AGREE Derivations using rule *Split*

Figure 5 shows a situation where both of the above cases occur: rule *TurnBlack* removes the dashed loops, preventing *Split* from being applied, and *Split* clones the triangle, creating three different matches for *TurnBlack*. Finally, Fig. 8 shows an example involving cloning and using a non-trivial embedding component ($TK1 \neq T(K1)$), where the two steps commute. In fact the two matches do not overlap, and thus are trivially parallel independent.

Summarizing, using AGREE rules, we have three new kinds of conflict:

clone-use (where **use** could be **delete** or **preserve** or **clone**): an item that is preserved/deleted/cloned by one rule is cloned by the other.

ctxdel-use (context deletion-use): an item used in one rule is specified for context-deletion by the embedding component TK of the other rule.

ctxclone-use (context clone-use): an item used in one rule is specified for context-cloning by the embedding component TK of the other rule[4].

5   The Church-Rosser Property for AGREE

This section is devoted to the main result of the paper, that is the identification of sufficient conditions for two AGREE rewrite steps to commute, according to Definition 5. Such conditions are identified in the next definition.

---

[4] We did not present examples of this kind of conflict, which can be avoided by requiring the embedding component to be included in $T(K)$.

Definition 6 (Parallel Independence in AGREE). *Let* **C** *be an adhesive category with a partial map classifier* (T, η). *Let* $\rho_i = (K_i \xrightarrow{l_i} L_i, K_i \xrightarrow{r_i} R_i, K_i \xrightarrow{t_i} T_{K,i})$, *for* $i \in \{1, 2\}$, *be two AGREE rules and let* $L_1 \xrightarrow{m_1} G$ *and* $L_2 \xrightarrow{m_2} G$ *be two matches for them to the same object* G. *Consider the corresponding AGREE rewriting steps* $G \Rightarrow_{\rho_1, m_1} H_1$ *and* $G \Rightarrow_{\rho_2, m_2} H_2$ *depicted in the following diagram.*[5]

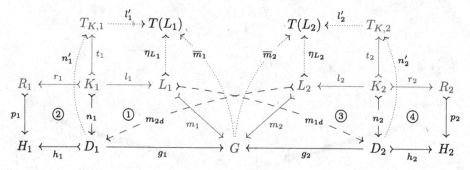

Then $G \Rightarrow_{\rho_1, m_1} H_1$ *and* $G \Rightarrow_{\rho_2, m_2} H_2$ *are parallel independent if the following are satisfied:*

1. *In the left diagram of (7) where the inner and the outer squares are built as pullbacks, the mediating morphism* $K_1 K_2 \to L_1 L_2$ *is an isomorphism.*
2. *The right diagram of (7) is a pullback for* $i \in \{1, 2\}$, *that is the image of* $T(L_1 L_2)$ *is reflected identically by* l'_i *to* $T_{K,i}$.

$$
\begin{array}{ccc}
K_1 K_2 \xrightarrow{\pi_2^K} K_2 & \qquad & T(L_1 L_2) \longrightarrow T_{K,i} \qquad (7) \\
\end{array}
$$

The main result is formulated as follows.

Theorem 1 (Local Church-Rosser). *If two AGREE rewrite steps are parallel independent, then they commute.*

As a first observation note that, unlike most related results for other algebraic approaches to GT, parallel independence does not require explicitly the existence

---

[5] For future reference we also depict the dashed arrows m_{1d} and m_{2d}, which are not mentioned in this definition.

of arrows m_{1d} and m_{2d}, which will be inferred in the proof from the other conditions. Nevertheless, note that the first condition can be seen as a direct translation in categorical terms of the classical *set-theoretical* definition of parallel independence (see [9]) requiring $m_1(L_1) \cap m_2(L_2) \subseteq m_1(l_1(K_1)) \cap m_2(l_2(K_2))$.

Since the conditions of Definition 6 are pretty technical, let us explain them by making reference to the specific case of graphs. The first condition guarantees that each item of G that is needed for the application of both rules (belongs to the intersection of the images of L_1 and L_2) is preserved by both rules (is in the image of both K_1 and K_2) and it is not cloned by any rule (it has only one inverse image in $K_1 K_2$). This forbids all delete-use and clone-use conflicts. Equivalently, if a rule duplicates or deletes an item of G, that item cannot be accessed by the other rule not even in a read-only way. For example, the application of rules **TurnGreen** and **Split** shown in Fig. 6 does not satisfy this condition because the pullback of $L3 \to G1$ and $L1 \to G1$ contains a single node, while the pullback of $K3 \to L3 \to G1$ and $K1 \to L1 \to G1$ contains three nodes, and thus they are not isomorphic.

For the second condition, remember from Sect. 2 that for any graph X, the partial map classifier $T(X)$ is made of a copy of X plus the $\star$-elements which, given any graph Y with a partial morphism to X, classify in a unique way the items of the context, i.e. the items of Y on which the morphism is not defined. Thus the second condition expresses a strong requirement on the embeddings $T_{K,i}$ of the two rules: they cannot modify (i.e. delete or duplicate) any item in the context of $L_1 L_2$. For example, this condition is not satisfied by the application of rules **TurnBlack** and **TurnGreen** to graph $G1$ in Fig. 7. In fact, in this case the pullback object of $L2 \to G1$ and $L3 \to G1$ is a single node (it is identical to $L2$), but $T(L2)$ is not reflected identically by $TK2 \to T(L2)$, because the embedding $TK2$ (see Fig. 4) does not contain the $\star$-loop on the left node.

Proof (of Theorem 1). We present the overview of the proof, which is detailed in the rest of the section. We focus on the application of ρ_2 and ρ_1 in this order, since the reverse order is symmetric. Consider Diagram (8), where for readability reasons we do not depict the embeddings of the rules and the partial maps classifiers, even if they are necessary for the constructions. Objects in plain math style exist by hypotheses, others (in bold) are introduced during the proof.

By Lemma 1, L_1 is reflected identically by $D_2 \text{-}g_2 \twoheadrightarrow G$ providing the mono $L_1 \rightarrowtail m_{1d} \twoheadrightarrow H_2$, which composed with $D_2 \rightarrowtail h_2 \twoheadrightarrow H_2$ becomes a match $L_1 \rightarrowtail m_{12} \twoheadrightarrow H_2$. By Construction 1 the AGREE rewrite step $H_2 \Rightarrow_{\rho_1, m_{12}} H_{12}$ generates objects D_{12} and H_{12} in the bottom line. Symmetrically, the AGREE rewrite step $H_1 \Rightarrow_{\rho_2, m_{21}} H_{21}$ generates the objects D_{21} and H_{21} in the right column. By Lemma 2, defining D as the pullback of square ⑤, K_1 is reflected identically by $D \text{-} d_1 \twoheadrightarrow D_1$ and R_1 is reflected identically by $D_{21} \text{-} g_{21} \twoheadrightarrow H_1$, providing monos $K_1 \rightarrowtail n_{1d} \twoheadrightarrow D$ and $R_1 \rightarrowtail p_{1d} \twoheadrightarrow D_{21}$. Lemma 3 shows that the only arrow $D \rightarrowtail d_{21} \twoheadrightarrow D_{21}$ that makes square ⑥ a pullback also makes the composed square ②+⑥ a pushout. It concludes by building H in ⑧ as the pushout object of $D_{21} \leftarrowtail D \rightarrowtail D_{12}$ (where the arrow $D \rightarrowtail D_{12}$ is built symmetrically to $D \rightarrowtail D_{21}$, making square ⑦ a pullback) and showing, by compositionality of pushouts, that H must be isomorphic to H_{12}. The result follows by symmetry.

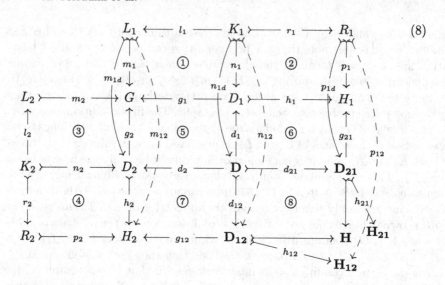

$$(8)$$

Lemma 1. *Consider the left diagram of (9). There is a unique (monic) arrow $m_{1d} : L_1 \to D_2$ making the top and the back-left faces commuting, and the top face a pullback. Thus L_1 is reflected identically by g_2.*

$$(9)$$

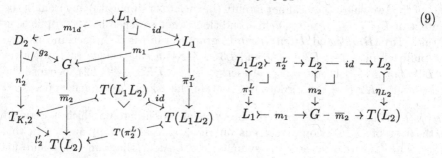

Proof. In the left cube, the front-left face is a pullback by construction of step $G \Rightarrow_{\rho_2, m_2} H_2$, the bottom face is a pullback by hypothesis (see (7)), and the back-right face is trivially a pullback. In addition the front-right face commutes: in fact on one hand we have $T(\pi_2^L) \circ \overline{\pi}_1^L = \varphi(\pi_1^L, \pi_2^L)$ by property (4) of partial maps classifiers, on the other hand the right diagram of (9) proves that $\overline{m}_2 \circ m_1 = \varphi(\pi_1^L, \pi_2^L)$. The statement follows by the decomposition property of pullbacks.

$$(10)$$

$$
\begin{array}{ccc}
T(L_1) \xleftarrow{\;l_1'\;} T_{K,1} & & T(L_2) \xleftarrow{\;l_2'\;} T_{K,2} \\
\eta_{L_1} \quad\uparrow \qquad \uparrow\; t_1 & & \eta_{L_2} \quad\uparrow \qquad \uparrow\; t_2 \\
\overline{m}_{12}\; L_1 \xleftarrow{\;l_1\;} K_1 \xrightarrow{\;r_1\;} R_1 & & \overline{m}_{21}\; L_2 \xleftarrow{\;l_2\;} K_2 \xrightarrow{\;r_2\;} R_2 \\
\downarrow m_{12} \quad n_{12} \downarrow \;\; n_{12}' \;\; \downarrow p_{12} & & \downarrow m_{21} \quad n_{21} \downarrow \;\; n_{21}' \;\; \downarrow p_{21} \\
H_2 \xleftarrow{\;g_{12}\;} D_{12} \xrightarrow{\;h_{12}\;} H_{12} & & H_1 \xleftarrow{\;g_{21}\;} D_{21} \xrightarrow{\;h_{21}\;} H_{21}
\end{array}
$$

Construction 1. *Arrow $m_{1d} : L_1 \to D_2$ of Lemma 1 composed with $h_2 : D_2 \to H_2$ (see (8)) provides a match $m_{12} = h_2 \circ m_{1d} : L_1 \to H_2$: it is mono because both m_{1d} and h_2 are, the latter because pushouts preserve monos in adhesive categories. The left diagram of (10) represents the resulting AGREE rewrite step $H_2 \Rightarrow_{\rho_1, m_{12}} H_{12}$. The right diagram of (10) represents the symmetric rewrite step $H_1 \Rightarrow_{\rho_2, m_{21}} H_{21}$, where $m_{21} = h_1 \circ m_{2d}$.*

The proofs of the next two lemmas are omitted for space constraints, and will appear in the full version of the paper.

Lemma 2. *Let $D_2 \blacktriangleleft d_2 - D - d_1 \blacktriangleright D_1$ be the pullback of $D_2 - g_2 \blacktriangleright G \blacktriangleleft g_1 - D_1$ (see square ⑤ of (8)), and consider the diagrams (11).*

1. *In the left cube, there is a unique (monic) arrow $n_{1d} : K_1 \to D$ making the top and the back-left faces commuting, and the top face a pullback. Thus K_1 is reflected identically by d_1.*
2. *In the right cube, there is a unique (monic) arrow $p_{1d} : R_1 \to D_{21}$ making the top and the back-left faces commuting, and the top face a pullback. Thus R_1 is reflected identically by g_{21}.*

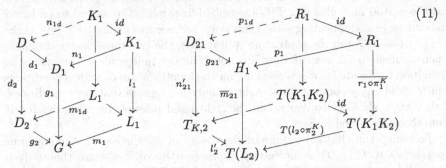

$$(11)$$

Lemma 3. *In the left diagram of (12) there is a unique arrow $d_{21} : D \to D_{21}$ making the top and the back-left faces commuting and the top face a pullback. Symmetrically, we get an arrow $d_{12} : D \to D_{12}$. Furthermore, the top face of the central diagram is a pushout. Now define $D_{21} \blacktriangleright H \blacktriangleleft D_{12}$ as the pushout of $D_{21} \blacktriangleleft d_{21} \cdot D \cdot d_{12} \blacktriangleright D_{12}$ (see square ⑧ of (8)). Then from the right diagram we infer that $H \cong H_{12}$.*

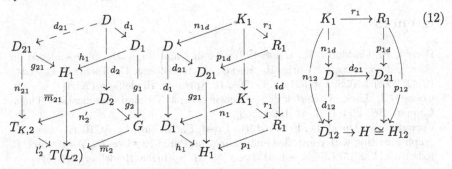

$$(12)$$

6 Conclusion and Related Works

In this paper we proposed sufficient conditions to ensure that two rewrite steps in the AGREE approach to GT commute. Unlike most of previous works on parallel independence [1,4,5,11–13], we consider an approach in which cloning is possible. Actually, general rules are considered also in the restricted version of the SqPO approach proposed in [7]: the exact relationship with those results is under investigation. The possibility of cloning makes the analysis of parallel independence more complex. Moreover, the fact that the embedding of cloned parts can be finely tuned in AGREE adds another layer of complexity: besides of conflicts that may arise from overlapping matches (as for classical approaches), new conflicts may arise from cloning or deletion of edges incident to the matched parts of the transformed graph.

The conditions for commutativity proposed in this paper are sufficient, but not necessary. It is easy to build a counterexample with two rules, that act as the identity transformation on a given graph G (the left- and right-hand sides are all identities on G), but differ in the embedding component in such a way that the second condition of Definiton 6 is not satisfied. For example, the first rule has the partial map classifier $T(G)$ as embedding, while the second has a larger embedding (e.g. duplicating some contextual arc). Since the first rule acts as the identity (both G and the context are preserved), the two rules clearly commute when applied to G, even if they are not "parallel independent" according to Definition 6. We are currently working on the identification of refined conditions which could enjoy completeness. A first analysis suggests that such conditions, if they exist, should also depend on the right-hand sides of the rules, differently from those identified in Sect. 5.

Following the classical outline of the theory of parallelism for the algebraic approaches to GT, other interesting topics worthy of study are the analysis of conditions for *sequential independence* for AGREE rewrite steps, and the definition of *parallel rules* allowing to model the simultaneous application of two rules to a state. Both topics look not obvious: the first one because AGREE rewrite steps are intrinsically non-symmetric (unlike, e.g., DPO or Reversible SqPO rewrite steps); the second because of the need of merging in some way the embedding components of the two rules.

References

1. Bonchi, F., Gadducci, F., Heindel, T.: Parallel and sequential independence for borrowed contexts. In: Ehrig, H., Heckel, R., Rozenberg, G., Taentzer, G. (eds.) ICGT 2008. LNCS, vol. 5214, pp. 226–241. Springer, Heidelberg (2008)
2. Cockett, J., Lack, S.: Restriction categories II: partial map classification. Theor. Comput. Sci. **294**(1–2), 61–102 (2003)
3. Corradini, A., Duval, D., Echahed, R., Prost, F., Ribeiro, L.: AGREE – algebraic graph rewriting with controlled embedding. In: Parisi-Presicce, F., Westfechtel, B. (eds.) ICGT 2015. LNCS, vol. 9151, pp. 35–51. Springer, Heidelberg (2015)

4. Corradini, A., Heindel, T., Hermann, F., König, B.: Sesqui-pushout rewriting. In: Corradini, A., Ehrig, H., Montanari, U., Ribeiro, L., Rozenberg, G. (eds.) ICGT 2006. LNCS, vol. 4178, pp. 30–45. Springer, Heidelberg (2006)
5. Corradini, A., Montanari, U., Rossi, F., Ehrig, H., Heckel, R., Löwe, M.: Algebraic approaches to graph transformation - part I: basic concepts and double pushout approach. In: Handbook of Graph Grammars and Computing by Graph Transformations. Foundations, vol. 1, pp. 163–246. World Scientific, Singapore (1997)
6. Danos, V., Feret, J., Fontana, W., Harmer, R., Hayman, J., Krivine, J., Thompson-Walsh, C.D., Winskel, G.: Graphs, rewriting and pathway reconstruction for rule-based models. In: FSTTCS 2012. LIPIcs, vol. 18, pp. 276–288. Schloss Dagstuhl - Leibniz-Zentrum fuer Informatik (2012)
7. Danos, V., Heindel, T., Honorato-Zimmer, R., Stucki, S.: Reversible sesqui-pushout rewriting. In: Giese, H., König, B. (eds.) ICGT 2014. LNCS, vol. 8571, pp. 161–176. Springer, Heidelberg (2014)
8. Ehrig, H., Kreowski, H.J., Montanari, U., Rozenberg, G. (eds.): Handbook of Graph Grammars and Computing by Graph Transformations. Concurrency, Parallelism and Distribution, vol. 3. World Scientific, Singapore (1999)
9. Ehrig, H.: Introduction to the algebraic theory of graph grammars (a survey). In: Claus, V., Ehrig, H., Rozenberg, G. (eds.) Graph-Grammars and Their Application to Computer Science and Biology. LNCS, vol. 73, pp. 1–69. Springer, Heidelberg (1979)
10. Ehrig, H., Golas, U., Hermann, F.: Categorical frameworks for graph transformation and HLR systems based on the DPO approach. Bulletin of the EATCS **102**, 111–121 (2010)
11. Ehrig, H., Habel, A., Lambers, L.: Parallelism and concurrency theorems for rules with nested application conditions. Electronic Communications of the EASST **26**, 1–23 (2010)
12. Ehrig, H., Löwe, M.: Parallel and distributed derivations in the single-pushout approach. Theor. Comput. Sci. **109**(1&2), 123–143 (1993)
13. Hermann, F., Corradini, A., Ehrig, H.: Analysis of permutation equivalence M-adhesive transformation systems with negative application conditions. Math. Struct. Comput. Sci. **24**(4), 1–47 (2014)
14. Lack, S., Sobociński, P.: Adhesive categories. In: Walukiewicz, I. (ed.) FOSSACS 2004. LNCS, vol. 2987, pp. 273–288. Springer, Heidelberg (2004)
15. Rosselló, F., Valiente, G.: Chemical graphs, chemical reaction graphs, and chemical graph transformation. Electr. Notes Theor. Comput. Sci. **127**(1), 157–166 (2005)
16. Taentzer, G., et al.: Generation of Sierpinski triangles: a case study for graph transformation tools. In: Schürr, A., Nagl, M., Zündorf, A. (eds.) AGTIVE 2007. LNCS, vol. 5088, pp. 514–539. Springer, Heidelberg (2008)

Model Checking Reconfigurable
Petri Nets with Maude

Julia Padberg$^{(\boxtimes)}$ and Alexander Schulz

Hamburg University of Applied Sciences, Hamburg, Germany
julia.padberg@haw-hamburg.de

Abstract. Model checking is a widely used technique to prove properties such as liveness, deadlock or safety for a given model. Here we introduce model checking of reconfigurable Petri nets. These are Petri nets with a set of rules for changing the net dynamically. We obtain model checking by converting reconfigurable Petri nets to specific Maude modules and using then the LTLR model checker of Maude. The main result of this paper is the correctness of this conversion. We show that the corresponding labelled transitions systems are bisimular. In an ongoing example reconfigurable Petri nets are used to model and to verify partial dynamic reconfiguration of field programmable gate arrays.

Keywords: Reconfigurable Petri nets · Rewrite logic · Maude · Model checking · Field programmable gate array · Dynamic partial reconfiguration

1 Introduction

Reconfigurable Petri nets – a family of formal modelling techniques – provide a powerful and intuitive formalism to model complex coordination and structural adaptation at run-time (e.g. mobile ad-hoc networks, communication spaces, ubiquitous computing). Their characteristic feature is the possibility to discriminate between different levels of change.

Model checking of reconfigurable Petri nets can be achieved by converting reconfigurable Petri nets into Maude specifications. Our main purpose is to guarantee that the reconfigurable Petri net and its conversion to a Maude specification are similar enough for the verification process to obtain valid results. The main theoretical contribution ensures the correctness of the conversion in terms of a bisimulation between the state space of a reconfigurable Petri net and the state space of corresponding Maude modules, see also [20]. Here, we give the main ideas and a sketch of the involved proofs. In general, a bisimulation relates state transition systems, that behave in the same way in the sense that one system simulates the other and vice versa. This is achieved by a bisimulation between the reachability graph for the reconfigurable Petri net and the search tree of the corresponding Maude modules. We have defined functions that convert syntactically all parts of the net such as places, transitions, the arcs or markings as well

© Springer International Publishing Switzerland 2016
R. Echahed and M. Minas (Eds.): ICGT 2016, LNCS 9761, pp. 54–70, 2016.
DOI: 10.1007/978-3-319-40530-8_4

as the set of net rules into the Maude modules. These functions are the basis for defining the relation between the corresponding labelled transition systems.

There are many proposals to use Petri nets to model the control logic for field programmable gate arrays (FPGAs). But they lack a formal foundation for the net's reconfiguration which models the partial dynamic reconfiguration of the FPGA (see Sect. 6). This gap can be closed using Petri nets to model FPGAs and using net transformations to model the dynamic reconfiguration of FPGAs. So, the paper's motivation and its ongoing example is to model dynamic partial reconfiguration of FPGA using reconfigurable Petri nets.

The paper is organized as follows: The next section deals with reconfigurable Petri nets and Sect. 3 illustrates our example. Section 4 introduces the rewriting logic Maude. Model checking reconfigurable Petri nets with Maude is given in Sect. 4. Correctness of the model checking approach is shown in Sect. 5. Finally we discuss related work in Sect. 6 and give some ideas concerning future work.

2   Reconfigurable Petri Nets

We use the algebraic approach to Petri nets, where the pre- and post-domain functions $pre, post : T \rightarrow P^{\oplus}$ map the transitions T to a multiset of places $P^{\oplus}$ given by the set of all linear sums over the set P. A marking is given by $m =\in P^{\oplus}$ with $m = \sum_{p \in P} k_p \cdot p$. The multiplicity of a single place p is given by $(\sum_{p \in P} k_p \cdot p)_{|p} = k_p$. The $\leq$ operator can be extended to linear sums: For $m_1, m_2 \in P^{\oplus}$ with $m_1 = \sum_{p \in P} k_p \cdot p$ and $m_2 = \sum_{p \in P} l_p \cdot p$ we have $m_1 \leq m_2$ if and only if $k_p \leq l_p$ for all $p \in P$. The operations "+" and "−" can be extended accordingly.

Definition 1 (Algebraic Approach to Petri nets). *A (marked) Petri net is given by $N = (P, T, pre, post, cap, p_{name}, t_{name}, m)$ where P is a set of places, T is a set of transitions. $pre : T \rightarrow P^{\oplus}$ maps a transition to its pre-domain and $post : T \rightarrow P^{\oplus}$ maps it to its post-domain. Moreover $cap : P \rightarrow \mathbb{N}_+^{\omega}$ assigns to each place a capacity (either a natural number or infinity ω), $p_{name} : P \rightarrow A_P$ is a label function mapping places to a name space, $t_{name} : T \rightarrow A_T$ is a label function mapping transitions to a name space and $m \in P^{\oplus}$ is the marking denoted by a multiset of places.*

A transition $t \in T$ is m-enabled for a marking $m \in P^{\oplus}$ if we have $pre(t) \leq m$ and $\forall p \in P : (m + post(t))_{|p} \leq cap(p)$. The follower marking m' – computed by $m' = m - pre(t) + post(t)$ – is the result of a firing step $m[t\rangle m'$.

If the marking is the focus we denote a net with its marking by (N, m).

Net morphisms are given as a pair of mappings for the places and the transitions preserving the structure, the decoration and the marking. Given two Petri nets $N_i = (P_i, T_i, pre_i, post_i, cap_i, p_{iname}, t_{iname}, m_i)$ for $i \in \{1, 2\}$ a net morphism $f : N_1 \rightarrow N_2$ is given by $f = (f_P : P_1 \rightarrow P_2, f_T : T_1 \rightarrow T_2)$, so that $pre_2 \circ f_T = f_P^{\oplus} \circ pre_1$ and $post_2 \circ f_T = f_P^{\oplus} \circ post_1$ and $m_1(p) \leq m_2(f_P(p))$ for all $p \in P_1$. Moreover, the morphism f is called strict if both f_P and f_T are injective and $m_1(p) = m_2(f_P(p))$ holds for all $p \in P_1$. A rule in the algebraic transformation approach is given by three nets called left-hand side L, interface K and right-hand side R, respectively, and a span of two strict net morphisms

$K \to L$ and $K \to R$. Then an occurrence morphism $o : L \to N$ is required that identifies the relevant parts of the left hand side in the given net N.

A transformation step $N \overset{(r,o)}{\Longrightarrow} M$ via rule r (see the commutative squares (1) and (2) in Fig. 1) can be constructed in two steps. Given a rule with an occurrence $o : L \to N$ the gluing condition has to be satisfied in order to apply a rule at a given occurrence. Its satisfaction requires that the deletion of a place implies the deletion of the adjacent transitions, and that the deleted place's marking does not contain more tokens than the corresponding place in L.

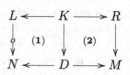

Fig. 1. Net transformation

Reconfigurable Petri nets exhibit dynamic behaviour using the token game of Petri nets and using net transformations by applying rules. So, a reconfigurable Petri net as in Definition 2 combines a net with a set of rules that modify the net [10,11].

Definition 2 (Reconfigurable Petri nets). *A reconfigurable Petri net $RN = (N, \mathcal{R})$ is given by a Petri net N and a set of rules $\mathcal{R}$.*

The labelled transition system for a reconfigurable Petri net LTS_{RPN} is based on the isomorphism classes of nets, where all reachable states are considered up to isomorphisms of marked nets. Isomorphisms are given by bijective mappings of places and transitions. The corresponding isomorphism classes are compatible with firing and transformation steps.

Definition 3 (Labelled Transition System for Reconfigurable Petri nets). *Given a reconfigurable Petri net $(N_0, \mathcal{R})$ the labelled transition system $LTS_{RPN} = (S_{RPN}, A_{RPN}, tr_{RPN})$ is based on the isomorphism classes of nets:*

1. ***Initial states:*** *$[(N_0, m_0)] \in S_{RPN}$*
   where m_0 is the marking of N_0 and $[(N_0, m_0)] = \{(N, m) \mid (N, m) \cong (N_0, m_0)\}$ is the isomorphism class containing (N_0, m_0)
2. ***Firing steps:*** *For $m[t\rangle m'$ in N with $(N, m) \in [(\overline{N}, \overline{m})] \in S_{RPN}$ we have:*
   $[(N, m')] \in S_{RPN}$, $t_{name}(t) \in A_{RPN}$ and $[(N, m)] \xrightarrow{t_{name}(t)} [(N, m')] \in tr_{RPN}$
3. ***Transformation steps:*** *For $(N, m) \overset{(r,o)}{\Longrightarrow} (N', m')$ with some rule $r = (r_{name}; L \leftarrow K \to R) \in \mathcal{R}$ and some occurrence $o : L \to N$ with $(N, m) \in [(\overline{N}, \overline{m})] \in S_{RPN}$ we have:*
   $[(N', m')] \in S_{RPN}$, $r_{name} \in A_{RPN}$ and $[(N, m)] \xrightarrow{r_{name}} [(N', m')] \in tr_{RPN}$
4. ***Finally:*** *$S_{RPN}, A_{RPN}, tr_{RPN}$ are the smallest sets satisfying the above conditions.*

3   Modelling Partial Dynamic Reconfiguration of FPGAs

In this section we give a small example for the use of reconfigurable Petri nets for modelling the control logic for FPGAs. The example is an extension of the example in [6] where Petri nets also have been used for the modelling, but the reconfiguration has been modelled merely informally. In Fig. 2 an industrial mixer of two components and water is illustrated. The water is heated before being fed into the mixer. The sensors x1, ..., x8 measure the corresponding fill level, x9 measures the water temperature and the actuators y1, ..., y9 control the valves, the heater and the mixer. The control logic has to determine in which order the components are fed into the mixer. Reconfiguration changes the control logic accordingly. In Fig. 3 the Petri net N_{mixer} describes the

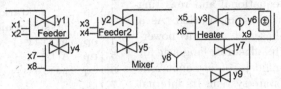

Fig. 2. Construction of the mixer

control logic, where both components and the hot water are added at the same time. This control logic can be translated into the binary coding of an FPGA (see e.g. [6]).

Dynamic partial reconfiguration of FPGAs allows changing the control logic, so that a different order for the components and the water can be employed. In Fig. 4 we present the rule $rule_1 : L \implies R$ where the interface K is indicated by node s in L and R with the same colour. It reconfigures the net by replacing the subnet that models the simultaneous feeding of both components and the water, by a subnet that models the sequential feeding, namely first the heated water, then both components. A second rule (omitted here) reconfigures the net by replacing the subnet that models the simultaneous feeding of both components and the water by a subnet that models the sequential feeding, namely first component 1 and the water, and at last component 2. There

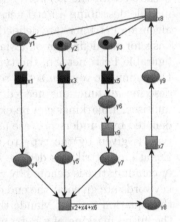

Fig. 3. Control logic N_{mixer}

are more rules as well as their inverse rules modelling the dynamic partial reconfiguration of the FPGA that implement the control logic of the mixer.

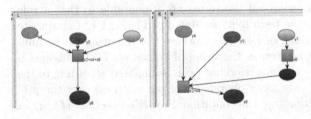

Fig. 4. $rule_1$: First water, then both components

The interaction of the system's control logic with the dynamic reconfiguration becomes much more complicated. Hence verification is of high importance. In this paper, we propose model checking of reconfigurable Petri

nets by translating the net and its rules to a Maude specification that subsequently can be used for model checking. In this example the absence of deadlocks, a property like "the initial state can be reached again" or a property like "the mixer never starts unless it is filled" are sensible requirements. In Sect. 4 we verify the absence of deadlocks using Maude.

4   Model Checking Reconfigurable Petri Nets with Maude

Maude is a high-level language supporting both equational and rewriting logic computation. As a base, it uses a powerful algebraic language for models of a concurrent state system. Its internal representation is given in [15] as a labelled rewrite theory. Implementations

```
1  sorts Places Transitions Markings.
2
3  op _ _ : Marking Marking → Marking
         [assoc comm] .
4  op initial : → Markings .
5  ops A B : → Markings .
6  eq initial = A .
7
8  rl [T] : A ⇒ B B.
```

Listing 1.1. Maude example

in Maude are based on one or many modules. Each module has types that are declared with the keyword "sorts". Subsequently we introduce Maude using a module describing a Petri net. Definitions of P/T nets, coloured Petri nets, and algebraic Petri nets are defined in [22] in a manner that makes Maude a suitable basis for the definition of a Maude net that models the net and rules of a reconfigurable Petri net. So, the types for a Petri net are given line 1 of Listing 1.1. Depending on a given set of sorts, the operators can be defined. The operators describe all functions needed to work with the defined types. For example, a multiset of markings can be expressed with a *whitespace*-functor. Place-holders, denoted by a underscore, are used for the types behind the colon, and the return type is given by the type to the right of the arrow, see line 3 of Listing 1.1. Equational attributes declare structural axioms. An operator being associative or commutative is denoted by with keywords such as "assoc" and "comm". These keywords are given at the end of line 3 for the multiset of places. The axioms are the equation logic of Maude that defines the operator's behavior. For example, the initial marking of a Petri net can be exemplified with the `initial` operator. The operators in line 3–5 describe the markings of the net in Fig. 5 and the equation in line 6 states the initial marking. The rewrite rule in line 8 describes the firing of transition T of the Petri net of Fig. 5. The rewrite rules replaces one multiset by another one, namely the pre-domain of T with its post-domain. As usual in a functional language, all the terms are immutable so that a rule can replace the term A with the term B B, see line 8 of Listing 1.1. The rewrite rule implements the token game of Petri nets, where the rewriting of the multiset A by B B in rule T can be seen as the firing of transition T. This is just a basic example. The firing in our conversion has been formulated according to the algebraic definition of Petri nets using operations of the multisets over the pre- and post-domain of a transition (see the conditional rewrite rule `crl [fire]` in Listing 1.3).

Maude's linear temporal logic for rewrite (LTLR) module can be used to test defined modules with LTL properties, such as deadlocks [3,12,13]. A first step to model checking of reconfigurable Petri nets using Maude has been given

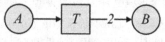

Fig. 5. Petri net

in [17]. The conversion of a net and a set of rules into a Maude modules used for (LTL) model checking with the module LTLR can be found in detail in [19]. The LTLR model-checking module contains all the usual operators, such as *true, false, conjunction, disjunction* and *negation*, and complex operators with the next-operator being written with $O \phi$ or the until-operator notated with $\psi U \phi$. Further, it supports release-operator statements, such as $\psi R \phi$ that are internally converted into $\neg(\neg\phi U \neg\psi)$. The future-operator written as $\Diamond \phi$ states that ϕ is possible in the future, and the global-operator written as $\Box \phi$ claims that ϕ is true in all states. The correctness of the LTLR model checker has been proven in [2]. In this section we sketch this approach to model checking reconfigurable Petri nets. The *ReConNet Model Checker* (rMC) (see [19]) defines Maude modules Net and Rules for a given reconfigurable Petri net. The modules contain the net and a set of rules as well as all mechanisms to fire a transition or to transform the net with a rule [19,20]. Together with the Maude modules RPN defining the firing behaviour, the Maude module PROP stating the properties to be verified and the Maude model checker LTLR-MODEL-CHECKER, these modules yield a rewrite theory that allows the verification of the linear temporal logic formulas over the properties implemented in module PROP. Listing 1.2 shows the net in Fig. 3 converted into the Maude module NET. Each net is modelled by the multisets of places and transitions. A place is defined as p(<label>|<identifier>|<capacity>). Transitions only consist of t(<label>|<identifier>). The pre- and post-domains are wrapped to a set by the pre- or post-operator. Finally, the initial marking is modelled as the multiset in Listing 1.2, line 13. The conditional rewriting rule for firing crl [fire] in Listing 1.3, line 1 uses the transition's pre-domain to determine if a transition is enabled and considers the capacity of each place in its post-domain.

```
1   mod NET is
2     including PROP .
3     including MODEL-CHECKER .
4
5     ops initial : -> Configuration .
6
7     eq initial =
8     net(
9       places{ p("y9" | 1517 | w) , p("y5" | 1518 | w) , p("
            y9" | 1519 | w) , p("y7" | 1520 | w) , p("y1" |
            1513 | w) , p("y8" | 1514 | w) , p("y3" | 1515 | w
            ) , p("y4" | 1516 | w) , p("y2" | 1512 | w) } ,
10      transitions{ t("x2+x4+x6" | 1599) : t("x6+x9" | 1526)
            : t("x1" | 1525) : t("x7" | 1527) : t("x8" |
            1522) : t("x5" | 1521) : t("x3" | 1524) } ,
11      pre{ (t("x2+x4+x6" | 1599) --> p("y5" | 1518 | w) , p
            ("y4" | 1516 | w) , p("y7" | 1520 | w)) , (t("x6+
            x9" | 1526) --> p("y9" | 1519 | w)) , (t("x1" |
            1525) --> p("y1" | 1513 | w)) , (t("x7" | 1527)
```

```
             —> p("y8" | 1514 | w)) , (t("x8" | 1522) —> p("
             y9" | 1517 | w)) , (t("x5" | 1521) —> p("y3" |
             1515 | w)) , (t("x3" | 1524) —> p("y2" | 1512 | w
             )) } ,
12     post{ (t("x2+x4+x6" | 1599) —> p("y8" | 1514 | w)) ,
             (t("x6+x9" | 1526) —> p("y7" | 1520 | w)) , (t("
             x1" | 1525) —> p("y4" | 1516 | w)) , (t("x7" |
             1527) —> p("y9" | 1517 | w)) , (t("x8" | 1522)
             —> p("y3" | 1515 | w) , p("y2" | 1512 | w) , p("
             y1" | 1513 | w)) , (t("x5" | 1521) —> p("y9" |
             1519 | w)) , (t("x3" | 1524) —> p("y5" | 1518 | w
             )) } ,
13     marking{ p("y1" | 1513 | w) ; p("y3" | 1515 | w) ; p
             ("y2" | 1512 | w) }
14     )
15         [...]
16 endm
```

Listing 1.2. Maude conversion of N_{mixer} from Fig. 3

```
1  crl [fire] :
2      net(P,
3          transitions{T : TRest},
4          pre{(T —> PreValue), MTupleRest1},
5          post{(T —> PostValue), MTupleRest2},
6          marking{PreValue ; M})
7              [...]
8      =>
9      net(P,
10         transitions{T : TRest},
11         pre{(T —> PreValue), MTupleRest1},
12         post{(T —> PostValue), MTupleRest2},
13         calc(((PreValue ; M) minus PreValue)
14                        plus PostValue))
15     [...]
16     if calc((PreValue ; M) plus PostValue) <=? PostValue
```

Listing 1.3. Firing as a rewrite rule in Maude module RPN

Listing 1.3 shows the firing of a transition, where each pre-domain condition is implemented in line 6. The subterm `PreValue; M` ensures that at least the pre-domain of the transition is part of the current marking. The if condition in line 16 in Listing 1.2 ensures the capacities using an operator `<=?` (called smallerAsCap) defined in the Maude module RPN.

```
1    crl [rule1 -PNML] :
2       net(
3       places{ p("y5" | Irule1044 | w) , p("y8" | Irule101298 | w
              ) , p("y4" | Irule1047 | w) , p("y7" | Irule1041 | w)
              , PRest } ,
4       transitions{ t("x2+x4+x6" | Irule1060) : TRest } ,
5       pre{ (t("x2+x4+x6" | Irule1060) --> p("y7" | Irule1041 | w
              ) , p("y4" | Irule1047 | w) , p("y5" | Irule1044 | w))
              , MTupleRest1 } ,
6       post{ (t("x2+x4+x6" | Irule1060) --> p("y8" | Irule101298
              | w)) , MTupleRest2 } ,
7       marking{ emptyMarking ; MRest }
8       )
9               [...]
10      =>
11      net(
12      places{ p("" | Aid1 | w) , p("y4" | Irule1047 | w) , p("y5
              " | Irule1044 | w) , p("y8" | Irule101298 | w) , p("y7
              " | Irule1041 | w) , PRest } ,
13      transitions{ t("x2+x4" | Aid2 ) : t("x6" | Aid3 ) : TRest
              } ,
14      pre{ (t("x2+x4" | Aid2 ) --> p("y4" | Irule1047 | w) , p("
              y5" | Irule1044 | w) , p("" | Aid1 | w)) , (t("x6" |
              Aid3 ) --> p("y7" | Irule1041 | w)) , MTupleRest1 } ,
15      post{ (t("x2+x4" | Aid2 ) --> p("y8" | Irule101298 | w)) ,
              (t("x6" | Aid3 ) --> p("" | Aid1 | w)) , MTupleRest2
              } ,
16      marking{ emptyMarking ; MRest }
17      )
18              [...]
19
20      if *** calculate new identifiers
21         AidRestNew := calculateAllIdentifiers /\
22         *** ∀p ∈ P_L which are deleted; prove if they
23         ***        are part of MRest (identity condition)
24         freeOfMarking( ( p(<label> | <identifier> |
25                            <capacity>) ) | MRest ) /\
26         *** ∀p ∈ P_L which are deleted; prove if there
27         *** is a related transition (dangling condition)
28         emptyNeighbourForPlace( p(<label> | <identifier>
29                                    | <capacity>) ,
30                       pre{ MTupleRest1 } ,
31                       post{ MTupleRest2 } ) /\
32         *** set new maximal identifier counter
33         NewMaxID := correctMaxID(MaxID | StepSize |
34                            |AidRestNew|) .
```

Listing 1.4. Rule $rule_1$ in Maude module RULES

The implementation of net rules as given in Fact 2 is illustrated in Listing 1.4. The rule application coincides with the pattern-matching algorithm of Maude, which ensures that the left-hand side is a subset of the current net state. If the conditions are successfully proven, the term describing the net is rewritten. Fact 1 states that the conditions `freeOfMarking` and `emptyNeighbourForPlace` ensure the satisfaction of the gluing condition (see page 3) in the current net. For more details see [20].

Fact 1 (Gluing Condition for Rewrite Rules). *Given a rule application with $r = (r_{name}, L \leftarrow K \rightarrow R)$ in a reconfigurable Petri net that satisfies the gluing condition. Then we have in the module RULES:*

- *`emptyNeighbourForPlace` ensures that p is not used in `pre{MTupleRest1}` or `post{MTupleRest2}`. So, a place p may be deleted only if there are no adjacent transitions that are not deleted.*
- *`freeOfMarking` ensures that for each deleted place $p \not\subseteq MRest$ holds.*

Fact 2 (Transformation Step in the Maude Module RULES). *For each transformation step $m \overset{(r,o)}{\Longrightarrow} m'$ with $r = (r_{name}, L \leftarrow K \rightarrow R)$ in a reconfigurable Petri net, there exists a rewriting rule in RULES so that*

- *there is a pattern match of the left-hand side ensuring that the left-hand side is a subset of the current net state*
- *the match satisfies the gluing condition of Fact 1.*

The implementation of this conversion is given by the ReConNet Model Checker (rMC) [19]. rMC is a Java-based tool that enables a user to convert a given reconfigurable Petri net[1] to the Maude modules introduced above. These Maude modules can be executed and analysed by the Maude interpreter. For the example in Sect. 3 the absence of deadlocks is expressed in Maude's notation using the LTL operators "finally" <> and "globally" []. The property `enabled` given in the Maude module `PROP` and states that at least the preconditions of one of the transitions or of one of the rules are met. So, the formula `[]<> enabled` asserts that the property enabled is globally finally true. The state `initial` corresponds to the reconfigurable Petri net $(N, \mathcal{R})$. Hence, `modelCheck(initial, []<> enabled)` is the operation that checks the formula `[]<> enabled` for the state `initial`. So, in Listing 1.5 we check that the reconfigurable Petri net $(N_{mixer}, \mathcal{R})$ is always enabled, i.e. there are no deadlocks.

A labelled transition system for the Maude module `NET` is defined by $LTS_{MNC} = (S_{MNC}, A_{MNC}, tr_{MNC})$, where S_{MNC} is a non-empty set that contains all states of a Maude breadth-first search tree. Maude's deduction rules are used to execute all rewrite rules, such as firing or transformation steps in the `RULES` module. A state $s \in S_{MNC}$ consists of a `Net` term as current state. A_{MNC} is defined by

---

[1] ReConNet (see [16]) is the tool for modelling and simulating reconfigurable Petri nets saving them as an extension of PNML.

$A_{MNC} = A_T \bigcup A_R$ and contains the labels of rewrite rules, such as the firing or the transformation. tr_{MNC} is defined as a set of transition relations that is based on $tr_{MNC} \subseteq S_{MNC} \times A_{MNC} \times S_{MNC}$. Therefore, two terms of S_{MNC} are related by a transition labelled with the name of the corresponding rewrite rule in A_{MNC}.

```
1        Maude 2.7 built: Aug  6 2014 22:54:44
2        Copyright 1997-2014 SRI International
3           Mon Sep 28 19:13:44 2015
4  ===================================================
5  rewrite in NET : modelCheck(initial , []<> enabled) .
6  rewrites: 17601 in 25ms cpu (48ms real) (704040 rewrites/
      second)
7  result Bool: true
```

Listing 1.5. Absence of deadlocks for $(N_{mixer}, \mathcal{R})$; proven by Maude

Definition 4 (Labelled Transition System for the Maude Module NET). *Given the Maude module NET, a labelled transition system $LTS_{MNC} = (S_{MNC}, A_{MNC}, tr_{MNC})$ is defined with respect to the term sets over the equation conditions of the Maude modules by:*

1. **Initial:** *initial $\in S_{MNC}$*
2. **Firing steps:** *If $s \in S_{MNC}$ and $s \to s'$ is a replacement for a rewrite rule [fire] of Listing 1.3 so that*

$$s = net(P,$$
$$transitions\{t(\overline{label}|identifier) : TRest\},$$
$$pre\{t(\overline{label}|identifier) --> PreValue, MTupleRest1\},$$
$$post\{t(\overline{label}|identifier) --> PostValue, MTupleRest2\},$$
$$marking\{PreValue; M\})$$

is used as left-hand side of Listing 1.3, then $s' \in S_{MNC}$, $t(\overline{label}) \in A_{MNC}$ and $s \xrightarrow{t(\overline{label})} s' \in tr_{MNC}$

3. **Transformation steps:** *If $s \in S_{MNC}$ and $s \to s'$ is a replacement for a rewrite rule [r_{name}] in the Maude module RULE and*

$$s = net(places \{ P_L, PRest\},$$
$$transitions\{T_L : TRest\},$$
$$pre\{Pre_L, MTupleRest1\},$$
$$post\{Post_L, MTupleRest2\},$$
$$marking\{M_L; M\})$$
$$rule(l(P_L, T_L, Pre_L, Post_L, M_L) , r(R))$$

then is:
$s' \in S_{MNC}$, $r \in A_{RPN}$ and $s \xrightarrow{r_{name}} s' \in tr_{MNC}$

4. **Finally:**
   $S_{MNC}, A_{MNC}, tr_{MNC}$ are the smallest sets satisfying the above conditions.

5   Correctness of Model Checking Approach

Model checking reconfigurable Petri nets using Maude is shown to be correct by proving a bisimulation between the corresponding labelled transition systems (LTS). Theorem 1 states a conversion of a reconfigurable Petri net $(N, \mathcal{R})$ into the corresponding Maude module NET. Then the LTS are calculated for both and in Theorem 2 these LTS are shown to be bisimular. The LTS_{RPN} is the reachability graph (up to isomorphism) of the reconfigurable Petri nets and is given by all reachable states using both firing and transformation steps (see Definition 3). The states are the isomorphism classes of marked Petri nets. LTS_{MNC} of a Maude module NET (see Definition 4) includes all rewriting rule applications of firing and transforming steps. Theorem 1 specifies the syntactical conversion for a given reconfigurable Petri net to the Maude modules. For the theorem itself, the following injective functions are used to convert all parts of a reconfigurable Petri net into a NET- and a RULES-module: *buildPlace* defines the conversion for places (see Lemma 1). The functions *buildTransition* (defining the conversion for transitions similar to *buildPlace*), *buildPre* (defining the conversion for each $pre(t)$ with $t \in T^{\oplus}$), *buildPost* (defining the conversion for each $post(t)$ with $t \in T^{\oplus}$ similar to *buildPre*), *buildNet* (defining the conversion of a net) and *buildRule* (defining the conversion of rules in $\mathcal{R}$) are constructed inductively as well, and can be found in [20]. Lemma 1 contains functions for the mapping of identifiers and capacities leading to conversion of places and transitions. The identifiers are defined as unique keys for nodes such as places or transitions and are used by the pre- and post-domain operations. The conversion of places is defined in Lemma 1. Each new place p' is converted to the Maude term p(<label>|<identifier>|<capacity>) using the identifier function and the place operator p.

Lemma 1 (buildPlace). *Let $N = (P, T, pre, post, cap, p_{name}, t_{name}, m)$ be a Petri net together with an injective identity function $id_P : P \rightarrow \mathbb{N}^+$, then there is an injective function $buildPlace : P^{\oplus} \rightarrow \mathbb{T}_{Places}$.*

Proof sketch: *buildPlace* is defined inductively over $|P|$ by:

- for $P = \emptyset$, $P^{\oplus} = \{0\}$ and $buildPlace(0) = \texttt{emptyPlace}$
- for $P' = P \uplus \{p'\}$ there is a $buildPlace' : P' \rightarrow \mathbb{T}_{Places}$ defined by
  $buildPlace'(s) = buildPlace(s)$ if $s \in P^{\oplus}$ and $buildPlace'(s) =$
  $buildPlace(s'), \texttt{p}(p_{name}(p'_1)|id_P(p'_1)|cap(p'_1)), ..., \texttt{p}(p_{name}(p'_k)|id_P(p'_k)|cap(p'_k))$
  with $p'_i = p'$ for $1 \leq i \leq k$ and $s = s' + k \cdot p', k \geq 1$ and $s' \in P^{\oplus}$

buildPlace is injective since id_p is injective and its inverse function $buildPlace^{-1}$ is defined accordingly [20].

Next, Theorem 1 states the conversion of one given reconfigurable Petri net into the Maude modules NET and RULES. The first part of the proof states that the functions *buildNet* and *buildRule* correspond to the initial state in Maude's term algebra. The Maude module NET comprises the **net**-operator, a set of rules defined by the **rule**-operator and some metadata. This metadata is used for example to ensure efficient use of identifiers. Moreover, the theorem states that the module RULES comprises rewrite rules for each rule in $\mathcal{R}$.

Theorem 1 (Syntactic conversion of a reconfigurable Petri net to Maude modules NET and RULES). *For each reconfigurable Petri net $(N, \mathcal{R})$, there are well-formed Maude modules NET and RULES.*

Proof sketch: Using *buildPlace* (see Lemma 1) *buildTransition*, *buildPre*, *buildPost* and *buildRule* (see [20]) each reconfigurable Petri net $(N, \mathcal{R})$ with $N = (P, T, pre, post, cap, p_{name}, t_{name}, m)$ and $\mathcal{R} = \{(r_{name_i}, L_i \leftarrow K_i \rightarrow R_i) | 1 \leq i \leq n\}$ yields the well-formed Maude module NET:

$$\text{eq initial} = buildNet(N)$$
$$buildRule(\mathcal{R})$$
$$\text{metadata}$$

Additionally, we have the Maude module RULES with the rewrite rules (see Fact 2), so that for each rule $r \in \mathcal{R}$ with $r = (r_{name_i}, L_i \leftarrow K_i \rightarrow R_i)$ there is:

```
crl [r_name] : net( places{buildPlaces(P_Li), PRest},
                    transitions{buildTransition(T_Li): TRest},
                    pre{buildPre(T_Li), MTupleRest1},
                    post{buildPost(T_Li), MTupleRest2},
                    marking{buildPlaces(M_Li); MRest})
              buildRule(R)
              metadata
              =>
         net( places{buildPlaces(P_Ri), PRest},
                    transitions{buildTransition(T_Ri): TRest},
                    pre{buildPre(T_Ri), MTupleRest1},
                    post{buildPost(T_Ri), MTupleRest2},
                    marking{buildPlaces(M_Ri); MRest})
              buildRule(R)
              new metadata
```

```
if *** for deleted places
    freeOfMarking(∀p ∈ P_Li| MRest)∧
    *** for places of deleted transitions
    emptyNeighbourForPlace(∀p ∈ P_Li \ P_Ri|
      pre{MTupleRest1} | post{MTupleRest2})∧
    calculate new metadata .
```

Listings 1.2 and 1.4 provide examples for both modules. Labelled transition systems are defined for reconfigurable Petri nets by LTS_{RPN} and for the corresponding Maude module NET by LTS_{MNC}. Both labelled transition systems are related by a surjective function map defined in Lemma 2. To distinguish the states of the respective labelled transition systems, the variables s for state in LTS_{MNC} and r for state in LTS_{RPN} are used. map relates a state $s \in LTS_{MNC}$ to a state $r \in LTS_{RPN}$. Note, the function map in Lemma 2 is not injective due to the isomorphism classes in Definition 3.

Lemma 2 (Surjective mapping of LTS_{MNC} to LTS_{RPN}). *Given a reconfigurable Petri net $(N_0, \mathcal{R})$ with $N_0 = (P_0, T_0, pre_0, post_0, p_{name0}, t_{name0}, cap_0, m_0)$ and the set of rules $\mathcal{R}$ together with the corresponding Maude modules NET and RULE as in Theorem 1. Then there is the surjective mapping $map : S_{MNC} \rightarrow S_{RPN}$ from the labelled transition system LTS_{MNC} in Definition 4 to the labelled transition system LTS_{RPN} in Definition 3 with $s = net(Places, Transitions, Pre, Post, Markings)$ | Rule Int Int IDPool by*
$map(s) = [(N, m)]$ *and*

- $P = \{p | p$ *is an atomic element in* $buildPlace^{-1}(Places)\}$
- $T = \{t | t$ *is an atomic element in* $buildTransition^{-1}(Transitions)\}$
- $pre : T \rightarrow P^{\oplus}$ *defined by* $pre(t) = buildPlace^{-1}(place)$; *if*

$$Transitions = transitions\{T : t(t_{name} \mid x)\} \ and$$
$$Pre = pre\{MT, (t(t_{name} \mid x) \rightarrow place)\}$$

- *post analogously.*
- $p_{name} : P \rightarrow A_P$ *defined by* $p_{name}(p) = label$; *if*

$$Places = places\{P, \ p(label \mid x \mid x)\}$$

- t_name *analogously.*
- $cap : P \rightarrow \mathbb{N}_+^\omega$ *defined by* $cap(p) = capacity$; *if*

$$Places = places\{P, \ p(str \mid x \mid capacity)\}$$

- m *is the atomic element in* $buildMarking^{-1}(Markings)$

Proof sketch: By induction.

- Given `initial` as defined above then $(N_0, m_0) \in map(\texttt{initial})$ and hence $r_0 = [(N_0, m_0)] = map(\texttt{initial})$.
- For each follower state $s_{n+1} \in S_{MNC}$ with $s_n \rightarrow s_{n+1} \in tr_{MNC}$ there is a $r_{n+1} \in S_{RPN}$ with $r_n \rightarrow r_{n+1} \in tr_{RPN}$ and $map(s_{n+1}) = r_{n+1}$:

  - **Firing:** Given $s_n \xrightarrow{t_{name}(t_s)} s_{n+1}$ and $map(s_n) = r_n = [(N, m)]$. Then there is also a step $r_n \xrightarrow{t_{name}(t_r)} r_{n+1}$ with $r_{n+1} = [(N, m')]$ in LTS_{RPN} since $pre^{\oplus}(t_r) \leq m$ as t_s is less or equal than the marking of s_n as `marking{PreValue ; M}` is rewritten by the rewrite rule `[fire]`. (For capacities an analogous argument holds.) If `calc(((PreValue ; M) minus PreValue) plus PostValue)` is rewritten by the rewrite rule `[fire]`, then the follower marking `Markings'` is given and the marking for r_{n+1} is calculated by $m' = (m \ominus pre^{\oplus}(t_r)) \oplus post^{\oplus}(t_r)$.

  - **Transformation:** Given $s_n \xrightarrow{r_{name}(r_s)} s_{n+1}$ then there is also a $r_n \xrightarrow{r_{name}(r_r)} r_{n+1}$ in LTS_{RPN}. If s_n can be rewritten by the rewrite rule `[r`$_{name}$`]`, then is the tem L a subterm of s_n. Hence, there is an occurrence $o : L \rightarrow N$ by r_r and $r_n \xrightarrow{r_{name}(r_r)} r_{n+1} \in tr_{RPN}$ as well as $r_{name}(r_s) = r_{name}(r_r)$. `freeOfMarking` applies for each deleted place $p \nsubseteq$ `MRest`, see Fact 2.
    `emptyNeighbourForPlace` holds for each deleted place p, since there is no occurrence in `Pre` and `Post`, Fact 2.

- Analogously, for each follower state $r_{n+1} \in S_{RPN}$ with $r_n \xrightarrow{l} r_{n+1} \in tr_{RPN}$ there is a $s_{n+1} \in S_{MNC}$ with $s_n \xrightarrow{l} s_{n+1} \in tr_{MNC}$ and $map(s_{n+1}) = r_{n+1}$.

The bisimulation between LTS_{RPN} and LTS_{MNC} is defined by the function map. Theorem 2 states the behavioural equivalence of both transition systems. For each pair $(s_n, r_n) \in map$ with $n \geq 0$ all outgoing actions are the same. The proof ensures that the reachable states s_{n+1} and r_{n+1} are again related by map.

Theorem 2 (Bisimulation). LTS_{RPN} and LTS_{MNC} are bisimilar.

Proof sketch: Given $s \in S_{MNC}$ and $r \in S_{RPN}$ with $map(s) = r = [N, m]$, we have:

- $s \rightarrow s'$: There is $map(s) = r$ and for each $a \in A_{MNC}$ we have $r \xrightarrow{a} r' \in tr_{RPN}$ because $s \xrightarrow{a} s' \in tr_{MNC}$ so that $map(s') = r'$, since map is well-defined (see Lemma 2).
- $r \rightarrow r'$: There is $map(s) = r$ and for each $a \in A_{MNC}$ we have $s \xrightarrow{a} s' \in tr_{MNC}$, due to $r \xrightarrow{a} r' \in tr_{RPN}$ so that $map(s') = r'$, since map is surjective (see Lemma 2).

Corollary 1 (LTS Properties are Preserved). For any LTL property ϕ we have:

$$LTS_{RPN} \models \phi \text{ iff } LTS_{MNC} \models \phi$$

Due to Theorem 3.1.5 and Theorem 7.6 in [3].

6   Conclusion

Related work concerns the translation of some modelling techniques into Maude. In [22] high-level Petri nets are modelled using Maude and the focus is on the soundness and correctness of the Maude structure. [7] shows a mapping for UML models to a Maude specification, where *AtoM* is used to convert the model into a Python-code representation that solves constraints inside the UML model. Closely related to our approach is [4], where Petri nets are converted into several Maude modules. [5] presents a graphical editor for CPNs, which uses Maude in the background to verify LTL properties. Specified Maude modules (similar to [22]) are defined, which contain one-step commands for the simulation. In [18] reference nets (a variant of the net-in-a-net approach) are used to model and decompose embedded systems.

Concerning the application to dynamic reconfigurable field programmable gate arrays (FPGAs) all the following approaches use Petri nets to model FPGAs, but merely have some informal mechanism to model its dynamic reconfiguration. [21] discusses how the FPGA architectures affect the implementation of Petri net specifications and shows how to obtain VHDL descriptions amenable to synthesis. [1] deals with the automatic translation of interpreted generalized Petri Nets with time into VHDL. [8] is concerned with an FPGA-based controller design to achieve simpler and affordable verification and validation. To model the interactions among processes both of state diagrams and Petri nets are used to model the concurrent processes. In [9] a Petri net variant called hierarchical configurable Petri nets modelling reconfigurable logic controllers are translated into Verilog language to be implemented in FPGAs. [14] proposes an approach analysis and testing of communication tasks of distributed control systems that uses timed colored Petri Nets for the simulation and performance estimation.

Summarizing, we have presented a correct model-checking technique for reconfigurable Petri nets. A first step towards the underlying conversion has been presented in [17]. In [20] these basis concepts have been extended, e.g. including capacities, gluing conditions, garbage collection etc. Here, we have sketched the improved conversion and have shown that it leads to a valid verification technique as the LTL properties are preserved by this conversion.

In [20] a preliminary evaluation our the approach to model checking reconfigurable Petri nets has been given based on another example. The performance of our approach based on Maude in version 2.7, including LTLR in version 1.0^2 has been compared to the established tool Charlie version 2.0^3. A Petri net with the same semantics (the same

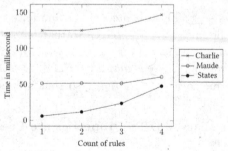

Fig. 6. Comparison

---

[2] http://maude.cs.illinois.edu/tools/tlr/, 11 March 2015.
[3] http://www-dssz.informatik.tu-cottbus.de/DSSZ/Software/Charlie, 11 March 2015.

state space), as the reconfigurable Petri net, has been used as an example. This net is used to perform a comparative analysis, including a transfer into a flat Petri net, where all transformations steps are modelled as part of the net.

Ongoing work is the introduction of control structures, as transformation units, negative application conditions and others into the tool RECONNET. These new features cause the need to adopt the conversion to Maude accordingly.

References

1. Andreu, D., Souquet, G., Gil, T.: Petri net based rapid prototyping of digital complex system. In: IEEE Computer Society Annual Symposium on VLSI, ISVLSI 2008, pp. 405–410. IEEE Computer Society (2008)
2. Bae, K., Meseguer, J.: A rewriting-based model checker for the linear temporal logic of rewriting. Electron. Notes Theoret. Comput. Sci. **290**, 19–36 (2012)
3. Baier, C., Katoen, J.: Principles of Model Checking. MIT Press, Cambridge (2008)
4. Barbosa, P., Barros, J.P., Ramalho, F., Gomes, L., Figueiredo, J., Moutinho, F., Costa, A., Aranha, A.: SysVeritas: a framework for verifying IOPT nets and execution semantics within embedded systems design. In: Camarinha-Matos, L.M. (ed.) Technological Innovation for Sustainability. IFIP AICT, vol. 349, pp. 256–265. Springer, Heidelberg (2011)
5. Boudiaf, N., Djebbar, A.: Towards an automatic translation of colored Petri nets to Maude language. Int. J. Comput. Sci. Eng. **3**(1), 253–258 (2009)
6. Bukowiec, A., Doligalski, M.: Petri net dynamic partial reconfiguration in FPGA. In: Moreno-Díaz, R., Pichler, F., Quesada-Arencibia, A. (eds.) EUROCAST 2013. LNCS, vol. 8111, pp. 436–443. Springer, Heidelberg (2013)
7. Chama, W., Elmansouri, R., Chaoui, A.: Using graph transformation and Maude to simulate and verify UML models. In: 2013 International Conference on Technological Advances in Electrical, Electronics and Computer Engineering (TAEECE), pp. 459–464 (2013)
8. Chen, C.K.: A Petri net design of FPGA-based controller for a class of nuclear I&C systems. Nucl. Eng. Des. **241**(7), 2597–2603 (2011)
9. Doligalski, M., Bukowiec, A.: Partial reconfiguration in the field of logic controllers design. Int. J. Electron. Telecommun. **59**(4), 351–356 (2013)
10. Ehrig, H., Hoffmann, K., Padberg, J., Prange, U., Ermel, C.: Independence of net transformations and token firing in reconfigurable place/transition systems. In: Kleijn, J., Yakovlev, A. (eds.) ICATPN 2007. LNCS, vol. 4546, pp. 104–123. Springer, Heidelberg (2007)
11. Ehrig, H., Padberg, J.: Graph grammars and Petri net transformations. In: Desel, J., Reisig, W., Rozenberg, G. (eds.) Lectures on Concurrency and Petri Nets. LNCS, vol. 3098, pp. 496–536. Springer, Heidelberg (2004)
12. Eker, S., Meseguer, J., Sridharanarayanan, A.: The Maude LTL model checker. Electr. Notes Theor. Comput. Sci. **71**, 162–187 (2002)
13. Eker, S., Meseguer, J., Sridharanarayanan, A.: The Maude LTL model checker and its implementation. In: Ball, T., Rajamani, S.K. (eds.) SPIN 2003. LNCS, vol. 2648, pp. 230–234. Springer, Heidelberg (2003)
14. Jamro, M., Rzonca, D., Rzasa, W.: Testing communication tasks in distributed control systems with sysml and timed colored Petri nets model. Comput. Ind. **71**, 77–87 (2015)

15. Meseguer, J.: A logical theory of concurrent objects. In: Yonezawa, A. (ed.) Proceedings of OOPSLA/ECOOP 1990, pp. 101–115. ACM (1990)
16. Padberg, J., Ede, M., Oelker, G., Hoffmann, K.: Reconnet: a tool for modeling and simulating with reconfigurable place/transition nets. ECEASST 54 (2012)
17. Padberg, J., Schulz, A.: Towards model checking reconfigurable Petri nets using Maude. ECEASST 68 (2014)
18. Richta, T., Janousek, V., Kočí, R.: Petri nets-based development of dynamically reconfigurable embedded systems. In: Moldt, D. (ed.) Petri Nets and Software Engineering (PNSE 2013), vol. 989, pp. 203–217. CEUR-WS.org (2013)
19. Schulz, A.: Converting reconfigurable Petri nets to Maude. Technical report, Cornell University (2014). http://arxiv.org/abs/1409.8404
20. Schulz, A.: Model checking of reconfigurable Petri Nets. Master's thesis, University of Applied Sciences Hamburg (2015). https://users.informatik.haw-hamburg.de/ubicomp/arbeiten/master/schulz.pdf
21. Soto, E., Pereira, M.: Implementing a Petri net specification in a FPGA using VHDL. In: Design of Embedded Control Systems, pp. 167–174. Springer, New York (2005)
22. Stehr, M.-O., Meseguer, J., Ölveczky, P.C.: Rewriting logic as a unifying framework for Petri nets. In: Ehrig, H., Juhás, G., Padberg, J., Rozenberg, G. (eds.) APN 2001. LNCS, vol. 2128, pp. 250–303. Springer, Heidelberg (2001)

Tools and Algorithms

A Software Package for Chemically Inspired Graph Transformation

Jakob L. Andersen[1,9(✉)], Christoph Flamm[2,8],
Daniel Merkle[1(✉)], and Peter F. Stadler[2,3,4,5,6,7]

[1] Department of Mathematics and Computer Science,
University of Southern Denmark, 5230 Odense, Denmark
{jlandersen,daniel}@imada.sdu.dk
[2] Institute for Theoretical Chemistry, University of Vienna, 1090 Wien, Austria
xtof@tbi.univie.ac.at
[3] Bioinformatics Group, Department of Computer Science,
and Interdisciplinary Center for Bioinformatics,
University of Leipzig, 04107 Leipzig, Germany
[4] Max Planck Institute for Mathematics in the Sciences, 04103 Leipzig, Germany
[5] Fraunhofer Institute for Cell Therapy and Immunology, 04103 Leipzig, Germany
[6] Center for Non-coding RNA in Technology and Health,
University of Copenhagen, 1870 Frederiksberg, Denmark
stadler@bioinf.uni-leipzig.de
[7] Santa Fe Institute, 1399 Hyde Park Rd, Santa Fe, NM 87501, USA
[8] Research Network Chemistry Meets Microbiology,
University of Vienna, 1090 Wien, Austria
[9] Earth-Life Science Institute, Tokyo Institute of Technology, Tokyo 152-8550, Japan
jlandersen@elsi.jp

Abstract. Chemical reaction networks can be automatically generated from graph grammar descriptions, where transformation rules model reaction patterns. Because a molecule graph is connected and reactions in general involve multiple molecules, the transformation must be performed on multisets of graphs. We present a general software package for this type of graph transformation system, which can be used for modelling chemical systems. The package contains a C++ library with algorithms for working with transformation rules in the Double Pushout formalism, e.g., composition of rules and a domain specific language for programming graph language generation. A Python interface makes these features easily accessible. The package also has extensive procedures for automatically visualising not only graphs and transformation rules, but also Double Pushout diagrams and graph languages in form of directed hypergraphs. The software is available as an open source package, and interactive examples can be found on the accompanying webpage.

Keywords: Double Pushout · Chemical graph transformation system · Graph grammar · Rule composition · Strategy framework

© Springer International Publishing Switzerland 2016
R. Echahed and M. Minas (Eds.): ICGT 2016, LNCS 9761, pp. 73–88, 2016.
DOI: 10.1007/978-3-319-40530-8_5

1   Introduction

It has been common practice in chemistry for more than a century to represent molecules as labelled graphs, with vertices representing atoms, and edges representing the chemical bonds between them [23]. It is natural, therefore, to formalize chemical reactions as graph transformations [6,11,13,20]. Many computational tools for graph transformation have been developed; some of them are either specific to chemistry [21] or at least provide special features for chemical systems [18]. General graph transformation tools, such as AGG [24], have also been used to modelling chemical systems [11].

Chemical graph transformation, however, differs in one crucial aspect from the usual setup in the graph transformation literature, where a single (usually connected) graph is rewritten, thus yielding a graph language. Chemical reactions in general involve multiple molecules. Chemical graph transformations therefore operate on *multisets* of graphs to produce a chemical "space" or "universe". A similar viewpoint was presented in [17], but here we let the basic graphs remain connected, and multisets of them are therefore dynamically constructed and taken apart in direct derivations.

Graph languages can be infinite. This is of course also true for chemical universes (which in general contain classical graph languages as subsets). In the case of chemistry, the best known infinite universes comprise polymers. The combinatorics of graphs makes is impossible in most cases to explore graph languages or chemical universes by means of a simple breadth-first search. This limitation can be overcome at least in part with the help of strategy languages that guide the rule applications. One such language has been developed for rewriting port graphs [12], implemented in the PORGY tool [5]. We have in previous work presented a similar strategy language [4] for transformation of multisets of graphs, which is based on partial application of transformation rules [2].

Here, we present the first part of the software package MedØlDatschgerl (in short: MØD), that contains a chemically inspired graph transformation system, based on the Double Pushout formalism [10]. It includes generic algorithms for composing transformation rules [2]. This feature can be used, e.g., to abstract reaction mechanisms, or whole pathways, into overall rules [3]. MØD also implements the strategy language [4] mentioned above. It facilitates the efficient generation of vast reaction networks under global constraints on the system. The underlying transformation system is not constrained to chemical systems. The package contains specialized functionalities for applications in chemistry, such as the capability to load graphs from SMILES strings [26]. This first version of MØD thus provides the main features of a chemical graph transformation system as described in [27].

The core of the package is a C++ 11 library that in turn makes use of the Boost Graph Library [22] to implement standard graph algorithms. Easy access to the library is provided by means of extensive Python bindings. In the following we use these to demonstrate the functionality of the package. The Python module provides additional features, such as embedded domain-specific languages for rule composition, and for exploration strategies. The package also provides comprehensive functionality for automatically visualising graphs, rules, Double Pushout diagrams, and hypergraphs of graph derivations, i.e., reaction networks. A LATEX package is additionally included that provides an easy mechanism for including visualisations directly in documents.

In Sect. 2 we first describe formal background for transforming multisets of graphs. Section 3 gives examples of how graph and rule objects can be used, e.g., to find morphisms with the help of the VF2 algorithms [8,9]. Sections 4 and 5 describes the interfaces for respectively rule composition and the strategy language. Section 6, finally, gives examples of the customisable figure generation functionality of the package, including the LATEX package.

The source code of MedØlDatschgerl as well as additional usage examples can be found at http://mod.imada.sdu.dk. A live version of the software can be accessed at http://mod.imada.sdu.dk/playground.html. This site also provides access to the large collection of examples.

2 Transformation of Multisets of Graphs

The graph transformation formalism we use is a variant of the Double Pushout (DPO) approach (e.g., see [10] more details). Given a category of graphs $\mathcal{C}$, a DPO rule is defined as a span $p = (L \xleftarrow{l} K \xrightarrow{r} R)$, where we call the graphs L, K, and R respectively the *left side*, *context*, and *right side* of the rule. A rule can be applied to a graph G using a match morphism $m \colon L \to G$ when the *dangling condition* and the *identification condition* are satisfied [10]. This results in a new graph H, where the copy of L has been replaced with a copy of R. We write such a direct derivation as $G \xRightarrow{p,m} H$, or simply as $G \xRightarrow{p} H$ or $G \Rightarrow H$ when the match or rule is unimportant. The graph transformation thus works in a category $\mathcal{C}$ of possibly disconnected graphs.

Let $\mathcal{C}'$ be the subcategory of $\mathcal{C}$ restricted to connected graphs. A graph $G \in \mathcal{C}$ will be identified with the multiset of its connected components. We use double curly brackets $\{\!\{\dots\}\!\}$ to denote the construction of multisets. Hence we write $G = \{\!\{g_1, g_2, \dots g_k\}\!\}$ for an arbitrary graph $G \in \mathcal{C}$ with not necessarily distinct connected components $g_i \in \mathcal{C}'$. For a set $\mathcal{G} \subseteq \mathcal{C}'$ of connected graphs and a graph $G = \{\!\{g_1, g_2, \dots, g_k\}\!\} \in \mathcal{C}$ we write $G \in^* \mathcal{G}$ whenever $g_i \in \mathcal{G}$ for all $i = 1, \dots, k$.

We define a graph grammar $\Gamma(\mathcal{G}, \mathcal{P})$ by a set of connected starting graphs $\mathcal{G} \subseteq \mathcal{C}'$, and a set of DPO rules $\mathcal{P}$ based on the category $\mathcal{C}$. The language of the grammar $L(\mathcal{G}, \mathcal{P})$ includes the starting graphs $\mathcal{G}$. Additional graphs in the language are constructed by iteratively finding direct derivations $G \xRightarrow{p} H$ with $p \in \mathcal{P}$ and $G, H \in \mathcal{C}$ such that $G \in^* L(\mathcal{G}, \mathcal{P})$. Each graph $h \in H$ is then defined to be in the language as well. A concise constructive definition of the language is thus $L(\mathcal{G}, \mathcal{P}) = \bigcup_{k=1}^{\infty} \mathcal{G}_k$ with $\mathcal{G}_1 = \mathcal{G}$ and

$$\mathcal{G}_{k+1} = \mathcal{G}_k \cup \bigcup_{p \in \mathcal{P}} \{h \in H \mid \exists G \in^* \mathcal{G}_k : G \xRightarrow{p} H\}$$

In MØD the objects of the category $\mathcal{C}$ are all undirected graphs without parallel edges and loops, and labelled on vertices and edges with text strings.

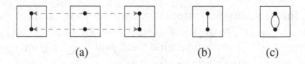

(a)   (b)   (c)

Fig. 1. The pushout object of (a) in the category of simple graphs is either not existing or is the graph depicted in (b) where the two edges are merged. For multigraphs the pushout object would be the graph depicted in (c).

The core algorithms can however be specialised for other label types. We also restrict the class of morphisms in $\mathcal{C}$ to be injective, i.e., they are restricted to graph monomorphisms. Note that this restriction implies that the identification condition of rule application is always fulfilled.

The choice of disallowing parallel edges is motivated by the aim of modelling of chemistry, where bonds between atoms are single entities. While a "double bond" consists of twice the amount of electrons than a "single bond", it does not in general behave as two single bonds. However, when parallel edges are disallowed a special situation arises when constructing pushouts. Consider the span in Fig. 1a. If parallel edges are allowed, the pushout object is the one shown in Fig. 1c. Without parallel edges we could identify the edges as shown in Fig. 1b. This approach was used in for example [7]. However, for chemistry this means that we must define how to add two bonds together, which is not meaningful. We therefore simply define that no pushout object exists for the span. A direct derivation with the Double Pushout approach thus additionally requires that the second pushout is defined.

The explicit use of multisets gives rise to a form of minimality of a derivation. If $\{\!\{g_a, g_b, g_b\}\!\} \xRightarrow{p,m} \{\!\{h_c, h_d\}\!\}$ is a valid derivation, for some rule p and match m, then the extended derivation $\{\!\{g_a, g_b, g_b, q\}\!\} \xRightarrow{p,m} \{\!\{h_c, h_d, q\}\!\}$ is also valid, even though q is not "used". We therefore say that a derivation $G \xRightarrow{p,m} H$ with the left-hand side $G = \{\!\{g_1, g_2, \ldots, g_n\}\!\}$ is *proper* if and only if

$$g_i \cap \operatorname{img}(m) \neq \emptyset, \forall 1 \leq i \leq n$$

That is, if all connected components of G are hit by the match. The algorithms in MØD only enumerate proper derivations.

3   Graphs and Rules

Graphs and rules are available as classes in the library. A rule $(L \xleftarrow{l} K \xrightarrow{r} R)$ can be loaded from a description in GML [16] format. As both l and r are monomorphisms the rule is represented without redundant information in GML by three sets corresponding somewhat to the graph fragments $L \backslash K$, K, and $R \backslash K$ (see Fig. 2 for details).

Graphs can similarly be loaded from GML descriptions, and molecule graphs can also be loaded using the SMILES format [26] where most hydrogen atoms are implicitly specified. A SMILES string is a pre-order recording of a depth-first traversal of the connected graph, where back-edges are replaced with pairs of integers.

Both input methods result in objects which internally stores the graph structure, where all labels are text strings. Figure 2 shows examples of graph and rule loading, using the Python interface of the software.

```
formaldehyde = graphGML("formaldehyde.gml")
caffeine = smiles("Cn1cnc2c1c(=O)n(c(=O)n2C)C")
ketoEnol = ruleGMLString("""rule [
    left [
        edge [ source 1 target 4 label "-" ]
        edge [ source 1 target 2 label "-" ]
        edge [ source 2 target 3 label "=" ]
        node [ id 3 label "O" ]
        node [ id 4 label "H" ]
    ]
    context [
        node [ id 1 label "C" ]
        node [ id 2 label "C" ]
    ]
    right [
        edge [ source 1 target 2 label "=" ]
        edge [ source 2 target 3 label "-" ]
        node [ id 3 label "O-" ]
        node [ id 4 label "H+" ]
    ]
]""")
```

Fig. 2. Creation of two graph objects and a transformation rule object in the Python interface. The (molecule) graph 'formaldehyde' is loaded from an external GML file, while the (molecule) graph 'caffeine' is loaded from a SMILES string [26], often used in cheminformatics. General labelled graphs can only be loaded from a GML description, and all graphs are internally stored simply as labelled adjacency lists. The DPO transformation rule 'ketoEnol' is loaded form an inline GML description. When the GML sections 'left', 'context', and 'right' are considered sets, they encode a rule $(L \leftarrow K \rightarrow R)$ with L = 'left' ∪ 'context', R = 'right' ∪ 'context', and K = 'context' ∪ ('left' ∩ 'right')'. Vertices and edges that change label are thus specified in both 'left' and 'right'. Note that in GML the endpoints of edges are described by 'source' and 'target', but for undirected graphs these tags have no particular meaning and may be exchanged. The graphs and rules are visualised in Fig. 5.

Graphs have methods for counting both monomorphisms and isomorphisms, e.g., for substructure search and for finding duplicate graphs. Counting the number of carbonyl groups in a molecule 'mol' can be done simply as

```
carbonyl = smiles("[C]=O") count = carbonyl.monomorphism(mol,
maxNumMatches=1337)
```

By default the 'monomorphism' method stops searching after the first morphism is found; alternative matches can be retrieved by setting the limit to a higher value.

Rule objects also have methods for counting monomorphisms and isomorphisms. A rule morphism $m\colon p_1 \to p_2$ on the rules $p_i = (L_i \xleftarrow{l_i} K_i \xrightarrow{r_i} R_i), i = 1, 2$ is a 3-tuple of graph morphisms $m_X\colon X_1 \to X_2, X \in \{L, K, R\}$ such that they commute with the morphisms in the rules. Finding an isomorphism between two rules can thus be used for detecting duplicate rules, while finding a monomorphism $m\colon p_1 \to p_2$ determines that p_1 is at least as general as p_2.

4   Composition of Transformation Rules

In [2,3] the concept of rule composition is described, where two rules $p_1 = (L_1 \leftarrow K_1 \to R_1), p_2 = (L_2 \leftarrow K_2 \to R_2)$ are composed along a common subgraph given by the span $R_1 \leftarrow D \to L_2$. Different types of rule composition can be defined by restricting the common subgraph and its relation to the two rules. MØD implements enumeration algorithms for several special cases that are motived and defined in [2,3]. The simplest case is to set D as the empty graph, denoted by the operator $\bullet_\emptyset$, to create a composed rule that implements the parallel application of two rules. In the most general case, denoted by $\bullet_\cap$, all common subgraphs of R_1 and L_2 are enumerated. In a more restricted setting R_1 is a subgraph of L_2, denoted by $\bullet_\subseteq$, or, symmetrically, L_2 is a subgraph of R_1, denoted by $\bullet_\supseteq$. When the subgraph requirement is relaxed to only hold for a subset of the connected components of the graphs we denoted it by $\bullet_\subseteq^c$ and $\bullet_\supseteq^c$.

The Python interface contains a mini-language for computing the result of rule composition expressions with these operators. The grammar for this language of expressions is shown in Fig. 3.

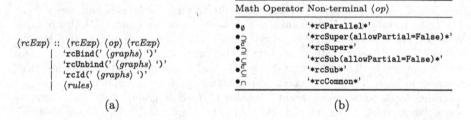

	Math Operator	Non-terminal $\langle op \rangle$
	$\bullet_\emptyset$	'*rcParallel*'
	$\bullet_\supseteq$	'*rcSuper(allowPartial=False)*'
$\langle rcExp \rangle$ :: $\langle rcExp \rangle$ $\langle op \rangle$ $\langle rcExp \rangle$	$\bullet_\supseteq^c$	'*rcSuper*'
| 'rcBind(' $\langle graphs \rangle$ ')'	$\bullet_\subseteq$	'*rcSub(allowPartial=False)*'
| 'rcUnbind(' $\langle graphs \rangle$ ')'	$\bullet_\subseteq^c$	'*rcSub*'
| 'rcId(' $\langle graphs \rangle$ ')'	$\bullet_\cap$	'*rcCommon*'
| $\langle rules \rangle$		
(a)		(b)

Fig. 3. Grammar for rule composition expressions in the Python interface, where $\langle graphs \rangle$ is a Python expression returning either a single graph or a collection of graphs. Similarly is $\langle rules \rangle$ a Python expression returning either a single rule or a collection of rules. The pseudo-operators $\langle op \rangle$ each correspond to a mathematical rule composition operator (see [2,3]). The three functions 'rcBind', 'rcUnbind', and 'rcId' refers to the construction of the respective rules $(\emptyset \leftarrow \emptyset \to G)$, $(G \leftarrow \emptyset \to \emptyset)$, and $(G \leftarrow G \to G)$ from a graph G.

Its implementation is realised using a series of global objects with suitable overloading of the multiplication operator. A rule composition expression can be passed to an evaluator, which will carry out the composition and discard duplicate results, as determined by checking isomorphism between rules. The result of each $\langle rcExp \rangle$ is coerced into a list of rules, and the operators consider all selections of rules from their arguments. That is, if 'P1' and 'P2' are two rule composition expressions, whose evaluation results in two corresponding lists of rules, P_1 and P_2. Then, for example, the evaluation of 'P1 *rcParallel* P2' results in the following list of rules:

$$\bigcup_{p_1 \in P_1} \bigcup_{p_2 \in P_2} p_1 \bullet_\emptyset p_2$$

Each of these rules encodes the parallel application of a rule from P_1 and a rule from P_2.

In the following Python code, for example, we compute the rules corresponding to the bottom span $(G \leftarrow D \rightarrow H)$ of a DPO diagram, arising from applying the rule $p = (L \leftarrow K \rightarrow R)$ to the multiset of connected graphs $G = \{\!\{g_1, g_2\}\!\}$.

```
exp = rcId(g1) *rcParallel* rcId(g2) *rcSuper(allowPartial=False)* p
rc = rcEvaluator(ruleList)  res = rc.eval(exp)
```

Here, the rule composition evaluator is given a list 'ruleList' of known rules that will be used for detecting isomorphic rules. Larger rule composition expressions, such as those found in [3], can similarly be directly written as Python code.

5   Exploration of Graph Languages Using Strategies

A breadth-first enumeration of the language of a graph grammar is not always desirable. For example, in chemical systems there are often constraints that can not be expressed easily in the underlying graph transformation rules. In [4] a strategy framework is introduced for the exploration of graph languages. It is a domain specific programming language that, like the rule composition expressions, is implemented in the Python interface, with the grammar shown in Fig. 4. The language computes on sets of graphs. Simplified, this means that each execution state is a set of connected graphs. An *addition strategy* adds further graphs to this state, and a *filter strategy* removes graphs from it. A *rule strategy* enumerates direct derivations based on the state, subject to acceptance by filters introduced by the *left-* and *right-predicate strategies*. Newly derived graphs are added to the state. Strategies can be sequentially composed with the '>>' operator, which can be extended to k-fold composition with the *repetition strategy*. A *parallel strategy* executes multiple strategies with the same input, and merges their output. During the execution of a program the discovered direct derivations are recorded as an annotated directed multi-hypergraph, which for chemical systems is a reaction network. For a full definition of the language see [4] or the MØD documentation.

A strategy expression must, similarly to a rule composition expression, be given to an evaluator which ensures that isomorphic graphs are represented by the same C++/Python object. After execution the evaluator contains the

⟨*strat*⟩ :: ⟨*strats*⟩ | ⟨*strat*⟩ '>>' ⟨*strat*⟩ | ⟨*rule*⟩
    | 'addSubset(' ⟨*graphs*⟩ ')' | 'addUniverse(' ⟨*graphs*⟩ ')'
    | 'filterSubset(' ⟨*filterPred*⟩ ')' | 'filterUniverse(' ⟨*filterPred*⟩ ')'
    | 'leftPredicate[' ⟨*derivationPred*⟩ '](' ⟨*strat*⟩ ')'
    | 'rightPredicate[' ⟨*derivationPred*⟩ '](' ⟨*strat*⟩ ')'
    | 'repeat' ['[' ⟨*int*⟩ ']'] '(' ⟨*strat*⟩ ')'
    | 'revive(' ⟨*strat*⟩ ')'

Fig. 4. Grammar for the domain specific language for guiding graph transformation, embedded in the Python interface of the software package. The non-terminal ⟨*strats*⟩ must be a collection of strategies, that becomes a **parallel** strategy from [4]. The production ⟨*strat*⟩ '>>' ⟨*strat*⟩ results in a sequence strategy.

generated derivation graph, which can be visualised or programmatically used for subsequent analysis.

The strategy language can for example be used for the simple breadth-first exploration of a grammar with a set of graphs 'startingGraphs' and a set of rules 'ruleSet', where exploration does not result in graphs above a certain size (42 vertices):

```
strat = (
      addSubset(startingGraphs)
   >> rightPredicate[
         lambda derivation: all(g.numVertices <= 42 for g in derivation.right)
   ](    repeat(ruleSet)    )
)
dg = dgRuleComp(startingGraphs, strat)
dg.calc()
```

The 'dg' object is the evaluator which afterwards contains the derivation graph. More examples can be found in [1,4] where complex chemical behaviour is incorporated into strategies. An abstract example can also be found in [4] where the puzzle game Catalan [14] is solved using exploration strategies.

6   Figure Generation

The software package includes elaborate functionality for automatically visualising graph, rules, derivation graphs, and derivations. The final rendering of figures is done using the TikZ [25] package for LaTeX, while the layouts for graphs are computed using Graphviz [15]. However, for molecule graphs it is possible to use the cheminformatics library Open Babel [19] for laying out molecules and reaction patterns in a more chemically familiar manner.

Visualisation starts by calling a 'print' method on the object in question. This generates files with LaTeX code and a graph description in Graphviz format. Special post-processing commands are additionally inserted into another file. Invoking the post-processor will then generate coordinates and compile the final layout. In addition, an aggregate summary document is compiled that includes all figures for easy overview. Figure 5 shows an example, where the wrapper script 'mod' provided by the package is used to automatically execute both a Python

```
p = GraphPrinter()
p.setMolDefault()
p.collapseHydrogens = False
formaldehyde.print(p)
p.edgesAsBonds = False
caffeine.print(p)
p.setReactionDefault()
ketoEnol.print(p)
```

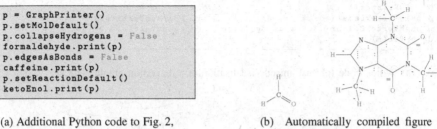

(a) Additional Python code to Fig. 2, for generating figures.

(b) Automatically compiled figure of the two graphs loaded in Fig. 2.

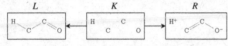

(c) Automatically compiled figure of the DPO rule loaded in Fig. 2.

Fig. 5. Example of automatic visualisation of graphs and rules, using the post-processor. The Python code is an extension of the code from Fig. 2, and can be executed using the provided 'mod' script that invokes both the Python interpreter 'python3' and the post-processor, 'mod_post'. Edges with special labels are as default rendered in a special chemical manner, as illustrated with the left graph of (b) (formaldehyde). In the right graph of (b) (caffeine) the edge labels are shown explicitly. Both graphs uses chemical colouring. The colouring of the transformation rule, (c) denote the differences between L, K, and R. (Color figure online)

script and subsequently the post-processor. The example also shows part of the functionality for chemical rendering options, such as atom-specific colouring, charges rendered in superscript, and collapsing of hydrogen vertices into their neighbours.

Derivation graphs can also be visualised automatically, where each vertex is depicted with a rendering of the graph it represents. The overall depiction can be customised to a high degree, e.g., by annotation or colouring of vertices and hyperedges using user-defined callback functions. Figure 6 illustrates part of this functionality.

Individual derivations of a derivation graph can be visualised in form of Double Pushout diagrams. The rendering of these diagrams can be customised similar to how rules and graph depictions can, e.g., to make the graphs have a more chemical feel. An example of derivation printing is illustrated in Fig. 7.

Composition of transformation rules is a core operation in the software, and for better understanding the operation we provide a mechanism for visualising individual compositions. An example of such a visualisation is shown in Fig. 8, where only the left and right graphs of two argument rules and the result rule are shown. The composition relation is shown as red dashed lines between the left graph of the first rule and the right graph of the second rule.

Including Figures in LATEX Documents. To make it easier to use illustrations of graphs and rules we have included a LATEX package in the software. It provides macros for automatically generating Python scripts that subsequently

```
p = DGPrinter()
p.pushVertexLabel(lambda g, dg: "|V| = %d" % g.numVertices)
p.pushVertexColour(lambda g, dg: "blue" if g.numVertices >= 16 else "")
dg.print(p)
```

(a) Python code for customised visualisation of a derivation graph   'dg'.

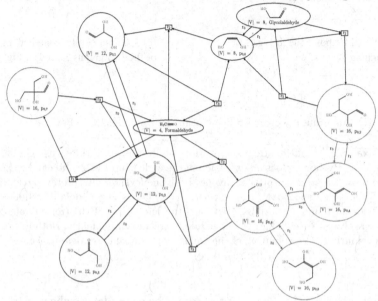

(b) Example of automatically laid out and rendered derivation graph with custom labelling and colour.

Fig. 6. Example of derivation graph printing. Each vertex is as default labelled with the name of the graph it represents, and a figure of the graph is embedded. Each hyperedge is as default labelled with the name of the rule used in the derivation the hyperedge represents. A general hyperedge is represented by a box, but for hyperedges with only 1 head and 1 tail the box is omitted, and a single labelled arc is rendered. (Color figure online)

generate figures and LaTeX code for inclusion into the original document. For example, the depictions in Fig. 5 are inserted with the following code.

```
\graphGML[collapse hydrogens=false][scale=0.4]{formaldehyde.gml}
\smiles[collapse hydrogens=false, edges as bonds=false][scale=0.4]
    {Cn1cnc2c1c(=0)n(c(=0)n2C)C}
\ruleGML{ketoEnol.gml}{\dpoRule[scale=0.4]}
```

Each '\graphGML' and '\smiles' macro expands into an '\includegraphics' for a specific PDF file, and a Python script is generated which can be executed to compile the needed files. The '\ruleGML' macro expands into

```
\dpoRule[scale=0.4]{fileL.pdf}{fileK.pdf}{fileR.pdf}
```

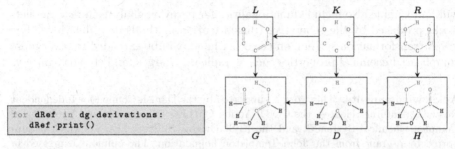

```
for dRef in dg.derivations:
    dRef.print()
```

(a) Python code for visualising all derivations in a derivation graph 'dg'.

(b) An automatically generated Double Pushout diagram.

Fig. 7. Example of visualisation of derivations. Each derivation from a derivation graph can be printed, with the same customisation options as for graphs and rules. Additional colouring is used to highlight the image of the rule into the lower span. (Color figure online)

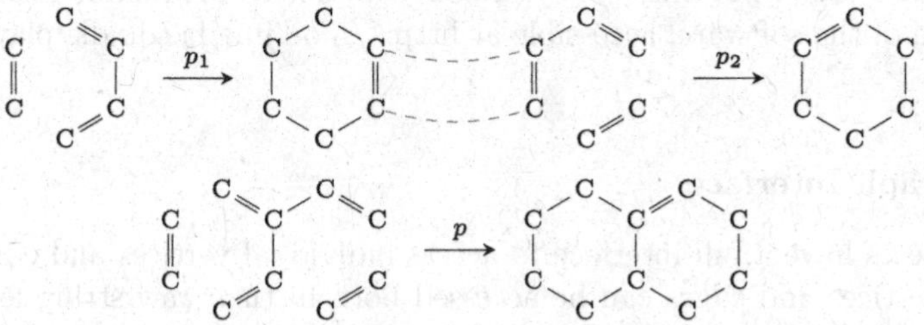

Fig. 8. Visualisation of the composition of two rules $p_i = (L_i \leftarrow K_i \rightarrow R_i, i = 1, 2$, along the a common subgraph of R_1 and L_2, indicated by the dashed red lines. Only the left and right graphs of both rules, and the resulting rule, are shown. The rendering can be customised in the same manner as the rendering for graphs and rules can. (Color figure online)

where the three PDF files depict the left side, context, and right side of the rule. The '\dpoRule' macro then expands into the final rule diagram with the PDF files included.

7 Summary

MedØlDatschgerl is a comprehensive software package for DPO graph transformation on multisets of undirected, labelled graphs. It can be used for generic, abstract graph models. By providing many features for handling chemical data it is particularly well-suited for modelling generative chemical systems. The package includes an elaborate system for automatically producing high-quality visualisations of graphs, rules, and DPO diagrams of direct derivations.

The first public version of MØD described here is intended as the foundation for a larger integrated package for graph-based cheminformatics. Future versions will for example also include functionalities for pathway analysis in reaction networks produced by the generative transformation methods described here. The graph transformation system, on the other hand, will be extended to cover more complicated chemical properties such as radicals, charges, and stereochemistry.

Acknowledgements. This work is supported by the Danish Council for Independent Research, Natural Sciences, the COST Action CM1304 "Emergence and Evolution of Complex Chemical Systems", and the ELSI Origins Network (EON), which is supported by a grant from the John Templeton Foundation. The opinions expressed in this publication are those of the authors and do not necessarily reflect the views of the John Templeton Foundation.

A   Examples

The following is a short list of examples that show how MedØlDatschgerl can be used via the Python interface. They are all available as modifiable script in the live version of the software, accessible at http://mod.imada.sdu.dk/playground.html.

A.1   Graph Interface

Graph objects have a full interface to access individual vertices and edges. The labels of vertices and edges can be accessed both in their raw string form, and as their chemical counterpart (if they have one).

```
g = graphDFS("[R]{x}C([O-])CC=O")
print("|V| =", g.numVertices)
print("|E| =", g.numEdges)
for v in g.vertices:
    print("v%d: label='%s'" % (v.id, v.stringLabel), end="")
    print("\tas molecule: atomId=%d, charge=%d" % (v.atomId, v.charge), end="")
    print("\tis oxygen?", v.atomId == AtomIds.Oxygen)
    print("\td(v) =", v.degree)
    for e in v.incidentEdges: print("\tneighbour:", e.target.id)
for e in g.edges:
    print("(v%d, v%d): label='%s'" % (e.source.id, e.target.id, e.stringLabel), end="")
    try:
        bt = str(e.bondType)
    except LogicError:
        bt = "Invalid"
    print("\tas molecule: bondType=%s" % bt, end="")
    print("\tis double bond?", e.bondType == BondType.Double)
```

A.2   Graph Morphisms

Graph objects have methods for finding morphisms with the VF2 algorithms for isomorphism and monomorphism. We can therefore easily detect isomorphic graphs, count automorphisms, and search for substructures.

```
mol1 = smiles("CC(C)CO")
mol2 = smiles("C(CC)CO")
# Check if there is just one isomorphism between the graphs:
isomorphic = mol1.isomorphism(mol2) == 1
print("Isomorphic?", isomorphic)
# Find the number of automorphisms in the graph,
# by explicitly enumerating all of them:
numAutomorphisms = mol1.isomorphism(mol1, maxNumMatches=1337)
print("|Aut(G)| =", numAutomorphisms)
```

```
# Let's count the number of methyl groups:
methyl = smiles("[CH3]")
# The symmetry of the group it self should not be counted,
# so find the size of the automorphism group of methyl.
numAutMethyl = methyl.isomorphism(methyl, maxNumMatches=1337)
print("|Aut(methyl)|", numAutMethyl)
# Now find the number of methyl matches,
numMono = methyl.monomorphism(mol1, maxNumMatches=1337)
print("#monomorphisms =", numMono)
# and divide by the symmetries of methyl.
print("#methyl groups =", numMono / numAutMethyl)
```

A.3   Rule Loading

Rules must be specified in GML format.

```
# A rule (L <- K -> R) is specified by three graph fragments:
# left, context, and right
destroyVertex = ruleGMLString('rule [   left    [  node [ id 1 label "A" ]    ]  ]')
createVertex = ruleGMLString( 'rule [   right   [  node [ id 1 label "A" ]    ]  ]')
identity = ruleGMLString(     'rule [   context [  node [ id 1 label "A" ]    ]  ]')
# A vertex/edge can change label:
labelChange = ruleGMLString("""rule [
    left   [   node [ id 1 label"A"]    edge [ source 1 target 2 label"A"]   ]
    # GML can have Python-style line comments too
    context [   node [ id 2 label"Q"]                                         ]
    right  [   node [ id 1 label"B"]    edge [ source 1 target 2 label"B"]   ]
]""")
# A chemical rule should probably not destroy and create vertices:
ketoEnol = ruleGMLString("""rule [
    left [
        edge [ source 1 target 4 label"-"]    edge [ source 1 target 2 label"-"]
        edge [ source 2 target 3 label"="]
        node [ id 3 label"O"]       node [ id 4 label"H"]
    ]
    context [
        node [ id 1 label"C"]   node [ id 2 label"C"]
    ]
    right [
        edge [ source 1 target 2 label"="]    edge [ source 2 target 3 label"-"]
        node [ id 3 label"O-"]       node [ id 4 label"H+"]
    ]
]""")
# Rules can be printed, but label changing edges are not visualised in K:
ketoEnol.print()
# Add with custom options, like graphs:
p1 = GraphPrinter()
p2 = GraphPrinter()
p1.disableAll()
p1.withTexttt = True
p1.withIndex = True
p2.setReactionDefault()
for p in inputRules: p.print(p1, p2)
# Be careful with printing options and non-existing implicit hydrogens:
p1.disableAll()
p1.edgesAsBonds = True
p2.setReactionDefault()
p2.simpleCarbons = True # !!
ketoEnol.print(p1, p2)
```

A.4   Rule Composition 1 — Unary Operators

Special rules can be constructed from graphs.

```
glycolaldehyde.print()
# A graph G can be used to construct special rules:
# (\emptyset <- \emptyset -> G)
bindExp = rcBind(glycolaldehyde)
# (G <- \emptyset -> \emptyset)
unbindExp = rcUnbind(glycolaldehyde)
# (G <- G -> G)
idExp = rcId(glycolaldehyde)
# These are really rule composition expressions that have to be evaluated:
rc = rcEvaluator(inputRules)
# Each expression results in a lists of rules:
bindRules = rc.eval(bindExp)
unbindRules = rc.eval(unbindExp)
idRules = rc.eval(idExp)
postSection("Bind Rules")
```

```
for p in bindRules: p.print()
postSection("Unbind Rules")
for p in unbindRules: p.print()
postSection("Id Rules")
for p in idRules: p.print()
```

A.5   Rule Composition 2 — Parallel Composition

A pair of rules can be merged to a new rule implementing the parallel transformation.

```
rc = rcEvaluator(inputRules)
# The special global object 'rcParallel' is used to make a pseudo-operator:
exp = rcId(formaldehyde) *rcParallel* rcUnbind(glycolaldehyde)
rules = rc.eval(exp)
for p in rules: p.print()
```

A.6   Rule Composition 3 — Supergraph Composition

A pair of rules can (maybe) be composed using a supergraph relation.

```
rc = rcEvaluator(inputRules)
exp = rcId(formaldehyde) *rcParallel* rcId(glycolaldehyde)
exp = exp *rcSuper* ketoEnol_F
rules = rc.eval(exp)
for p in rules: p.print()
```

A.7   Reaction Networks 1 — Rule Application

Transformation rules (reaction patterns) can be applied to graphs (molecules) to create new graphs (molecules). The transformations (reactions) implicitly form a directed (multi-)hypergraph (chemical reaction network).

```
# Reaction networks are expaned using a strategy:
strat = (    # A molecule can be active or passive during evaluation.
            addUniverse(formaldehyde) # passive
            >> addSubset(glycolaldehyde) # active
            # Aach reaction must have a least 1 active educt.
            >> inputRules    )
# We call a reaction network a 'derivation graph'.
dg = dgRuleComp(inputGraphs, strat)
dg.calc()
# They can also be visualised.
dg.print()
```

A.8   Reaction Networks 2 — Repetition

A sub-strategy can be repeated.

```
strat = (    addUniverse(formaldehyde)
            >> addSubset(glycolaldehyde)
            # Iterate the rule application 4 times.
            >> repeat[4](inputRules)    )
dg = dgRuleComp(inputGraphs, strat)
dg.calc()
dg.print()
```

References

1. Andersen, J.L., Andersen, T., Flamm, C., Hanczyc, M.M., Merkle, D., Stadler, P.F.: Navigating the chemical space of HCN polymerization and hydrolysis: guiding graph grammars by mass spectrometry data. Entropy 15(10), 4066–4083 (2013)
2. Andersen, J.L., Flamm, C., Merkle, D., Stadler, P.F.: Inferring chemical reaction patterns using rule composition in graph grammars. J. Syst. Chem. 4(1), 4 (2013)
3. Andersen, J.L., Flamm, C., Merkle, D., Stadler, P.F.: 50 shades of rule composition. In: Fages, F., Piazza, C. (eds.) FMMB 2014. LNCS, vol. 8738, pp. 117–135. Springer, Heidelberg (2014)
4. Andersen, J.L., Flamm, C., Merkle, D., Stadler, P.F.: Generic strategies for chemical space exploration. Int. J. Comput. Biol. Drug Des. 7(2/3), 225–258 (2014). TR: http://arxiv.org/abs/1302.4006
5. Andrei, O., Fernández, M., Kirchner, H., Melançon, G., Namet, O., Pinaud, B.: PORGY: strategy driven interactive transformation of graphs. In: Proceedings of the 6th International Workshop on Computing with Terms and Graphs (TERM-GRAPH 2011). Electronic Proceedings in Theoretical Computer Science, vol. 48, pp. 54–68 (2011)
6. Benkö, G., Flamm, C., Stadler, P.F.: A graph-based toy model of chemistry. J. Chem. Inf. Comput. Sci. 43(4), 1085–1093 (2003)
7. Braatz, B., Golas, U., Soboll, T.: How to delete categorically - two pushout complement constructions. J. Symb. Comput. 46(3), 246–271 (2011). Applied and Computational Category Theory
8. Cordella, L., Foggia, P., Sansone, C., Vento, M.: A (sub) graph isomorphism algorithm for matching large graphs. IEEE Trans. Pattern Anal. Mach. Intell. 26(10), 1367 (2004)
9. Cordella, L.P., Foggia, P., Sansone, C., Vento, M.: An improved algorithm for matching large graphs. In: Proceedings of the 3rd IAPR-TC15 Workshop on Graph-based Representations in Pattern Recognition, pp. 149–159 (2001)
10. Corradini, A., Montanari, U., Rossi, F., Ehrig, H., Heckel, R., Löwe, M.: Algebraic approaches to graph transformation - Part I: Basic concepts and double pushout approach. In: Rozenberg, G. (ed.) Handbook of Graph Grammars and Computing by Graph Transformation. Chapter 3, pp. 163–245. World Scientific, Singapore (1997)
11. Ehrig, K., Heckel, R., Lajios, G.: Molecular analysis of metabolic pathway with graph transformation. In: Corradini, A., Ehrig, H., Montanari, U., Ribeiro, L., Rozenberg, G. (eds.) ICGT 2006. LNCS, vol. 4178, pp. 107–121. Springer, Heidelberg (2006)
12. Fernández, M., Kirchner, H., Namet, O.: A strategy language for graph rewriting. In: Vidal, G. (ed.) LOPSTR 2011. LNCS, vol. 7225, pp. 173–188. Springer, Heidelberg (2012)
13. Flamm, C., Ullrich, A., Ekker, H., Mann, M., Högerl, D., Rohrschneider, M., Sauer, S., Scheuermann, G., Klemm, K., Hofacker, I.L., Stadler, P.F.: Evolution of metabolic networks: a computational framework. J. Syst. Chem. 1(4), 4 (2010)
14. Increpare games: Catalan (2011). http://www.increpare.com/2011/01/catalan/
15. Gansner, E.R., North, S.C.: An open graph visualization system and its applications to software engineering. Softw. Pract. Exp. 30(11), 1203–1233 (2000)
16. Himsolt, M.: GML: a portable graph file format. http://www.fim.uni-passau.de/fileadmin/files/lehrstuhl/brandenburg/projekte/gml/gml-technical-report.pdf

17. Kreowski, H.J., Kuske, S.: Graph multiset transformation: a new framework for massively parallel computation inspired by DNA computing. Nat. Comput. **10**(2), 961–986 (2011)
18. Mann, M., Ekker, H., Flamm, C.: The graph grammar library - a generic framework for chemical graph rewrite systems. In: Duddy, K., Kappel, G. (eds.) ICMB 2013. LNCS, vol. 7909, pp. 52–53. Springer, Heidelberg (2013)
19. O'Boyle, N.M., Banck, M., James, C.A., Morley, C., Vandermeersch, T., Hutchison, G.R.: Open Babel: an open chemical toolbox. J. Cheminformatics **3**, 33 (2011)
20. Rosselló, F., Valiente, G.: Analysis of metabolic pathways by graph transformation. In: Ehrig, H., Engels, G., Parisi-Presicce, F., Rozenberg, G. (eds.) ICGT 2004. LNCS, vol. 3256, pp. 70–82. Springer, Heidelberg (2004)
21. Rosselló, F., Valiente, G.: Chemical graphs, chemical reaction graphs, and chemical graph transformation. Electron. Notes Theor. Comput. Sci. **127**(1), 157–166 (2005). Proceedings of the International Workshop on Graph-Based Tools (GraBaTs 2004) Graph-Based Tools 2004
22. Siek, J.G., Lee, L.Q., Lumsdaine, A.: Boost Graph Library: The User Guide and Reference Manual. Pearson Education, Upper Saddle River (2001). http://www.boost.org/libs/graph/
23. Sylvester, J.J.: On an application of the new atomic theory to the graphical representation of the invari- ants and covariants of binary quantics, with three appendices. Am. J. Math. **1**(1), 64–128 (1878)
24. Taentzer, G.: AGG: A graph transformation environment for modeling and validation of software. In: Pfaltz, J.L., Nagl, M., Böhlen, B. (eds.) AGTIVE 2003. LNCS, vol. 3062, pp. 446–453. Springer, Heidelberg (2004)
25. Tantau, T.: The TikZ and PGF Packages (2013). http://sourceforge.net/projects/pgf/
26. Weininger, D.: SMILES, a chemical language and information system. 1. Introduction to methodology and encoding rules. J. Chem. Inf. Comput. Sci. **28**(1), 31–36 (1988)
27. Yadav, M.K., Kelley, B.P., Silverman, S.M.: The potential of a chemical graph transformation system. In: Ehrig, H., Engels, G., Parisi-Presicce, F., Rozenberg, G. (eds.) ICGT 2004. LNCS, vol. 3256, pp. 83–95. Springer, Heidelberg (2004)

A Tool Environment for Managing Families of Model Transformation Rules

Daniel Strüber[✉] and Stefan Schulz

Philipps-Universität Marburg, Marburg, Germany
{strueber,schulzs}@informatik.uni-marburg.de

Abstract. Model transformation systems often contain families of rules that are substantially similar to each other. Variability-based rules are a recent approach to express such families of rules in a compact representation, enabling the convenient editing of multiple rule variants at once. On the downside, this approach gives rises to distinct maintenance drawbacks: Users are required to view and edit presence conditions. The complexity and size of the resulting rules may impair their readability.

In this paper, we propose to facilitate the editing of variability-based rules through suitable tool support. Inspired by the paradigms of *filtered editing* and *virtual seperation of concerns*, we present a tool environment that offers editable views for variants expressed in a variability-based rule. We demonstrate that our tool environment is helpful to address the identified issues, rendering variability-based rules a highly feasible reuse approach.

1 Introduction

Model transformation is a key enabling technology for Model-Driven Engineering. Algebraic graph transformation is one of the main paradigms in this field, enabling a high-level, declarative specification based on graph rewriting rules [1]. Non-trivial graph transformation systems often contain rules that are substantially similar to each other. Such rules may share a large bulk of intended actions, while differing only marginally, leading to a large amount of pattern duplications.

Several approaches can be used to capture such *families of rules* while avoiding pattern duplication. Many of these approaches embody a composition-based paradigm: rule variants are assembled from fragmentary building blocks. In the case of rule inheritance [2], the implementation of a rule family comprises a hierarchy of a base rule with sub-rules. Rule refinement [3] extends this concept by supporting multiple base rules and the capability to modify super-rules. While these approaches clearly avoid pattern duplication, they may entail managing a large number of interrelated fragments. Their semantics are often intricate; a scheduling mechanism may be required to handle conflicts during composition.

Inspired by product line engineering approaches [4], we propose *variability-based (VB) rules*, an annotative approach to managing families of rules. The key idea is to encode a family of rules as one VB rule. Portions of this VB rule are

© Springer International Publishing Switzerland 2016
R. Echahed and M. Minas (Eds.): ICGT 2016, LNCS 9761, pp. 89–101, 2016.
DOI: 10.1007/978-3-319-40530-8_6

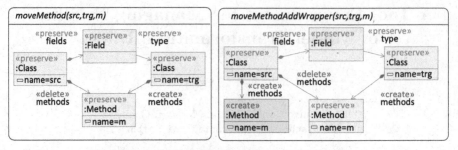

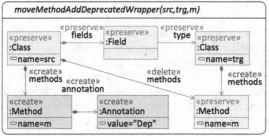

Fig. 1. Three variants of the *move method* refactoring.

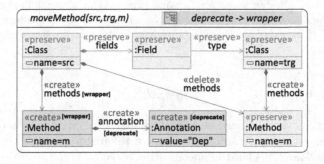

Fig. 2. Variability-based rule expressing the same three variants.

annotated with presence conditions to assign them to a subset of the encoded rules. The portion common to all rules, called the *base rule*, is not annotated.

Example. Consider a family of three in-place transformation rules. The rules, shown in Fig. 1, express variants of the *move method* refactoring for class models. The first rule specifies the relocation of a method between two classes. The second one additionally creates a wrapper method of the same name in the source class. The third one adds an annotation to mark the wrapper method as deprecated.

These three rule variants can be expressed using the VB rule shown in Fig. 2. Several elements are annotated with presence conditions over the literals **wrapper** and **deprecate**. The variants are obtained by configuring the rule, i.e., binding these literals to *true* or *false* and removing elements whose presence condition evaluates to *false*. Configuration {*wrapper=false; deprecate=false*} yields the base rule, a rule isomorphic to rule *moveMethod* in Fig. 1. The rules induced by the configurations {*wrapper=true; deprecate=false*} and {*wrapper=true; deprecate=true*} produce the additional variants. To avoid the illegal configuration {*wrapper=false; deprecate=true*}, the rule has a constraint called *variability model*, shown in the title bar, requiring `wrapper` to be true if `deprecate` is true.

This example highlights several maintainability benefits of VB rules: (i) During maintenance, all included variants are viewed and edited at once. While evidently convenient, this editing style may also be less error-prone: Fixing the same bug in multiple rules manually may lead to residual bugs not being considered. (ii) The representation is more compact, in terms of the number of rules, the total number of rule elements and the used amount of space. While compactness does not necessarily equal better readability, it is still an explicit goal of compositional approaches [3]. (iii) In contrast to compositional approaches, no additional mechanism is needed to glue fragments together, adding to the compactness of the specification. The structure of each variant is directly present; maintainers are not required to obtain a mental representation by assembling fragments.

Conversely, the example also illustrates a set of drawbacks of VB rules. (i) The use of presence conditions creates a "noisy" or "cluttered" impression, impairing readability. To make matters worse, these presence conditions are required to be edited manually, a tedious and potentially error-prone process. (ii) The rule size in terms of *average* number of elements per rule is greater. A detrimental effect of diagram size on readability is reported in [5]. (iii) To understand individual rule variants, developers are required to identify and focus on selected portions, posing a high cognitive effort. While color-coding would be helpful to mitigate this issue, it is at least complicated if not unavailable due to existing color-coding.

In this work, we address the following research question: **How can the efficient viewing and editing of variability-based rules be facilitated?** Our key idea is to provide dynamic representations suitable to the task at hand rather than one static representation – an idea inspired by the paradigms of *filtered editing* [6] and *virtual separation of concerns* [7]. We propose a tool environment that offers views on rule variants selected by the user. These views are helpful to mitigate the identified drawbacks by (i) removing the need to read and edit presence conditions, (ii) being smaller in size, and (iii) reducing the cognitive effort in deriving mental representations. In addition, we provide support for converting a legacy rule set into a VB rule with little manual effort. The basic concepts of VB rules and their automatic creation have been introduced elsewhere [8–10].

We have implemented our tool environment on top of Henshin [11], a model transformation language based on algebraic graph transformations. Lifting the concepts proposed in this work to other languages and paradigms is desirable, but left to future work. The tool and a description of its use can be found at https://www.uni-marburg.de/fb12/swt/forschung/software/varhenshin/.

2   Variability-Based Rules

In this section, we briefly revisit the main concepts of variability-based rules. We assume the reader to be familiar with double-pushout graph transformation rules, such as those shown in Fig. 1. The underlying graph kind may include typing and attributes since these concepts are orthogonal to variability. We further use the concept of subrule, a rule that can be embedded into a larger rule in an injective manner. A detailed account of these concepts is given in [10].

Definition 1 (Variability-based (VB) rule). *Given a set of atomic terms V, called* variability points, *a VB rule $\check{r} = (r, S, vm, pc)$ consists of a rule r, a set S of subrules of r, a propositional term $vm \in \mathcal{L}_V$ and a function $pc : S \cup \{r\} \to \mathcal{L}_V$, where $\mathcal{L}_V$ is the set of propositional terms over V. Term vm is called* variability model. *Function pc defines* presence conditions *for subrules s.t. $pc(r)$ is true and $\forall s \subseteq s' : pc(s') \implies pc(s)$. The* base rule *is the intersection of all subrules.*

Figure 2 shows a VB rule over variability points {*wrapper, deprecate*}. The rule is shown in a compact representation where subrules are not shown explicitly, but denoted using element presence conditions. Rule r is the entire rule, ignoring annotations. S comprises a subrule for each propositional term over V. Each subrule contains those elements whose presence conditions are implied by its own presence condition. For instance, subrule s with $pc(s) = wrapper \land \neg deprecate$ contains all elements annotated with *wrapper* and without annotations, but not those annotated with *deprecate*. The variability model is *deprecate $\to$ wrapper*.

Definition 2 (Configuration). *Let a VB rule $\check{r} = (r, S, vm, pc)$ over V be given. A* configuration *is a total function $c : V \to \{true, false\}$. A configuration c* satisfies *a term $t \in \mathcal{L}_V$ if t evaluates to true when each variable v in t is substituted by $c(v)$. A configuration c is* valid *if c satisfies vm.*

In the example, {*wrapper=true; deprecate=false*} is a valid configuration, satisfying the presence condition *wrapper*, but not the presence condition *deprecate*.

Definition 3 (Rule variant). *For a valid configuration c, there exists a unique set of subrules $S_c \subseteq S$ s.t. $\forall s \in S : s \in S_c$ iff c satisfies $pc(s)$. Gluing together all elements contained in one of these subrules yields a rule r_c, called* rule variant *induced by c.*

The example VB rule can be used to produce three variants; details are provided in the previous description of the example. Categorically, the gluing can be expressed as a consecutive multi-pullback and multi-pushout construction [10].

There are two main application scenarios for VB rules. First, a specific user intention may lead to the selection and application of one particular rule variant. For instance, in the example, the user may configure the rule so that it produces a wrapper method. Such an *external* configuration process leads to an individual rule being applied in the classic way. Second, all rules in a rule set may be applied simultaneously. Such rule sets are found in batch transformation scenarios, such as translation or migration suites. In this case, configurations can be set *internally* by the transformation engine. This approach allows to consider the base rule of all variants at once, leading to considerable performance savings [8].

3   Main Features

In this section, we present the main features of our tool environment. The design of these features is informed by *Cognitive Dimension* [12] (CD), a framework of usability dimensions for visual programming environments. First, we give an overview of the features, relating each to the CD framework. Second, we exemplify the use of these features from the user perspective.

- **View specific rule variants**: Each variant expressed in a VB rule corresponds to a configuration, a binding of all variability points to *true* or *false*. To view specific variants, we provide a *live configuration* feature: The user performs a partial or total binding of variability points, leading to immediate feedback. Irrelevant rule elements can be either turned invisible or toned down. The former option helps the user during the comprehension of individual variants. The latter one facilitates the comparison of variants.

  This feature addresses several cognitive dimensions: The *visibility* of rule variants is increased. The need for *hard mental operations* is reduced by shielding users from the cognitive effort of deriving variants. Notational *diffuseness* is reduced as fewer different symbols are needed to capture variability.

- **Edit rule variants**: A crucial issue of editing VB rules is the requirement to have users edit presence conditions, a tedious and error-prone process. We provide features to mitigate this issue: When creating a new element, a presence condition corresponding to the currently selected configuration is assigned automatically. We also support the reassigning of elements to different variants by moving them to a more general or specific configuration (i.e., one where more or less variability points are unbound).

  By lifting the abstraction level from editing presence conditions to moving elements between variants, we aim to reduce *error-proneness*. The capability to move multiple elements also reduces *viscosity*, the resistance to change.

- **Explore relationships between rule variants**: We provide multiple features to support exploring multiple variants and their interrelations. First, a *favorites* feature allows rapid switching between variants. Second, a *quick access* feature provides instant access to distinguished variants such as the base rule and the maximum rule. Third, an *auto-completion feature* reduces the configuration effort of by inferring certain open bindings automatically.

  These features are key to increasing *role-expressiveness*, the ease of understanding *"how each component [...] relates to the whole"* [12].

- **Manage variability points**: We provide a dedicated viewer component for the management of variability points. Using this viewer, variability points can be created and deleted. To ensure consistency, presence conditions referring to the deleted variability point can be updated automatically.

This dedicated component inceases the *visibility* of variability management.

- **Sanitize legacy rule sets**: Legacy rule sets may exhibit a high degree of pattern duplication, notably, if they were devised in a copy-and-paste manner. To sanitize such rule sets, we provide *clone detection* and *merge refactoring* features. Clone detection allows identifying cloned portions in a set of rules. These portions may serve as input to merge refactoring, a feature that creates VB rules automatically, including an optimization to preserve layout information from the input rules. We present this technique in [9].

  This feature shields from *premature commitment*: VB rules do not have to be devised from scratch. Users may develop rules in an ad-hoc manner and decide to use VB rules later, while retaining key layout information.

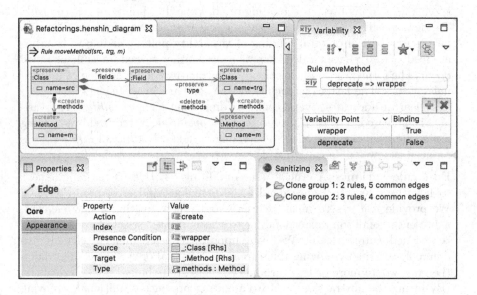

Fig. 3. Our tool environment from the user perspective.

User Perspective. Our tool environment is integrated with the default user interface of Henshin. The main components of this user interface, a graphical editor and its attached *properties* view, are shown in the left of Fig. 3. As custom components, we provide the *variability* and *sanitizing* views, shown in the right. The variability view comprises features for the definition and configuration of variability points. The variant produced from the current configuration is displayed in the editor. The sanitizing view can be used to sanitize legacy transformations.

Variability view. The variability view allows variability points in a rule to be created and deleted. To view and edit variants individually, the user configures the rule by setting the bindings for these variability points. Three literals are supported: *true*, *false*, and *unbound*. Per default, each variability point is *unbound*, yielding the maximal rule, all elements regardless of their annotations. Configurations are validated against the given variability model, *deprecate* → *wrapper* in this case. Invalid configurations and rules lead to error messages being displayed.

To navigate variants efficiently, frequently used configurations can be saved as favorites using the ☆ button in the toolbar. The star appears in yellow if a favored configuration is currently active. Each configuration has a user-specified name. In Fig. 4, the user has created two favorites, *WrapperWithDeprecate* and *WrapperWithoutDeprecate*, the latter one being active. Upon selection, the configuration is loaded and shown in the table at the bottom of the view.

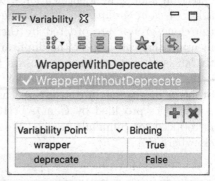

Fig. 4. Variability view with favorites menu.

A *view mode* feature allows to access distinguished variants rapidly. In the *maximum rule* mode, represented by the ⬚ icon, all elements included in the rule are shown regardless of the configuration. In the *variant mode* (⬚), elements absent in the current configuration are concealed. In the *base rule* mode (⬚), elements with a non-empty presence condition are concealed.

To further improve the handling of variability, the view allows the users to choose a *concealing strategy*, depicted in Fig. 5. First, elements can be turned invisible. This avoids a cluttered representation of the rule and lets users focus on the variant at hand. On the other hand, to allow the comparison of a variant with the full rule, users may choose to have the elements toned down instead.

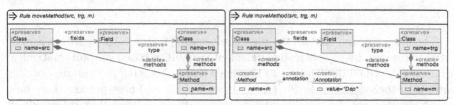

(a) Hiding unrelated elements.   (b) Toning down unrelated elements.

Fig. 5. Concealing strategies.

Using the ⬚ button, users can select an *editing mode* to define which variants are affected by edits to the rule. The supported options are: all variants, variants included in the selected configuration, or variants associated to the current view mode. In particular, the editing mode determines which presence condition is assigned during the addition or deletion of elements to a rule.

Sanitizing view. The sanitizing view, shown in the lower right of Fig. 6, supports two operations for sanitizing legacy rule sets: clone detection and merging. Clone detection allows the identification of duplications in the rule set. The result is a list of *clone groups*. To display the most relevant clone groups prominently, the clone groups are ordered by their size, i.e., the number of common elements. Users can inspect the duplication interactively; when a rule is selected, the affected portions in the rule are focused and highlighted. Internally, clone detection aims to identify isomorphic sub-graphs, a computationally expensive problem in general. To ensure reasonable response times, our approach uses a heuristics provided by ConQAT [13], a clone detection technique originally introduced for Simulink models. We discuss our adaptation of this technique elsewhere [14].

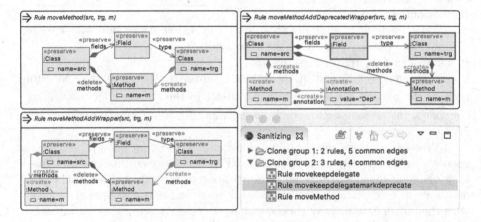

Fig. 6. Sanitizing view.

The *merge* button enables the merge refactoring feature. An algorithm is executed to construct a semantically equivalent variability-based rule automatically, using the identified duplication as base rule and annotating the variant-specific parts with their rule names [9]. The user can inspect and post-process the refactoring result using the viewing and editing features. In case that the result is not satisfactory, the process can be undone.

Context menus. Additional context menu entries allow to manage variability at the level of individual elements. Multiple nodes, edges and attributes can be selected and moved to a different configuration simultaneously.

4   Architecture and Implementation

In this section, we describe the architecture of our tool and our design principles during the implementation.

We give an overview of the architecture in Fig. 7. The novel features presented in this work are encapsulated by *Variability UI*, an integrating layer on top of the *UI*, *Clone Detection*, *Merging* and *Variability* extensions for the *Henshin* language core. To combine the Henshin UI with the variability implementation first introduced in [8], the Variability UI provides the variability view and its editor integration. Clone detection and merging are made

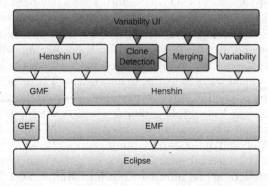

Fig. 7. Architecture.

available to users in the *Sanitizing View*. The merging component acts as a bridge between the clone detection and variability extensions: It enables the conversion of rules affected by cloning to variability-based rules. *GMF*, *GEF*, *EMF* and *Eclipse* are featured as underlying frameworks. The Henshin language is based on an EMF meta-model. The Henshin UI comprises a GMF-based editor to enable the visual viewing and editing of transformations.

The main design goal in our architectural design was non-intrusiveness: Changing the Henshin core and UI should be avoided where possible. The rationale for this design goal was to keep the Henshin language core, its UI and analysis functionality as simple as possible. Variability is deployed as a drop-in language extension, orthogonal to additional extensions, such as the existing support for Triple Graph Grammars [15] or possible future support for uncertainty [16]. Including multiple of these extensions might lead to feature interactions that need to be addressed explicitly by the designers of the extensions.

To define language extensions in a non-intrusive way, the Henshin meta-model provides a flexible annotation mechanism. Any element contained in a transformation may be annotated with key-value pairs of strings. The language extension at hand is responsible for processing these annotations. This concept allowed us to implement variability-based rules in the variability extension alone, without modifying the language core.

5   Related Work

Model Transformation Reuse. There are two groups of reuse approaches for model transformations. The first group focuses on *intra-transformation* reuse [17], the reuse of artifacts within the same transformation. In this group, many approaches are composition-based: Rule refinement, the modularization of a rule into a set of fragmentary rules, has been implemented in the eMoflon tool [18]. The modularization of a graph transformation rules into multiple *aspects* is another compositional approach [19]. A feature-based composition approach for the reuse of ATL transformations has been proposed by Sijtema [20]. Many of these approaches do not provide an automated tool to split a legacy rule into a set of composition fragments, an issue that might be addressed by applying a general-purpose splitting tool [21,22]. An important annotative approach is rule amalgamation. While this approach allows the specification of mandatory and optional parts in a rule, in contrast to VB rules, the optional parts are matched and applied as often as possible. VB rules provide the capability to assign one element to multiple variants, which is not supported in amalgamation. Amalgamation has been implemented in the AToM3 meta-modeling tool [23] and the eMoflon Triple Graph Grammar tool [24].

The second group focuses on *inter-transformation* reuse [17], the reuse of artifacts across multiple transformations. VCT [25] is a comprehensive toolkit that allows to accommodate variability in a chain of multiple transformations and to compose larger transformations from smaller ones. Cuadrado et al. [26] have introduced a component model to orchestrate the reuse of model transformations across multiple different modeling languages. Their Bentō [27] tool provides support for generic rules for ATL transformations. These generic rules are typed over *concepts*, abstract meta-models. To consider a new scenario, concepts are instantiated by binding them to the types of the required meta-model. To increase the flexibility of this approach, de Lara et al. propose an extension that accounts for heterogeneity between concepts and meta-models [28]. Criado et al. [29] propose to reuse existing transformations by annotating them. These works are orthogonal to ours as they address a different reuse scenario.

Implementation Approaches to Software Product Lines. We adopted the distinction of annotative and composition-based mechanisms from software product line (SPL) engineering [4], where it refers to different approaches to implementing a SPL. An important composition-based approach is *Feature-Oriented Programming* [30], in which a SPL is developed by dividing its specification into features and implementing each feature as a separate module. The AHEAD [31] tool made this approach applicable for Java. An example for an annotative approach are *#ifdef directives*: Portions of the source code are annotated with variability conditions and optionally removed during compilation. *Virtual separation of concerns* (VSoC) is a paradigm aiming to combine the benefits of both approaches by means of tool support [7]. In the CIDE tool [4], users are provided custom views, visual representations, and variability-aware type checks. Based on the VSoC paradigm, Walkingshaw et al. [32] propose an editing model for variational software based on an isolation principle: Edits to a view shall only affect the variants associated with this view. We adopt this principle in one of the editing modes of our tool. In a related work of Schwägerl et al. [33], the scope of variants affected by an edit is set using a separate configuration.

The FeatureIDE [34] framework integrates many of these approaches and makes them available during the entire development cycle. Its aim is to establish a *uniformity principle* of managing variability consistently in all design, implementation, and documentation artifacts. The integration of our approach into this framework is a promising avenue for future work.

Usability-oriented Model Transformation. As we aim to improve the maintainability of complex rules, our work is related to usability-oriented model transformation, a field of research addressed in [35]. Based on the premise that users may prefer mature model editors to experimental transformation tools, the authors provide a new modeling language that can be instantiated in any given model editor and mapped back to a host transformation language. This work is complementary to ours since we aim to contribute to the maturity of model transformation tools instead of replacing them.

6 Future Work and Additional Improvement

The most important task left to future work is a user study to validate the claim that our tool environment improves usability during editing. Such a study would substantiate our anecdotic evidence that the development of rule families without a dedicated reuse concept is a highly inconvenient and error-prone task. Furthermore, we are eliciting future usability improvements. First, the visibility of distinguished variants, such as the base rule, can be further increased by providing a "hot corner" feature. Implementing such a feature proves to be challenging due to limitations of the underlying editor framework. Second, relationships between variability points are currently expressed using a logical formula. In product line engineering, dedicated formalism have emerged to capture variability, the most important one being feature models [36]. Combining feature models with VB rules seems a promising avenue for future work.

References

1. Ehrig, H., Ehrig, K., Prange, U., Taentzer, G.: Fundamentals of Algebraic Graph Transformation. Springer, Heidelberg (2006)
2. Wimmer, M., Kappel, G., Kusel, A., Retschitzegger, W., Schönböck, J., Schwinger, W., Kolovos, D.S., Paige, R.F., Lauder, M., Schürr, A., Wagelaar, D.: Surveying rule inheritance in model-to-model transformation languages. J. Object Technol. **11**(2), 3:1–3:46 (2012)
3. Anjorin, A., Saller, K., Lochau, M., Schürr, A.: Modularizing triple graph grammars using rule refinement. In: Gnesi, S., Rensink, A. (eds.) FASE 2014 (ETAPS). LNCS, vol. 8411, pp. 340–354. Springer, Heidelberg (2014)
4. Kästner, C., Apel, S., Kuhlemann, M.: Granularity in software product lines. In: Proceedings of the International Conference on Software Engineering (ICSE). ACM, pp. 311–320 (2008)
5. Störrle, H.: On the impact of layout quality to understanding UML diagrams: size matters. In: Dingel, J., Schulte, W., Ramos, I., Abrahão, S., Insfran, E. (eds.) MODELS 2014. LNCS, vol. 8767, pp. 518–534. Springer, Heidelberg (2014)

6. Sarnak, N., Bernstein, R.L., Kruskal, V.: Creation and maintenance of multiple versions. In: SCM. Berichte des German Chapter of the ACM, vol. 30, pp. 264–275. Teubner (1988)
7. Kästner, C.: Virtual separation of concerns, Ph.D. dissertation, University of Magdeburg (2010)
8. Strüber, D., Rubin, J., Chechik, M., Taentzer, G.: A variability-based approach to reusable and efficient model transformations. In: Egyed, A., Schaefer, I. (eds.) FASE 2015. LNCS, vol. 9033, pp. 283–298. Springer, Heidelberg (2015)
9. Strüber, D., Rubin, J., Arendt, T., Chechik, M., Taentzer, G., Plöger, J.: Rule-Merger: automatic construction of variability-based model transformation rules. In: Stevens, P., Wąsowski, A. (eds.) FASE 2016. LNCS, vol. 9633, pp. 122–140. Springer, Heidelberg (2016). doi:10.1007/978-3-662-49665-7_8
10. Strüber, D.: Model-driven engineering in the large: Refactoring techniques for models and model transformation systems, Ph.D. dissertation, Philipps-Universität Marburg (2016)
11. Arendt, T., Biermann, E., Jurack, S., Krause, C., Taentzer, G.: Henshin: advanced concepts and tools for in-place EMF model transformations. In: Rouquette, N., Haugen, Ø., Petriu, D.C. (eds.) MODELS 2010, Part I. LNCS, vol. 6394, pp. 121–135. Springer, Heidelberg (2010)
12. Green, T.R.G., Petre, M.: Usability analysis of visual programming environments: a 'cognitive dimensions' framework. J. Vis. Lang. Comput. **7**(2), 131–174 (1996)
13. Deissenboeck, F., Hummel, B., Juergens, E., Pfaehler, M., Schaetz, B.: Model clone detection in practice. In: International Workshop on Software Clones, pp. 57–64. ACM (2010)
14. Strüber, D., Plöger, J., Acreţoaie, V.: Clone detection for graph-based model transformation languages. In: International Conference on Model Transformation (ICMT). Springer, 2016
15. Hermann, F., Gottmann, S., Nachtigall, N., Braatz, B., Morelli, G., Pierre, A., Engel, T.: On an automated translation of satellite procedures using triple graph grammars. In: Duddy, K., Kappel, G. (eds.) ICMB 2013. LNCS, vol. 7909, pp. 50–51. Springer, Heidelberg (2013)
16. Famelis, M.: Managing design-time uncertainty in software models, Ph.D. dissertation, University of Toronto (2016)
17. Kusel, A., Schönböck, J., Wimmer, M., Kappel, G., Retschitzegger, W., Schwinger, W.: Reuse in model-to-model transformation languages: are we there yet? J. Softw. Syst. Model. **14**, 1–36 (2013)
18. Kulcsár, G., Leblebici, E., Anjorin, A.: A solution to the FIXML case study using triple graph grammars and eMoflon. In: TTC@STAF, pp. 71–75 (2014)
19. Machado, R., Foss, L., Ribeiro, L.: Aspects for graph grammars. Electron. Commun. EASST **18** (2009)
20. Sijtema, M.: Introducing variability rules in ATL for managing variability in MDE-based product lines. MtATL **10**, 39–49 (2010)
21. Strüber, D., Selter, M., Taentzer, G.: Tool support for clustering large meta-models. In: BigMDE Workshop on the Scalability of Model-Driven Engineering. ACM Digital Library, pp. 7.1–7.4 (2013)
22. Strüber, D., Lukaszczyk, M., Taentzer, G.: Tool support for model splitting using information retrieval and model crawling techniques. In: BigMDE: Workshop on Scalability in Model Driven Engineering, pp. 44–47. CEUR-WS.org (2014)
23. de Lara, J., Ermel, C., Taentzer, G., Ehrig, K.: Parallel graph transformation for model simulation applied to timed transition petri nets. Electron. Notes Theor. Comput. Sci. **109**, 17–29 (2004)

24. Leblebici, E., Anjorin, A., Schürr, A., Taentzer, G.: Multi-amalgamated triple graph grammars. In: Parisi-Presicce, F., Westfechtel, B. (eds.) ICGT 2015. LNCS, vol. 9151, pp. 87–103. Springer, Heidelberg (2015)
25. Basso, F.P., Pillat, R.M., Oliveira, T.C., Becker, L.B.: Supporting large scale model transformation reuse. In: ACM SIGPLAN Notices, vol. 49(3), pp. 169–178. ACM (2013)
26. Sánchez Cuadrado, J., Guerra, E., de Lara, J.: A component model for model transformations. IEEE Trans. Softw. Eng. **40**(11), 1042–1060 (2014)
27. Cuadrado, J.S., Guerra, E., de Lara, J.: Reusable model transformation components with bentō. In: Kolovos, D., Wimmer, M. (eds.) ICMT 2015. LNCS, vol. 9152, pp. 59–65. Springer, Heidelberg (2015)
28. de Lara, J., Guerra, E.: Towards the flexible reuse of model transformations: a formal approach based on graph transformation. J. Logical Algebraic Methods Program. **83**(5), 427–458 (2014)
29. Criado, J., Martínez, S., Iribarne, L., Cabot, J.: Enabling the reuse of stored model transformations through annotations. In: Kolovos, D., Wimmer, M. (eds.) ICMT 2015. LNCS, vol. 9152, pp. 43–58. Springer, Heidelberg (2015)
30. Prehofer, C.: Feature-oriented programming: a fresh look at objects. In: Akşit, M., Matsuoka, S. (eds.) ECOOP 1997. LNCS, vol. 1241, pp. 419–433. Springer, Heidelberg (1997)
31. Batory, D.: Feature-oriented programming and the AHEAD tool suite. In: International Conference on Software Engineering (ICSE), pp. 702–703. IEEE Computer Society (2004)
32. Walkingshaw, E., Ostermann, K.: Projectional editing of variational software. In: Generative Programming: Concepts and Experiences, pp. 29–38. ACM (2014)
33. Schwägerl, F., Buchmann, T., Westfechtel, B.: SuperMod: a model-driven tool that combines version control and software product line engineering. In: International Conference on Software Paradigm Trends. SCITEPRESS, pp. 5–18 (2015)
34. Thüm, T., Kästner, C., Benduhn, F., Meinicke, J., Saake, G., Leich, T.: FeatureIDE: an extensible framework for feature-oriented software development. Sci. Comput. Program. **79**, 70–85 (2014)
35. Acretoaie, V., Störrle, H., Strüber, D.: Transparent model transformation: turning your favourite model editor into a transformation tool. In: Kolovos, D., Wimmer, M. (eds.) ICMT 2015. LNCS, vol. 9152, pp. 121–130. Springer, Heidelberg (2015)
36. Kang, K.C., Cohen, S.G., Hess, J.A., Novak, W.E., Peterson, A.S.: Feature-oriented domain analysis (FODA) feasibility study. Technical report, DTIC Document (1990)

Compiling Graph Programs to C

Christopher Bak and Detlef Plump[✉]

The University of York, York, UK
{cb574,detlef.plump}@york.ac.uk

Abstract. We show how to generate efficient C code for a high-level domain-specific language for graphs. The experimental language GP 2 is based on graph transformation rules and aims to facilitate formal reasoning on programs. Implementing graph programs is challenging because rule matching is expensive in general. GP 2 addresses this problem by providing *rooted* rules which under mild conditions can be matched in constant time. Using a search plan, our compiler generates C code for matching rooted graph transformation rules. We present run-time experiments with our implementation in a case study on checking graphs for two-colourability: on grid graphs of up to 100,000 nodes, the compiled GP 2 program is as fast as the tailor-made C program given by Sedgewick.

1 Introduction

GP 2 is an experimental domain-specific language for graphs whose basic command is the application of graph transformation rules. The language has a simple syntax and semantics to support formal reasoning on programs (see [14] for a Hoare-logic approach to verifying graph programs). GP 2's initial implementation is an interpreter running in one of two modes, either fully exploring the non-determinism inherent to transformation rules or attempting to produce a single result [3]. In this paper, we report on a compiler for GP 2 which translates programs directly into efficient C code.

The bottleneck for generating fast code for graph transformation rules is the cost of graph matching. In general, to match the left-hand graph L of a rule within a host graph G requires time $\text{size}(G)^{\text{size}(L)}$ (which is polynomial since L is fixed). As a consequence, linear-time imperative programs operating on graphs may be slowed down to polynomial time when they are recast as rule-based graph programs. To speed up graph matching, GP 2 allows to distinguish some nodes in rules and host graphs as so-called roots, and to match roots in rules with roots in host graphs. This concept goes back to Dörr [7] and was also studied by Dodds and Plump [6].

Our compiler, described in Sect. 3, translates GP 2 source code directly into C code, bridging the large gap between graph transformation rules and C. We use a *search plan* to generate code for graph matching, deconstructing each matching step into a sequence of primitive matching operations from which structured

---

C. Bak—This author's work was supported by an EPSRC Doctoral Training Grant.

R. Echahed and M. Minas (Eds.): ICGT 2016, LNCS 9761, pp. 102–117, 2016.
DOI: 10.1007/978-3-319-40530-8_7

code is generated. The code generated to evaluate rule conditions is interleaved in the matching code such that conditions are evaluated as soon as their parameters are assigned values, to rule out invalid matches at the first opportunity. Another non-standard aspect of the compiler is that programs are analysed to establish when the state (host graph) needs to be recorded for potential backtracking at runtime. Backtracking is required by GP 2's transaction-like branching constructs if-then-else and try-then-else which may contain arbitrary subprograms as guards.

In [4] we identified *fast* rules, a large class of conditional rooted rules, and proved that they can be applied in constant time if host graphs have a bounded node degree (an assumption often satisfied in practice). In Sect. 4, we demonstrate the practicality of rooted graph programs with fast rules in a case study on graph colouring: we give a GP 2 program that 2-colours host graphs in linear time. We show that on grid graphs of up to 100,000 nodes, the compiled GP 2 program matches the speed of Sedgewick's tailor-made implementation in C [17]. In this way, users get the best of both worlds: they can write visual, high-level graph programs with the performance of a relatively low-level language.

2 The Graph Programming Language GP 2

GP 2 is the successor to the graph programming language GP [12]. This section gives a brief introduction to GP 2. The original language definition is [13], an up-to-date version is given in the PhD thesis of the first author [2].

2.1 Conditional Rule Schemata

GP 2's principal programming constructs are conditional rule schemata (abbreviated to *rule schemata* or, when the context is clear, *rules*). Rule schemata extend standard graph transformation rules[1] with expressions in labels and with application conditions. Figure 1 shows the declaration of a conditional rule schema rule. The numbered nodes are the *interface* nodes. Nodes that are in the left-hand side but not in the interface are deleted by the rule. Similarly, nodes that are in the right-hand side but not in the interface are added.

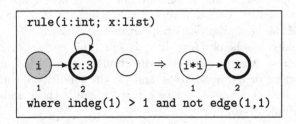

Fig. 1. Declaration of a conditional rule schema

---

[1] In the double-pushout approach with injective matching.

The top line of the declaration states the name of the rule schema and lists the variables that are used in the labels and in the condition. All variables occurring in the right-hand side and in the condition must also occur in the left-hand side because their values a runtime are determined by matching the left-hand side with a subgraph of the host graph.

Each variable is declared with a type which is either `int`, `char`, `string`, `atom` or `list`. Types form a subtype hierarchy in which integers and character strings are basic types, both of which are atoms. Atoms in turn are considered as lists of length one. Labels in host graphs are variable-free expressions containing only constructor operations such as list or string concatenation. Lists are constructed by the colon operator which denotes list concatenation.[2] String concatenation is represented by a dot.

To avoid ambiguity in variable assignments when constructing a mapping between the left-hand graph of a rule schema and a host graph, we require that expressions in the left graph are *simple*: they (1) contain no arithmetic operators, (2) contain at most one occurrence of a list variable, and (3) do not contain string expressions with more than one occurrence of a string variable. Labels in the right-hand side of a rule schema may contain arithmetic expressions.

The labels of nodes and edges can be *marked* with colours from a fixed set, in addition to a dashed mark for edges only. Marked items match only host graph items with the same mark. There is a special mark **any** that matches arbitrary host graph marks. Nodes with thick borders are *root nodes*. Their purpose is to speed up graph matching, discussed in more detail in the next section.

The programmer can specify a textual condition to add further control to where the rule is applicable, declared by the keyword **where** followed by a boolean expression. GP 2 offers a number of predicates for querying the host graph. For example, the predicate `indeg(1) > 1` in Fig. 1 ensures that node 1 is only matched to suitable host graph nodes with more than one incoming edge.

2.2 Fast Rule Schemata

The idea of *rooted* graph transformation [4] is to equip both rule and host graphs with root nodes which support efficient graph matching. Root nodes in rules must match compatible root nodes in the host graph. In this way, the search for a match is localised to the neighbourhood of the host graph's root nodes. It is possible to identify a class of rooted rule schemata that are applicable in constant time if the host graph satisfies certain restrictions.

A conditional rule schema $\langle L \Rightarrow R, c \rangle$ is *fast* if (1) each node in L is reachable from some root (disregarding edge directions), (2) neither L nor R contain repeated list, string or atom variables, and (3) the condition c contains neither an **edge** predicate nor a test $e_1 = e_2$ or $e_1 != e_2$ where both e_1 and e_2 contain a list, string or atom variable.

---

[2] Not to be confused with Haskell's colon operator which adds an element to the beginning of a list.

The first condition ensures that matches can only occur in the neighbourhood of roots. The other conditions rule out linear-time operations, such as copying lists or strings in host graph labels of unbounded length. In [4] it is shown that fast rule schemata can be matched in constant time if there are upper bounds on the maximal node degree and the number of roots in host graphs. The remaining steps of rule application, namely checking the dangling condition and the application condition, removing items from $L - K$, adding items from $R - K$, and relabelling nodes, are achievable in constant time.

2.3   Programs

GP 2 programs consist of a finite number of rule schema declarations and a main command sequence which controls their application order. Execution starts at the top-level procedure `Main`. The user may declare other named procedures, which consist of a mandatory command sequence and optional local rule and procedure declarations. Recursive procedures are not allowed.

The control constructs are: application of a set of conditional rule schemata $\{r_1, \ldots, r_n\}$, where one of the applicable schemata in the set is non-deterministically chosen; sequential composition $P; Q$ of programs P and Q; as-long-as-possible iteration $P!$ of a program P; and conditional branching statements if C then P else Q and try C then P else Q, where C, P and Q are arbitrary command sequences. The meaning of these constructs is formalised with a small-step operational semantics [2].

We just discuss the branching statements. To execute if C then P else Q on a graph G, first C is executed on G. If this produces a graph, then this result is thrown away and P is executed on G. Alternatively, if C fails on G, then Q is executed on G. In this way, graph programs can be used to test a possibly complex condition on a graph without destroying the graph. If one wants to continue with the graph resulting from C, the command try C then P else Q can be used. It first executes C on G and, if this fails, executes Q on G. However, if C produces a graph H, then P is executed on H rather than on G.

3   The GP 2 Compiler

The language is implemented with a compiler, written in C, that translates GP 2 source code to C code. The generated code is executed with the support of a runtime library containing the data structures and operations for graphs and morphisms. We describe how we convert high-level, non-deterministic and rule-based programs into deterministic, imperative programs in C.

3.1   Rule Application

Implementing a graph matching algorithm in the context of graph transformation systems is a well-researched problem. A frequently-used technique is the

```
bool match_rule(morphism *m) {
   return match_n0;
}

bool match_n0(morphism *m) {
   for(root nodes N of the host graph) {
      if(N is not a valid match for n0) continue;
      else {
         flag N as matched;
         update morphism;
         if(match_e0) return true;
      }
   }
   return false;
}

bool match_e0(morphism *m) {
   for(outedges E of match(n0)) {
      if(E is not a valid match for e0) continue;
      else {
         flag E as matched;
         update morphism;
         return true;
      }
   }
   return false;
}
```

Fig. 2. Skeleton of the rule matching code.

search plan, a decomposition of the matching problem into a sequence of primitive matching operations [7]. The compiler supports operations to match an isolated node, to match an edge incident to an already-matched node, and to match a node incident to an already-matched edge. A search plan is constructed by an undirected depth-first traversal of the left-hand side of a rule. When a node or edge is first visited, an operation to match that item is appended to the current search plan. Every iteration of the depth-first search starts at a root node, if one exists, to ensure that the initial "find node" operation is as cheap as possible. If all root nodes have been visited, it starts at an arbitrary unexplored node.

The generated code is a nested chain of matching functions corresponding to the search plan operations. The top-level function is named match_R for rule R. The pseudocode in Fig. 2 illustrates this structure for a rule that matches a root node with a looping edge.

Four checks are made to test if a host graph item h is a valid match for a particular rule item. They are listed below.

1. h is flagged as matched (note that we use injective matching).
2. The rule item is not marked **any** and h's mark is not equal to the rule item's mark.
3. h is not structurally compatible with respect to the rule and the current partial morphism.
4. h's label cannot match the expression of the rule item's label.

The third check differs for nodes and edges. Host graph nodes are ruled out if their degrees are too small. For example, a rule node with two outgoing edges cannot match a host graph node with only one outgoing edge. Host graph edges are checked for source and target consistency. For example, if the target of a rule edge is already matched, the host edge's target must correspond with that node.

To evaluate rule conditions, the compiler writes a function for each predicate and a function to evaluate the whole condition. The predicate functions modify the values of global boolean variables that are queried by the condition evaluator. The condition is checked directly after each call to a predicate function. If the condition is true or all variables in the condition have not been assigned values, matching continues. Otherwise, the match fails and the current matcher returns false, triggering a backtrack. At runtime, the predicate functions are called as soon as they are needed. For example, the function to check the predicate `indeg(1) = indeg(2)` is called immediately after rule node 1 is matched and immediately after rule node 2 is matched. This is done in order to detect an invalid match as soon as possible. To make this possible, a complex data structure is used at compile time to represent conditional rule schemata. The data structure links nodes and variables in the rule to each condition predicate querying that node or variable.

A rule schema contains complete information on the behaviour of the rule, including which items are added, which items are deleted, which items are relabelled, and which variables are required in updated labels. The rule is analysed at compile time to generate code to apply the rule given a morphism. Host graph modifications are performed in the following order to prevent conflicts and dangling edges: delete edges, relabel edges, delete nodes, relabel nodes, add nodes, add edges. The appropriate host nodes, host edges and values of variables are pulled from morphism data structures populated during the matching step.

3.2 Program Analysis for Graph Backtracking

The semantics of GP 2's loop and conditional branching commands require the host graph to be backtracked to a previous state in certain circumstances. For example, the `if-then-else` statement throws away the graph obtained by executing the condition before taking the **then** or **else** branch. Therefore there needs to be a mechanism to preserve older host graph states. We achieve this by maintaining a stack of changes made to the host graph. This is more space-efficient than storing multiple copies of the host graph. This concept is taken from the implementation of the first version of the GP language [11]. At compile time the program text is analysed to determine which portions of the program require recording of the host graph state. This analysis is quite subtle. For instance, a condition that requires graph backtracking in an `if-then-else` statement may not require graph backtracking in a `try-then-else` statement. We omit the details for lack of space. The first author's PhD thesis [2] describes the program analysis in detail, including the algorithm used by the compiler.

3.3   Program Translation

The main function of the generated C program is responsible for calling the matching and application functions as designated by the command sequence of the GP 2 program. Executing the program amounts to applying a sequence of rules. The code generator writes a short code fragment for each rule call and translates each control construct into an equivalent C control construct. The runtime code is supported by a number of global variables, including the host graphs and morphisms. A global boolean variable *success*, initialised to true, stores the outcome of a computation to support the control flow of the program.

A standard rule call generates code trying to match the rule. If a match is found, the code calls the rule application function and sets the success flag to true. If not, control passes to failure code. Certain classes of rules allow simpler code to be generated. For example, a rule with an empty left-hand side does not generate code to call a matching function. The failure code is context-sensitive. If there is a failure at the top level, the program is terminated after reporting to the user and freeing memory. Failure in a condition guard sets the global success flag to false so that control goes to the else branch of the conditional statement. Failure in a loop sets the success flag to false and calls the function `undoChanges` (described below) to restore the host graph to the state at the start of the most recent loop iteration.

Figure 3 summarises the translation of some GP 2 control constructs to C. The rule set call {R1, R2} is tackled by applying the rules in textual order until either one rule matches or they all fail. The do-while loop is used to exit the rule set if a rule matches before the last rule has been reached. The condition of a branching statement is executed in a do-while loop: if failure occurs before the last command of the condition, the break statement is used to exit the condition, and control is assumed by the then/else branch. GP 2's loop translates directly to a C while loop. One subtlety is the looped command sequence, where the line `if(!success) break;` is printed after the code for all commands except the last. A second subtlety is that success is set to true after exiting a loop because GP 2's semantic rules state that a loop cannot fail. Command sequences are handled by generating the code for each command in the designated order. When a procedure call is encountered in the program text, the code generator inlines the command sequence of the procedure at the point of the call.

Restore points (the variables named *rp* in Fig. 3) are created and assigned to the top of the graph change stack when graph backtracking is required. The function *undoChanges* restores a previous host graph state by popping and undoing changes from the stack until the restore point is reached. The function *discardChanges* pops

Command	Generated Code
{R1, R2}	```do { if(matchR1(M_R1)) { <success code> break; } if(matchR2(M_R2)) <success code> else <failure code> } while(false)```
if C then P else Q	```int rp = <top of GCS>; do C while(false); undoChanges(host, rp); if(success) P else Q;```
try C then P else Q	```int rp = <top of GCS>; do C while(false); if(success) P else { undoChanges(host, rp); Q }```
(P; Q)!	```int rp = <top of GCS>; while(success) { P if(!success) break; Q if(success) discardChanges(rp); } success = true;```

Fig. 3. C code for GP 2 control constructs

the changes but does not undo them. It is only called at the end of a successful loop iteration to prevent a failure in a future loop iteration from causing the host graph to roll back beyond the start of its preceding iteration. Each restore point has a unique identifier to facilitate multiple graph backtracking points.

The compiler respects the formal semantics of GP 2 (given in [13] and in updated form in [2]) in that any output graph of the generated code is admissible by the semantics. Similarly, a program run ending in failure is possible only if the semantics allows it. We did not formally prove this kind of soundness—that would be a tremendous project far beyond the scope of this work. Also, there is no guarantee that a program run terminates if a terminating execution path exists (this would require a breadth-first strategy which is impractical).

3.4   Runtime Library

The runtime library is a collection of data structures and operations used by the generated code during rule matching, rule application and host graph backtracking. As aforementioned, graph backtracking is performed by a graph change stack. We describe the other core data structures of the runtime library.

The host graph structure stores node and edge structures in dynamic arrays. Free lists are used to prevent fragmentation. Nodes and edges are uniquely identified by their indices in these arrays. The graph structure also stores the node count, the edge count, and a linked list of root node identifiers for fast access to the root nodes in the host graph. A node structure contains the node's identifier, its label, its degrees, references to its inedges and outedges, a root flag, and a matched flag. An edge structure contains the edge's identifier, its label, the identifiers of its source and target, and a matched flag.

A label is represented as a structure containing the mark (an enumerated type), a pointer to the list and the length of the list. GP 2 lists are represented internally as doubly-linked lists. Each element of the list stores a type marker and a union of integers and strings, equivalent to GP 2's atom type. Lists are stored centrally in a hash table to prevent unnecessary and space-consuming duplication of lists for large host graphs with repeated labels.

The morphism data structure needs to capture the node-to-node and edge-to-edge mappings, and the assignment mapping variables to their values. Thus the data structure used to represent morphisms contains the following four substructures: (1) an array of host node identifiers, (2) an array of host edge identifiers, (3) an array of assignments, and (4) a stack of variable identifiers. At compile time each node, edge and variable in a rule is identified with an index of its array in the morphism, allowing quick access to the appropriate elements. The stack is used to record assignment indices in the order in which the variables are assigned values. This is needed because the variables encountered at runtime are not guaranteed to agree with the compile-time order.

4   Case Study: 2-Colouring

Vertex colouring has many applications [18] and is among the most frequently considered graph problems. We focus on 2-colourability: a graph is *2-colourable*, or *bipartite*, if one of two colours can be assigned to each node such that the source and target of each non-loop edge have different colours.

Figure 4 shows a rooted GP 2 program for 2-colouring. The input is a connected, unmarked and unrooted graph G. If G is bipartite, the output is a valid 2-colouring of G. Otherwise, the output is G. The edges in this program are *bidirectional* edges, graphically denoted by lines without an arrowhead. Such a rule matches a host graph edge incident to two suitable nodes independent of the edge's direction. (This is syntactic sugar: a rule with one bidirectional edge is equivalent to a rule set containing two rules with the edge pointing in different directions.) The rules colour_red and joined_blues are omitted, which are the "inverted" versions of the rules colour_blue and joined_reds with respect to

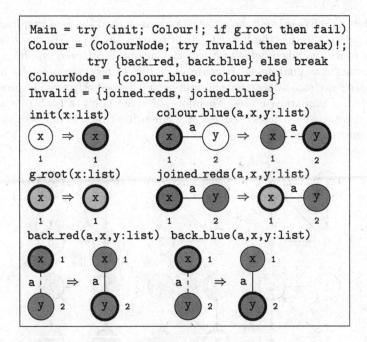

Fig. 4. The program 2colouring (Colour figure online)

the node marks. In particular, the right-hand side of joined_blues also has a grey root node.

At its core, 2colouring is an undirected depth-first traversal in which the source node is chosen non-deterministically. The root node represents the current position in the traversal. The rule init prepares the search by matching an arbitrary host graph node, making it the root node, and colouring it red. Each iteration of the Colour! loop does the following:

1. ColourNode: move the root node to an adjacent uncoloured node and colour it with the opposite colour. Dash the edge connecting the current root node to the previous one.
2. try Invalid else break: check if the current root node is adjacent to any nodes with the same colour. If so, mark the root node grey and break the inner loop.
3. Repeat steps (1) and (2) until no more rules are applicable.
4. try {back_red, back_blue} else break: move the root along a dashed edge and undash the edge. If this is not possible, break the outer loop.

Observe that the dashed edges act as a "trail of breadcrumbs" to facilitate backtracking. If the 2-colourability is violated at any point during the computation, the root node is marked grey, which acts as a flag for non-bipartiteness. Once the Colour! exits, the remainder of the program (if g_root then fail) checks if the root node is grey. If the root node is grey, then the fail command causes the `try-then-else` to take the `else` branch, and the host graph assumes its state before entering the branch, which returns the input graph. Otherwise, the `then` branch is taken, which returns the current 2-coloured graph.

Termination is guaranteed because each rule either decreases the number of unmarked nodes or decreases the number of dashed edges while preserving the number of unmarked nodes. Therefore, at some point, a back rule will fail because there exist no dashed edges, or a colouring rule will fail because there exist no unmarked nodes.

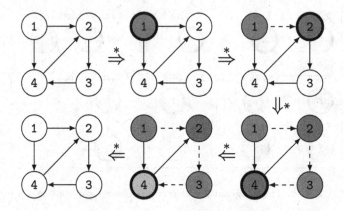

Fig. 5. Example run of `2colouring` (Colour figure online)

Figure 5 shows the execution of `2colouring` on the host graph in the upper-left of the diagram. This graph is clearly not 2-colourable. The rule `init` colours node 1 red. The rule `colour_blue` nondeterministically matches the edge $1 \rightarrow 2$. It roots node 2, colours it blue and dashes the edge. The colouring rules are applied twice more to give the lower-right graph. At this point the rule `joined_blues` matches the edge $4 \rightarrow 2$. This colours the root node grey. The inner loop breaks, and control passes to `Backtrack`. Both back rules fail because neither match a grey root node. This causes the outer loop to break. Finally, `g_root` succeeds, causing the `try` statement to fail and return the original graph.

The following result, proved in [2], assumes that input graphs are unmarked and connected.

Proposition 1 (Time Complexity of `2colouring`). *On graphs with bounded node degree, the running time of `2colouring` is linear in the size of graphs. On unrestricted graphs, the running time is quadratic in the size of graphs.*

Here "size of graphs" refers to the number of nodes and edges in host graphs. The result is independent of the size of host graph labels.

5   Performance

To experimentally validate the time complexity of 2colouring, and to test the performance of the language implementation, we ran the generated C code for 2colouring against an adaptation of Sedgewick's hand-crafted C program for 2-colouring [17].

We chose two classes of input graphs. The first class is *square grids* (abbreviated *grids*), which are suitable because: (1) grids are 2-colourable. This guarantees that both programs perform the same computation, namely matching and colouring every node in the graph; (2) grids have bounded node degree, which tests the linear complexity of 2colouring; (3) it is relatively simple to generate large grids. The second class is *star graphs*, used to test the performance on graphs of unbounded degree. A star graph consists of a central node with k outgoing edges. The targets of these outgoing edges themselves have a single outgoing edge. Star graphs share properties (1) and (3) of grid graphs. Examples can be seen in Fig. 6.

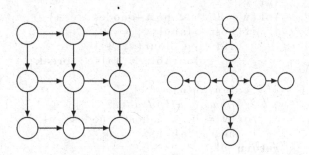

Fig. 6. Examples of a square grid graph and a star graph

5.1   2-Colouring in C

This section describes a C implementation of 2-colouring based on the code in Sedgewick's textbook *Algorithms in C* [17] which uses an adjacency list data structure for host graphs. For a graph with n nodes, an adjacency list is a node-indexed array containing n linked lists. An edge $i \to j$ is represented by the presence of j in the ith linked list, and vice versa if the graph is undirected. For our purposes there is no requirement to implement a graph data structure that supports the complete GP 2 feature set. Instead, we exploit some of the properties of the algorithms and host graphs we wish to execute in order to develop a minimal graph data structure.

We adapt Sedgewick's adjacency-list data structure and functions for host graphs. The main graph structure stores counts of the number of nodes and edges, a node-indexed array of adjacency lists, and a node-indexed array of integer node labels. Adjacency lists are represented internally by linked lists, where each list element stores the node identifier of a target of one of its outgoing edges.

```
1   bool dfsColour(int node, int colour) {
2     Link *l = NULL;
3     int new_colour = colour == 1 ? 2 : 1;
4     host->label[node] = new_colour;
5     for(l = host->adj[node]; l != NULL;
6         l = l->next)
7       if(host->label[l->id] == 0) {
8         if(!dfsColour(l->id, new_colour))
9           return false;
10        }
11      else if(host->label[l->id] != colour)
12        return false;
13    return true;
14  }
15
16  int main(int argc, char **argv) {
17    host = buildHostGraph(argv[1]);
18    bool colourable = true;
19    int v;
20    for(v = 0; v < host->nodes; v++)
21      if(host->label[v] == 0)
22        if(!dfsColour(v, 1)) {
23          colourable = false; break;
24        }
25    if(!colourable)
26      // Unmark all nodes.
27      for(v = 0; v < host->nodes; v++)
28        host->label[v] = 0;
29    return 0;
30  }
```

Fig. 7. DFS 2-colouring in C

At runtime, the GP 2 compiler's host graph parser is used to read the host graph text file and construct the graph data structure. This minimises the gap between the handwritten C code and the code generated from the GP 2 compiler, so that the comparison between the performance of the actual computations on the host graph is as fair as possible.

The C algorithm for 2-colouring is given in Fig. 7. Code for error checking, host graph building, and declaration of global variables is omitted. The program takes a single command line argument: the file path of the host graph. The function **buildHostGraph** initialises and adds edges to the graph (via the global Graph pointer *host*) through the GP 2 host graph parser.

Nodes are labelled 0, 1 or 2. Node labels are initialised to 0, representing an uncoloured and unvisited node. 1 and 2 represent the two colours of the algorithm. The function **dfsColour** is called recursively on all uncoloured nodes of the host graph. It is passed a node v and a colour c as its argument. It colours v

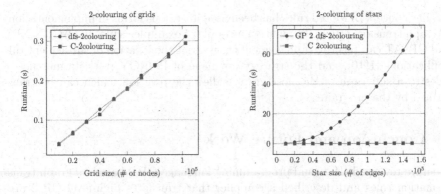

Fig. 8. Plots of the runtimes of the 2-colouring programs in GP 2 and C

with the contrasting colour c', and goes through v's adjacency list. If an adjacent node is uncoloured, `dfsColour` is called on that node. If an adjacent node is also coloured c', the function returns false, which will propagate through its parent calls and to the main function. If main detects a failure (line 25), it sets the label of all nodes to 0 and exits. Otherwise, the coloured graph is returned.

Figure 8 show the comparison of runtimes of both programs. There is almost no difference between the time it takes for either program to 2-colour grids, a remarkable result considering the compiled GP 2 code explicitly performs (rooted) subgraph matching at each step, while the tailored C program navigates a simple pointer structure. However, the star graph plot makes it clear that tailored C code is not limited by bounds on node degree. The compiled GP 2 code displays quadratic time complexity because it searches the outgoing edge list of the central node in the same order for every rule match.

6 Related Work

There exist a number of tools and languages for programming with graph transformation rules, including AGG [15], GROOVE [9] and PORGY [8]. We highlight three implementations with code generation. PROGRES [16] generates efficient Modula-2 or C code from transformation specifications. The code generator is more complex than that of GP 2 because it must handle sophisticated language features, for example arbitrary path expressions in rules and derived attributes. Programs in GrGEN.NET [10] are compiled to highly-optimised .NET assemblies for high performance execution. The code generator of the model transformation tool GReAT [19] has some similarity to that of GP 2: both generate pattern matching code that searches the host graph with user-declared root points to prune the search space. However, there are some differences because of the different feature sets of the languages. For example, GReAT's code generator must handle passing matched objects from one rule to another, while GP 2's code generator must handle application conditions during graph matching.

The concept of rooted rules has been used in various forms in implementations of graph transformation. To mention a couple of examples, rules in GrGEN.NET and GReAT can return graph elements to restrict the location of subsequent rule applications [1,10], and the strategy language of PORGY restricts matches of rules to a subgraph of the host graph called the position which can be transformed by the program [8].

7   Conclusion and Future Work

We have reviewed the visual programming language GP 2 based on graph transformation rules and described a compiler that translates high-level GP 2 programs to C code. A novel aspect of our implementation is generating search plans at compile time and using them to systematically generate structured and readable C code. Another distinctive feature is the static analysis of programs to determine if code needs to be generated to facilitate the recording of the host graph state. Using the compiler, we show that the generated C code for a depth-first 2-colouring program performs as quickly as a handcrafted C program also based on depth-first search on a class of host graphs with bounded node degree. These initial results are good, but more case studies ought to be investigated to further demonstrate the efficiency of the generated code, in particular programs that transform the host graph structurally, such as a reduction program to identify membership in a specific graph class.

A limitation of the GP 2 implementation is that it makes little effort to optimise rule matching for rules without root nodes. One method of speeding up matching is to compute optimal search plans at runtime based on an analysis of the host graph. This has been implemented in GrGEN.NET [5]. Another approach is to optimise rule matching at compile time. An example of such an optimisation is transforming a looped rule call to code that finds all matches in the host graph and performs the modifications in one step, which in general is more efficient than finding one match and starting a new search for the next match. This requires some care because pairs of matches could be in conflict.

References

1. Agrawal, A., Karsai, G., Neema, S., Shi, F., Vizhanyo, A.: The design of a language for model transformations. Softw. Syst. Model. **5**(3), 261–288 (2006)
2. Bak, C.: GP 2: Efficient Implementation of a Graph Programming Language. Ph.D. thesis, Department of Computer Science, The University of York (2015)
3. Bak, C., Faulkner, G., Plump, D., Runciman, C.: A reference interpreter for the graph programming language GP 2. In: Proceedings of Graphs as Models (GaM 2015), vol. 181 of Electronic Proceedings in Theoretical Computer Science, pp. 48–64 (2015)
4. Bak, C., Plump, D.: Rooted graph programs. In: Proceedings of International Workshop on Graph-Based Tools (GraBaTs 2012), vol. 54 of Electronic Communications of the EASST (2012)

5. Veit Batz, G., Kroll, M., Geiß, R.: A first experimental evaluation of search plan driven graph pattern matching. In: Schürr, A., Nagl, M., Zündorf, A. (eds.) AGTIVE 2007. LNCS, vol. 5088, pp. 471–486. Springer, Heidelberg (2008)
6. Dodds, M., Plump, D.: Graph transformation in constant time. In: Corradini, A., Ehrig, H., Montanari, U., Ribeiro, L., Rozenberg, G. (eds.) ICGT 2006. LNCS, vol. 4178, pp. 367–382. Springer, Heidelberg (2006)
7. Dörr, H. (ed.): Efficient Graph Rewriting and Its Implementation. LNCS, vol. 922. Springer, Heidelberg (1995)
8. Fernández, M., Kirchner, H., Mackie, I., Pinaud, B.: Visual modelling of complex systems: towards an abstract machine for PORGY. In: Beckmann, A., Csuhaj-Varjú, E., Meer, K. (eds.) CiE 2014. LNCS, vol. 8493, pp. 183–193. Springer, Heidelberg (2014)
9. Ghamarian, A.H., de Mol, M., Rensink, A., Zambon, E., Zimakova, M.: Modelling and analysis using GROOVE. Int. J. Softw. Tools Technol. Transf. 14(1), 15–40 (2012)
10. Jakumeit, E., Buchwald, S., Kroll, M.: GrGen.NET - the expressive, convenient and fast graph rewrite system. Softw. Tools Technol. Transf. 12(3–4), 263–271 (2010)
11. Manning, G., Plump, D.: The GP programming system. In: Proceedings of Graph Transformation and Visual Modelling Techniques (GT-VMT 2008), vol. 10 of Electronic Communications of the EASST (2008)
12. Plump, D.: The graph programming language GP. In: Bozapalidis, S., Rahonis, G. (eds.) CAI 2009. LNCS, vol. 5725, pp. 99–122. Springer, Heidelberg (2009)
13. Plump, D.: The design of GP 2. In: Proceedings of International Workshop on Reduction Strategies in Rewriting and Programming (WRS 2011), vol. 82 of Electronic Proceedings in Theoretical Computer Science, pp. 1–16 (2012)
14. Poskitt, C.M., Plump, D.: Hoare-style verification of graph programs. Fundamenta Informaticae 118(1–2), 135–175 (2012)
15. Runge, O., Ermel, C., Taentzer, G.: AGG 2.0 – new features for specifying and analyzing algebraic graph transformations. In: Schürr, A., Varró, D., Varró, G. (eds.) AGTIVE 2011. LNCS, vol. 7233, pp. 81–88. Springer, Heidelberg (2012).
16. Schürr, A., Winter, A., Zündorf, A.: The PROGRES approach: language and environment. In: Ehrig, H., Kreowski, H.J., Rozenberg, G. (eds.) Handbook of Graph Grammars and Computing by Graph Transformation, pp. 487–550. World Scientific, Singapore (1999)
17. Sedgewick, R.: Algorithms in C: Part 5: Graph Algorithms. Addison-Wesley, New York (2002)
18. Skiena, S.S.: The Algorithm Design Manual, 2nd edn. Springer, London (2008)
19. Vizhanyo, A., Agrawal, A., Shi, F.: Towards generation of efficient transformations. In: Karsai, G., Visser, E. (eds.) GPCE 2004. LNCS, vol. 3286, pp. 298–316. Springer, Heidelberg (2004)

An Algorithm for the Critical Pair Analysis of Amalgamated Graph Transformations

Kristopher Born[✉] and Gabriele Taentzer

Philipps-Universität Marburg, Marburg, Germany
{born,taentzer}@informatik.uni-marburg.de

Abstract. Graph transformation has been shown to be well suited as formal foundation for model transformations. While simple model changes may be specified by simple transformation rules, this is usually not sufficient for more complex changes. In these situations, the concept of amalgamated transformation has been increasingly often used to model *for each* loops of rule applications which coincide in common core actions. Such a loop can be specified by a kernel rule and a set of extending multi-rules forming an interaction scheme.

The Critical Pair Analysis (CPA) can be used to show local confluence of graph transformation systems. Each critical pair reports on a potential conflict between two rules. It has been shown recently that the generally infinite set of critical pairs for interaction schemes can be reduced to a finite set of non-redundant pairs being sufficient to show local confluence of the transformation system. Building on this basic result, we present an algorithm that is able to compute all non-redundant critical pairs for two given interaction schemes. The algorithm is implemented for Henshin, a model transformation environment based on graph transformation concepts.

1 Introduction

In model-based software development, models play a primary role w.r.t. requirements elicitation, software design and software validation. Model changes can be well specified as model transformations. If several developers work concurrently on the same model, they may run into conflicts that have to be resolved. For the execution of several model changes, a specific order may be necessary due to causal dependencies. To analyze such conflicts and dependencies as early as possible, critical pair analysis (CPA) [8,18] has been used. This analysis allows to check transformation rules for potential conflicts and dependencies at specification time, i.e., before run time. A critical pair describes a minimal conflicting situation that may occur in the transformation system. If every critical pair can be resolved by finitely many transformation steps, the system is locally confluent. Potential dependencies between rules can be discovered by inverting the first

This work was partially funded the German Research Foundation, Priority Program SPP 1593 "Design for Future - Managed Software Evolution".

R. Echahed and M. Minas (Eds.): ICGT 2016, LNCS 9761, pp. 118–134, 2016.
DOI: 10.1007/978-3-319-40530-8_8

rule and using it as input to the CPA, together with the second rule. In that case, local confluence of critical pairs show how resulting models of dependent transformations can be reached alternatively.

Conflicts as well as dependencies of model transformations have been analyzed by the CPA for several different applications as, e.g., finding conflicts and dependencies in functional requirement specifications of software systems [11], analyzing conflicts and dependencies of model refactorings [17] as well as in aspect-oriented modeling [16], and using conflict and dependency results to find the right order of edit operations for reporting model differences on an application-specific abstraction level [13].

While simple model changes can be well specified using simple rules, this is usually not sufficient for more complex model changes. Amalgamated graph transformation has proven to be suitable for specifying core actions equipped with a number of optional or context-dependent actions. Considered applications are, e.g., an interpreter semantics for statecharts [4], automatic model migration [15], and the specification of complex model edit operations [13]. A typical example of such complex changes are model refactorings where, e.g., equal attributes in subclasses are pulled up to one attribute in their super class.

Collaborative working developers are interested in understanding when model changes can be applied in parallel and when they are a potential source for conflicts. Being in conflict, it would be interesting to understand if and how these conflicts can be resolved. Hence, the notions of parallel independence, conflict and conflict resolution have to be lifted to amalgamated graph transformation.

An amalgamated transformation is specified by a interaction scheme containing a kernel rule and a set of extending multi-rules. While the kernel rule is intended to be matched exactly once, each multi-rule is matched as often as possible – in the general case, a fixed, but arbitrary number of times. An amalgamated rule over an interaction scheme contains at least the kernel rule and arbitrary many copies of multi-rules overlapping at the kernel rule. Hence, an interaction scheme specifies infinitely many amalgamated rules in general.

Applying the CPA to analyze conflicts and dependencies between interaction schemes confronts us with the problem to check infinitely many rule pairs and therefore, critical pairs. [22] shows that a finite set of CPs is enough to show local confluence of the overall transformation system. This result is proven for algebraic graph transformation [8]. Model transformations that are based on the Eclipse Modeling Framework (EMF) have been formally based on graph transformation in [5].

To apply the CPA to amalgamated transformations in practice, we need an algorithm that implements it. The main challenge is to find out an effective termination criterion when enumerating pairs of amalgamated rules and their critical pairs. The main contributions of this paper are the following:

1. An algorithm for the CPA of pairs of interaction schemes. We argue for the correctness of the presented algorithm w.r.t. the underlying theory. In particular, we focus on the termination of this algorithm.

2. An implementation of the algorithm within the model transformation tool Henshin based on EMF models.
3. First tests of the algorithm: We report on the CPA of an example transformation system, focussing on termination issues.

The paper is organized as follows: The main concepts of amalgamated graph transformation are recalled in Sect. 2. The main ideas for the CPA for amalgamated graph transformation are summarized in Sect. 3. Thereafter, we present our algorithm and argue for its correctness in Sect. 4.

2   Amalgamated Transformations

In the context of graph transformation, amalgamated transformation has been introduced to perform a kernel action exactly once and context-dependent actions as often as possible. In this section, we consider amalgamated transformations based on EMF [21] and use model refactorings as running example. The formal basis is given by amalgamated graph transformation as presented in [9] and the conflict analysis in [22]. Since the subtle differences do not play a role throughout this paper, we use the notions model and graph as synonyms in the following.

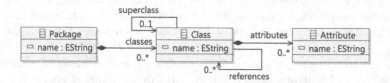

Fig. 1. Meta-model for class models

EMF is a common and widely-used open source technology in model-based software development. It extends Eclipse by modeling facilities and allows for defining (meta-)models and modeling languages by means of structured data models. An EMF-model can formally be considered as an instance graph with a prominent containment hierarchy.

Example 1 (Simple class models). In the running example, we consider selected refactorings of simple class models. A simple class model consists of a package being the container of all classes. A class is named and may have any number of attributes just given by their names. Classes may be related in two ways: A class may have a superclass and any number of references to other classes. The meta-model for simple class models is shown in Fig. 1.

A very simple instance model to this meta-model is shown in Fig. 2; it represents two classes "List" and "Stack" where the stack is inheriting from the list. Class "List" has an attribute called "first". Since this design is not optimal, it will be refactored later on.

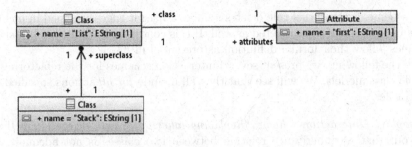

Fig. 2. Example instance model

In the context of EMF, refactorings are specified by model transformations. See e.g. [3,14]. In the following, we consider rule-based model transformations formulated in Henshin [2], a model transformation language based on graph transformation concepts. In Henshin, rules may be depicted in an integrated form annotating each model element node and reference edge by a change action. Nodes and edges that have to exist but are not changed during transformation are annotated with << preserve >> while others may be deleted or created dependent on their annotations. In addition, rule may have application conditions. In negative application conditions, nodes and edges may be forbidden meaning that they must not occur in the specified form for applying the rule. In contrast, positive application conditions may require model elements.

Example 2 (Deleting an empty class). A class which does not have any attribute or reference and which is not involved in class inheritance relations is called empty class. Since an empty class does not have any effect on other classes, it can be deleted. This refactoring is specified by a simple Henshin rule as shown in Fig. 3. Note that the class may only be deleted if no context edge is adjacent, i.e., the *dangling condition* has to hold.

When performing model refactorings, a restructuring action is often accompanied by update actions on all involved model elements. For example, pulling up an attribute to a superclass implies the deletion of such an attribute from all subclasses. Such *for all* actions are specified by additional multi-rules comprising the basic rule (also called kernel rule). The overall rule (with optionally contained

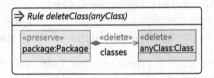

Fig. 3. Refactoring rule "Delete Class" (DC)

multi-rules) is also called *interaction scheme*; its semantics is defined by a set of rules (see below). A rule without any multi-rule is a special case of interaction scheme consisting of just one rule. In the following, an interaction scheme is represented in an integrated way, i.e., all multi-rules are represented in one diagram overlapping in the kernel rule. Note that - given an interaction scheme - the kernel rule always performs a subset of actions specified in a multi-rule of that scheme. If the kernel rule deletes a node, adjacent edges specified in including multi-rule have to be deleted as well. If this condition is fulfilled, interaction schemes follow their formal definition as presented in [9].

In the following, we present several interaction schemes for the refactoring of simple class models. We will see that they all include *for all* actions specified by multi-rules.

Example 3 (Interaction scheme "Replacing inheritance with delegation"). If we find out that an inheritance relation between two classes is not adequate as, e.g., pointed out in Example 1 for class "Stack" inheriting from class "List", the inheritance relation might be replaced by a reference. Formerly inherited attributes have to be copied in that case. This refactoring can be specified by a kernel rule just replacing the superclass reference while the extending multi-rule copies all attributes. Figure 4 shows the corresponding specification in Henshin. All cascaded nodes and adjacent edges are in the multi-rule only while all other nodes and edges are also contained in the kernel rule.

Given an interaction scheme, i.e. a kernel rule with multi-rules, its semantics consists of an infinite set of simple rules called *amalgamated rules*. Each rule of this set consists of the kernel rule extended by 0, 1, 2 or more copies of its multi-rules. For each multi-rule, the exact number of copies depends on the number of different matches found in the instance graph the interaction scheme is applied to. It is assumed that all multi-rule matches overlap in the match of

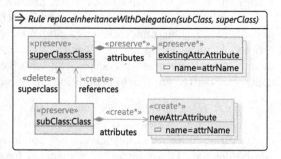

Fig. 4. Interaction scheme for refactoring "Replace Inheritance With Delegation" (RIWD)

their common kernel rule. In the following, we show example amalgamated rules for the refactoring "Replace Inheritance With Delegation".

Example 4 (Amalgamated rules and their application). Given the interaction scheme "replaceInheritanceWithDelegation" as in Fig. 4, Fig. 5 shows three amalgamated rules as concrete examples using 0, 1 or 2 copies of the multi-rule.

Considering the instance model in Fig. 2, the inheritance between classes "List" and "Stack" shall be replaced by a delegation. Hence, we apply the refactoring "Replace Inheritance With Delegation" here. Since class "List" has one attribute, the multi-rule is applied exactly once which means that amalgamated rule riwd_1 is selected for application. The result is the model in Fig. 6. The effect is that the inheritance relation between classes "List" and "Stack" is replaced by a reference and an attribute with name "first" is added to class "Stack".

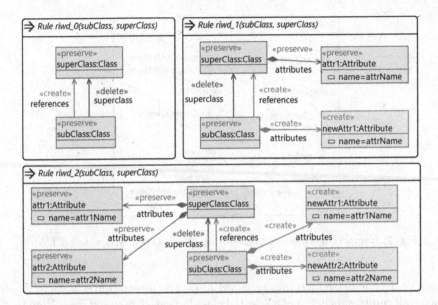

Fig. 5. Amalgamated rules which replace inheritance with delegation for classes with 0 to 2 attributes

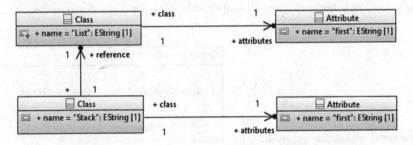

Fig. 6. Example instance model after refactoring

After having specified one refactoring we consider three further refactoring specifications below. They are used to investigate selected refactoring conflicts and their potential resolutions below. They all use multi-rules.

Example 5 (Interaction scheme "Push down attribute"). An attribute of a given superclass may be pushed down to all its subclasses. This refactoring is needed if the modeled attribute shall be modified in its subclasses in different ways. This refactoring is the opposite of "Pull up attribute" which is not considered in detail in this paper. The diagram in Fig. 7 shows the specifying interaction scheme. The kernel rule pushes down an attribute to one subclass while the multi-rule pushes down the attribute to all further subclasses.

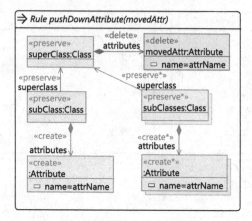

Fig. 7. Interaction scheme for refactoring "Push Down Attribute" (PDA)

Example 6 (Further refactoring specifications). The interaction scheme in Fig. 8 deletes all empty subclasses of a selected class indicated as superClass. Note that model nodes may only be deleted if they do not leave any edges dangling. This means for a class that it must not have attributes, references or subclasses. I.e., the interaction scheme deletes all empty subclasses of a given superclass.

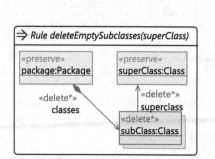

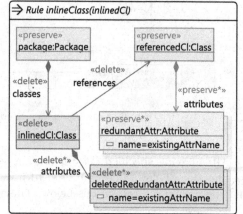

Fig. 8. Interaction scheme for refactoring "Delete Empty Subclasses" (DES)

Fig. 9. Interaction scheme for refactoring "Inline Class" (IC)

Another class refactoring is the inlining of classes shown in Fig. 9. If all the attributes of a class A have corresponding attributes (with the same names) in a referenced class B then class A can be inlined into class B. This means that class A and all its attributes are deleted. Again, the dangling condition checks if there are no further attributes (with different names), adjacent references or inheritance relations. In that case, the inlining must not take place.

3   Critical Pair Analysis

The critical pair analysis (CPA) is a well-known technique to analyze potential conflicts and dependencies of transformation systems. It has first been introduced for term rewriting and later generalized to graph transformation [8,18]. A critical pair describes a minimal conflicting situation that may occur in the transformation system. It is well-known that if all critical pairs can be shown to be strictly confluent, the transformation system is locally confluent. This means that each pair of direct transformation steps can be resolved by arbitrary many steps to a common graph. The notion of strict cofluence means that the jointly preserved part of a critical pair is also preserved by its resolution [19].

This theory has been extended to amalgamated graph transformation in [22]. Here, we have to face the problem that an interaction scheme generally describes infinite many rules and therefore, also infinite many critical pairs may exist for a given interaction scheme. In [22], we show that it is enough to check finitely many critical pairs to decide for local confluence. The key observation is that, from a certain number n of multi-rule copies, critical pairs over amalgamated rules do not lead to new kinds of conflict resolutions, i.e., all larger critical pairs are redundant to smaller ones. Up to now, however, there does not exist a construction to determine this number n. As main contribution of this paper, we present an algorithm for the CPA of amalgamated transformations below. As a prerequisite, the main definitions are recalled and illustrated at the running example here.

Two transformations are *conflicting* if (1) one transformation deletes a graph element the other uses (delete/use conflict), (2) one transformation produces a graph element the other forbids (produce/forbid conflict), or (3) one transformation changes an attribute the other uses (change/use conflict). A *critical pair*, short CP, consists of two conflicting transformations $G \overset{r1,m1}{\Longrightarrow} H1$ and $G \overset{r2,m2}{\Longrightarrow} H2$ applying rules $r1, r2$ at matches $m1, m2$ such that G is minimal. If rules $r1$ and $r2$ do not have application conditions, G is just an overlap graph of their left-hand sides. For rules with negative application conditions (NACs), also slightly larger graphs have to be considered taking parts of their NACs into account as well.

Example 7 (Critical pair). Applying refactorings "Delete Empty Subclass" and "Replace Inheritance With Delegation" in parallel may lead to conflicts. Figure 10 shows a CP over corresponding amalgamated rules, each one applying exactly one multi-rule copy. In this case, exactly one empty subclass is deleted and one attribute is copied to a referring class. This CP shows a delete/use conflict since a subclass that is deleted cannot be changed to be a delegating class. This is a potential conflict that may occur during transformations. It may be resolved by inlining the delegating class on the right yielding the model graph

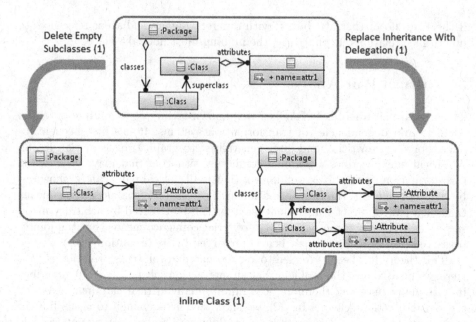

Fig. 10. Critical pair applying refactorings DES and RIWD with one multi-rule copy each

on the left. Note that the potential conflict shown here becomes concrete when all the variables are instantiated by concrete values.

Given two critical pairs $cp_s = (ts1 : G_s \implies H_{s1}, ts2 : G_s \implies H_{s2})$ and $cp_l = (tl1 : G_l \implies H_{l1}, tl2 : G_l \implies H_{l2})$ of set $CP(is1, is2)$ where G_s is part of G_l. cp_l is an extension of cp_s if H_{s1} is part of H_{l1} and H_{s2} is part of H_{l2} and moreover, the same interaction schemes $is1$ and $is2$ are applied. Then these CPs are considered to be *redundant* if their transformations $ts1$ and $tl1$ (as well as $ts2$ and $tl2$) allow for equivalent partial matches only, considering all rules of a given transformation system. Given a transformation $G \implies H$ two partial matches m and m' to a graph H are *equivalent* if each pair of isomorphic range elements has the same history, i.e., both are newly created or both do already exist. Due to this definition, partial matches are considered equivalent if they differ only in range elements stemming from different multi-rule copies. If the dangling condition is set for a rule, the equivalence check comprises the satisfaction check of this condition as well.

Example 8 (Redundancy of critical pairs). Considering the critical pairs in Figs. 10 and 11, we can notice that the CP in Fig. 10 applies one multi-rule copy on each side while the one in Fig. 11 applies two copies on each side. Basically, the same potential delete/use conflict is reported: A subclass that is deleted cannot be changed to a delegating class. However, the contexts are different. Although this is the case, the conflict resolution for the larger CP can be similar to the one for the smaller CP. Inlining the delegating class (with two attributes now) on the

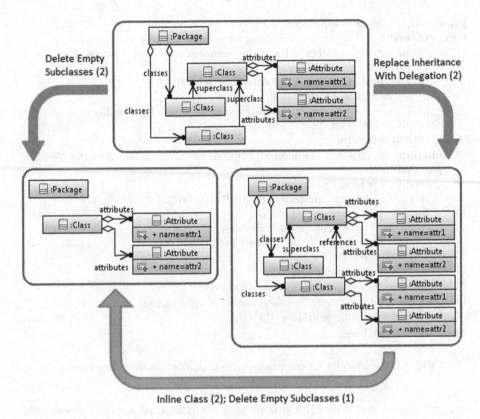

Fig. 11. Critical pair applying refactorings DES and RIWD with two multi-rule copies each

right followed by deleting the remaining empty subclass yields the model graph on the left. Any other interaction scheme is not applicable on the left or right. Comparing all the partial matches that exist in both cases and check whether they are equivalent w.r.t. the above definition, we find out that this is the case for all interaction schemes except of DES. Since the right graph in Fig. 11 still contains a generalization relation, some new partial matches can be found here. Hence, the CP in Fig. 11 is not redundant to the one in Fig. 10 although a very similar conflict is reported.

4   Algorithm for the Critical Pair Analysis

In the following, we present the core algorithm for computing all relevant critical pairs between two interaction schemes. As shown in [22], a finite set of critical pairs is enough to decide for local confluence of a given transformation system. The computation is performed for increasingly larger amalgamated rules. The maximal number of multi-rules applied define the level of computation. The

Input: *is1,is2*: Input interaction schemes
Output: *resultCps*: Output set containing all non-redundant critical pairs

```
 1: function COMPUTECRITICALPAIRS(InteractionScheme is1, InteractionScheme
       is2): CpaResult
 2:     rulePairHandler = new RulePairHandler(is1,is2);
 3:     resultCps = new CpaResult();
 4:     levelCps = analyseLevelForNonRedundantCps(0, rulePairHandler);
 5:     resultCps.add(levelCps);
 6:     return resultCps;
 7: function ANALYSELEVELFORNONREDUNDANTCPS(int level, RulePairHandler
       rph): CpaResult
 8:     resultCps = new CpaResult();
 9:     currentRulePairs = rph.getRulePairsOfLevel(level);
10:     for rulePair : currentRulePairs do
11:         currentCps = computeCps(rulePair);
12:         newCps = extractNonRedundantCps(currentCps,level);
13:         rulePair.setCps(newCps);
14:         resultCps.add(newCps);
15:     if resultCps.size() != 0 || level == 0 then
16:         resultOfNextLevel = analyseLevelForNonRedundantCps(level+1);
17:         resultCps.add(resultOfNextLevel);
18:     return resultCps;
```

Fig. 12. Pseudocode for computing critical pairs of interaction schemes.

stop criterion is met if all critical pairs that are computed for the current level turn out to be redundant to critical pairs of lower levels. The main algorithm is presented in Fig. 12.

Class *RulePairHandler* is a container for rule pairs and their critical pairs and the associated partial matches of the transformation system. Function *getRule-PairsOfLevel* computes all pairs of amalgamated rules where each rule has at most as many multi-rule copies as *level* prescribes. Given a concrete rule pair, function *computeCps* computes all critical pairs of this rule pair. Each critical pair is reported by a minimal model with two matches of participating rules. After having computed all critical pairs of a given level, function *extractNonRe-dundantCps* filters out all those critical pairs that are not redundant to already existing ones computed in lower levels. The identification of redundant critical pairs is achieved by comparing the partial matches of the whole transformation system for each new critical pair against the already known.

Correctness. Given two interaction schemes *is1* and *is2*, we have to show that COMPUTECRITICALPAIRS yields the set of all non-redundant CPs between these two rules schemes. The main design decision is here that CPs are computed level-wise starting with level 0. All CPs of level n are computed if $n \leq 1$ or new non-redundant CPs have been computed for level $n-1$. Level one has to be considered anytime due to the fact that level zero doesn't involve the amalgamations at all. Given level n, function ANALYSELEVELFORNONREDUNDANTCPS computes all

```
 1: function EXTRACTNONREDUNDANTCPS(cpaResult cps): cpaResult
 2:     nonRedundantCps = new cpaResult();
 3:     for cp : cps do
 4:         tl1 = cp.getTransitionOfR1();
 5:         pMaTl1 = findAllNonEquivPartialMatches(tl1);
 6:         tl2 = cp.getTransitionOfR2();
 7:         pMaTl2 = findAllNonEquivPartialMatches(tl2);
 8:         rulePairs = rph.getReducedRulePairs(cp.getRulePair());
 9:         for rulePair : rulePairs do
10:             reducedCps = rulePair.getCriticalPairs();
11:             for reducedCp : reducedCps do
12:                 isExtension = isExtension(cp, reducedCp);
13:                 if isExtension && (pMaTl1.size()>0 || pMaTl2.size()>0) then
14:                     ts1 = reducedCp.getTransformation1();
15:                     alreadyKnownpMaTl1 = extractEquivParMatches(ts1, tl1);
16:                     pMaTl1.removeAll(alreadyKnownpMaTl1);
17:                     ts2 = reducedCp.getTransformation2();
18:                     alreadyKnownpMaTl2 = extractEquivParMatches(ts2, tl2);
19:                     pMaTl2.removeAll(alreadyKnownpMaTl2);
20:         if pMaTl1.size()>0 || pMaTl2.size()>0 then
21:             nonRedundantCps.addResult(cp);
22:     return nonRedundantCps;
```

Fig. 13. Pseudocode of function *extractNonRedundantCps*.

non-redundant CPs of rule pairs of that level. All rule pairs for that level where each rule has at most n multi-rule copies, are collected in *currentRulePairs*. All their CPs are collected in *currentCps*. Function EXTRACTNONREDUNDANTCPS directly implements the check for non-redundant CPs based on the definition given in [22] which is informally recalled above. If any rule pair of a level yields new non-redundant CPs, the set of resulting CPs becomes non-empty and the next level has to be considered. Finally, the non-redundant CPs of all levels are joined to the set *resultCps*.

In the algorithm in Fig. 13 we extract all critical pairs that are not redundant to existing ones. First of all we compute all partial matches of the two involved transformations *tl1* and *tl2* resulting in *pMaTl1* and *pMaTl2*. Thereafter we take the rule pair of *cp* and collect all critical pairs of reduced rule pairs in *reducedCps*. A rule pair is reduced if it applies at least one multi-rule less than the original rule pair (on the left or right-hand side). Then we check if the current critical pair *cp* is an extension of a reduced one, consisting of the transformations *ts1* and *ts2*. If this is the case, we have to further check if it is even redundant. For this check all the partial matches on *ts1* and *tl1* as well as *ts2* and *tl2* have to be compared. All partial matches on *ts1* (*ts2*) that are equivalent to partial matches on *tl1* (*tl2*) are removed from the original set of partial matches *pMaTl1* (*pMaTl2*). If there are non-equivalent partial matches left in *pMaTl1* or *pMaTl2*, the current critical pair *cp* is considered non-redundant.

The central question to be answered here is: Does this algorithm terminate? The answer should be yes due to the main result in [22]. The proof of this result contains the following key idea: For each pair of interaction schemes, there are two finite numbers c and d such that all rule pairs of amalgamated rules with at most c and d multi-rule copies yield redundant CPs only. If we take $max(c, d)$ as current level, there would not be any further non-redundant conflict found. The key idea for termination is that new CPs do not provide new non-equivalent partial matches for the rules of our transformation system. An extreme over-approximation can go like this: Given a transformation system with interaction schemes, let $|L| = x$ be the number of graph elements of the left-hand side of the largest (multi-)rule. There are at most $|\mathcal{P}(L)| = 2^x$ different subsets of elements, i.e., domains for partial matches, over L. The different ranges are not interesting in detail. We only check if range elements are preserved or newly created. If two partial matches with the same domain are equal w.r.t. this range classification, they are considered equivalent. Hence, the largest number of non-equivalent partial matches is $2^{2x} = 4^x$. Although this number is extremely high, it tells us that there is an upper limit for non-equivalent partial matches independent of the result graphs $H1$ and $H2$ occurring in concrete CPs. Usually, the number of non-equivalent partial matches is much smaller since element types, attributes and graph structures have to be taken into account as well. Moreover, partial matches cannot exist if sub-matches do not exist as well. The following examples produce numbers of partial matches being smaller than 200 for any CP result graph although the extreme over-approximation yields 4^9. The value nine is based on the interaction scheme "Push down attribute", which has the most model elements in its left-hand side compared to the other ones in the transformation system.

Example 9 (Algorithm run). To illustrate the algorithm, we consider an example run now: Given the interaction schemes for refactorings DES and RIWD in Figs. 8 and 4, all non-redundant CPs with a DES rule as first and a RIWD rule as second shall be computed. As pointed out above, the CPs of levels 0 and 1 always have to be computed. For all pairs of kernel and multi-rules of participating interaction schemes, the number of CPs found are shown in Table 1. Moreover, it shows the numbers of CPs for all pairs of amalgamated rules of levels 2 and 3. This is needed since the check by *extractNonRedundantCps* finds out that RIWD and IC have new non-equivalent partial matches to CP graphs of rule pairs of level 1, as shown in Table 2. Therefore, there are non-redundant CPs on level 1 and hence, level 2 has to be considered. Note that the table entries always show the number of non-redundant CPs as well as the number of all CPs. One example CP of level 2 is shown in Fig. 11. In contrast to CPs of level 1, an inheritance relation may remain after applying refactoring RIWD which leads to new non-equivalent partial matches of refactorings DES and PDA. Therefore, new non-redundant CPs occur on level 2 and the algorithm does not yet terminate. Hence, amalgamated rules of level 3 have to be checked for non-redundant CPs as well. As explained at Table 2, it turns out that new non-equivalent partial matches

Table 1. Numbers of non-redundant/all critical pairs between refactorings DES (first rule) and RIWD (second rule)

CPs 1./2.	RIWD (0)	RIWD (1)	RIWD (2)	RIWD (3)
DES(0)	0/0	0/0	0/0	0/0
DES(1)	1/1	1/1	1/1	0/1
DES(2)	2/2	2/2	0/2	0/2
DES(3)	0/3	0/3	0/3	0/3

Table 2. Numbers of non-equivalent partial matches to the left and right result graphs of CPs between refactorings DES and RIWD

# Part. matches	RIWD (0)	RIWD (1)	RIWD (2)	RIWD (3)
DES(0)	-	-	-	-
DES(1)	15/28	20/84	10/0	0/0
DES(2)	0/4	0/2	0/0	0/0
DES(3)	0/0	0/0	0/0	0/0

do not occur and therefore, all newly found CPs are redundant leading to the termination of the algorithm for the considered interaction schemes.

Given all the CPs between refactorings DES and RIWD, Table 2 shows the numbers of non-equivalent matches to the left and right result graphs of each CP. (Remember that a CP consists of two conflicting transformations both starting from the same graph and resulting in two graphs. The left one is the result after applying refactoring DES while the right one is obtained by applying refactoring RIWD.) We see that on level 3, there are no further non-equivalent matches discovered. Hence, the computation of non-redundant CPs terminates after level 3 as stated above.

In the following, we summarize the number of non-redundant CPs found in our transformation system and show for each pair of interaction schemes how many levels of amalgamation have to be considered for the CPA until the termination criterion is fulfilled. We consider two variants of our transformation system: The first one contains all presented interaction schemes without PDA while the second one includes PDA. Tables 3 and 4 show the results. In both tables, we see that the number of levels needed is moderate. Often, the consideration of levels 0 and 1 is already enough. Tables 3 and 4 do not only differ in the numbers of rows and columns but also w.r.t. their entries. As an example, the CPA for DES and RIWD needs 3 or 4 levels, resp. The tables also show the numbers of non-redundant CPs found for pairs of interaction schemes which is 7 for pairs (DES,RIWD) as well as (RIWD,DES) on level 3 and 13 on level 4. The differences arise due to the fact that PDA causes new kinds of partial matches. These additional conflicts have to be taken into account for future confluence check. A more detailed view of the results can be found at [12].

The presented algorithm for the CPA of interaction schemes has been prototypically implemented for rules specified in Henshin. It relies on the CPA implementation for basic rules as presented in [6]. The current CPA implementation for interaction schemes supports the conflict detection only (i.e. does not support the detection of dependencies yet). Furthermore, rules with application conditions are not supported yet. These limitations are easy to erase which will be done in the near future.

Table 3. Level of termination and number of non-redundant critical pairs with four interaction schemes

Level of amalg /# non-red. CPs	DC	RIWD	DES	IC
DC	1/0	1/0	1/2	1/0
RIWD	1/0	2/1	3/7	1/0
DES	1/0	3/7	1/0	1/0
IC	1/0	1/0	2/4	1/0

Table 4. Level of termination and number of non-redundant critical pairs with five interaction schemes

Level of amalg ./# non-red. CPs	DC	RIWD	DES	IC	PDA
DC	1/0	1/0	1/2	1/0	1/0
RIWD	1/0	2/1	4/13	1/0	2/10
DES	1/0	4/13	1/0	1/0	2/3
IC	1/0	1/0	2/4	1/0	1/0
PDA	1/0	3/4	3/8	3/4	2/6

5   Related Work and Conclusion

Multi-objects and other variants that match graph parts as often as possible have been considered in several graph transformation approaches: in tool environments such as PROGRES [20] and Fujaba [1] as well as in conceptual approaches by Grönmo [10] and Drewes et al. [7]. These tools and approaches, however, do not support the critical pair analysis (CPA) for graph transformation systems expressing such variability.

While a basic graph transformation approach is taken in [22] to develop the necessary theory for the CPA for amalgamated graph transformation, we switch to model transformation based on the Eclipse Modeling Framework (EMF) and Henshin here. EMF models have typed, attributed graphs as conceptual basis while Henshin is based on graph transformation concepts. Hence, we have developed the CPA for the amalgamated transformation of typed, attributed graphs here. However, we do not yet consider application conditions for rules.

The main contribution of this paper is an algorithm for computing all non-redundant critical pairs of two given interaction schemes. It shows that the CPA for pairs of simple rules can be reused. The key idea is to compute all critical pairs for pairs of small amalgamated rules. This computation stops at level n when all pairs of rules with at most n copies of multi-rules yield redundant critical pairs only. We have implemented this algorithm in Henshin. First tests with a set of refactoring interaction schemes have shown that the CPA for interaction schemes is performed in a reasonable amount of time. An extensive evaluation is planned for future work.

References

1. The Fujaba tool suite. www.fujaba.de
2. Arendt, T., Biermann, E., Jurack, S., Krause, C., Taentzer, G.: Henshin: advanced concepts and tools for in-place EMF model transformations. In: Petriu, D.C., Rouquette, N., Haugen, Ø. (eds.) MODELS 2010, Part I. LNCS, vol. 6394, pp. 121–135. Springer, Heidelberg (2010)
3. Arendt, T., Taentzer, G.: A tool environment for quality assurance based on the eclipse modeling framework. Autom. Softw. Eng. **20**(2), 141–184 (2013)

4. Biermann, E., Ehrig, H., Ermel, C., Golas, U., Taentzer, G.: Parallel independence of amalgamated graph transformations applied to model transformation. In:
   Engels, G., Lewerentz, C., Schäfer, W., Schürr, A., Westfechtel, B. (eds.) Nagl
   Festschrift. LNCS, vol. 5765, pp. 121–140. Springer, Heidelberg (2010)
5. Biermann, E., Ermel, C., Taentzer, G.: Formal foundation of consistent EMF model
   transformations by algebraic graph transformation. Softw. Syst. Model. 11(2), 227–
   250 (2012). http://dx.doi.org/10.1007/s10270-011-0199-7
6. Born, K., Arendt, T., Heß, F., Taentzer, G.: Analyzing conflicts and dependencies
   of rule-based transformations in Henshin. In: Egyed, A., Schaefer, I. (eds.) FASE
   2015. LNCS, vol. 9033, pp. 165–168. Springer, Heidelberg (2015)
7. Drewes, F., Hoffmann, B., Janssens, D., Minas, M.: Adaptive star grammars and
   their languages. Theor. Comput. Sci. 411(34–36), 3090–3109 (2010)
8. Ehrig, H., Ehrig, K., Prange, U., Taentzer, G.: Fundamentals of Algebraic Graph
   Transformation. Monographs in Theoretical Computer Science. Springer, New
   York (2006)
9. Golas, U., Habel, A., Ehrig, H.: Multi-amalgamation of rules with application
   conditions in M-adhesive categories. Math. Struct. Comput. Sci. 24(4) (2014)
10. Grønmo, R., Krogdahl, S., Møller-Pedersen, B.: A collection operator for graph
    transformation. In: Paige, R.F. (ed.) ICMT 2009. LNCS, vol. 5563, pp. 67–82.
    Springer, Heidelberg (2009)
11. Hausmann, J.H., Heckel, R., Taentzer, G.: Detection of conflicting functional
    requirements in a use case-driven approach: a static analysis technique based on
    graph transformation. In: Proceedings of the 22rd International Conference on
    Software Engineering, ICSE 2002, Orlando, Florida, USA, 19–25 May, pp. 105–
    115. ACM (2002)
12. More details on the results. http://www.uni-marburg.de/fb12/swt/cpa_amal
13. Kehrer, T., Kelter, U., Taentzer, G.: Consistency-preserving edit scripts in model
    versioning. In: 2013 28th IEEE/ACM International Conference on Automated Soft-
    ware Engineering, ASE 2013, Silicon Valley, CA, USA, 11–15 November, pp. 191–
    201. IEEE (2013)
14. Kolovos, D.S., Paige, R.F., Polack, F., Rose, L.M.: Update transformations in the
    small with the epsilon wizard language. J. Object Technol. 6(9), 53–69 (2007).
    http://dx.doi.org/10.5381/jot.2007.6.9.a3
15. Mantz, F., Taentzer, G., Lamo, Y., Wolter, U.: Co-evolving meta-models and their
    instance models: a formal approach based on graph transformation. Sci. Comput.
    Program. 104, 2–43 (2015)
16. Mehner-Heindl, K., Monga, M., Taentzer, G.: Analysis of aspect-oriented mod-
    els using graph transformation systems. In: Moreira, A., Chitchyan, R., Araújo,
    J., Rashid, A. (eds.) Aspect-Oriented Requirements Engineering, pp. 243–270.
    Springer, New York (2013)
17. Mens, T., Taentzer, G., Runge, O.: Analysing refactoring dependencies using graph
    transformation. Softw. Syst. Model. 6(3), 269–285 (2007)
18. Plump, D.: Critical pairs in term graph rewriting. In: Privara, I., Ružička, P.,
    Rovan, B. (eds.) MFCS 1994. LNCS, vol. 841, pp. 556–566. Springer, Heidelberg
    (1994)
19. Plump, D.: On termination of graph rewriting. In: Nagl, M. (ed.) WG 1995. LNCS,
    vol. 1017, pp. 88–100. Springer, Heidelberg (1995)
20. Schürr, A., Winter, A., Zündorf, A.: The PROGRES approach: language and envi-
    ronment. In: Ehrig, H., Engels, G., Kreowski, H.J., Rozenberg, G. (eds.) Handbook
    of Graph Grammars and Computing by Graph Transformation. Applications, Lan-
    guages and Tools, vol. 2, pp. 487–550. World Scientific (1999)

21. Steinberg, D., Budinsky, F., Patenostro, M., Merks, E.: EMF: Eclipse Modeling Framework, 2nd edn. Pearson Eduction, London (2009)
22. Taentzer, G., Golas, U.: Towards local confluence analysis for amalgamated graph transformation. In: Parisi-Presicce, F., Westfechtel, B. (eds.) ICGT 2015. LNCS, vol. 9151, pp. 69–86. Springer, Heidelberg (2015). https://opus4.kobv.de/opus4-zib/frontdoor/index/index/docId/5494

Queries

Rete Network Slicing for Model Queries

Zoltán Ujhelyi[2]([✉]), Gábor Bergmann[1], and Dániel Varró[1]

[1] Department of Measurement and Information Systems, MTA-BME Lendület
Research Group on Cyber-Physical Systems,
Budapest University of Technology and Economics,
Magyar Tudósok Krt. 2, Budapest 1117, Hungary
{bergmann,varro}@mit.bme.hu
[2] IncQuery Labs Ltd., Bocskai út 77-79, Budapest 1113, Hungary
zoltan.ujhelyi@incquerylabs.com

Abstract. Declarative model queries captured by graph patterns are
frequently used in model driven engineering tools for the validation of
well-formedness constraint or the calculation of various model metrics.
However, their high level nature might make it hard to understand all
corner cases of complex queries. When debugging erroneous patterns, a
common task is to identify which conditions or constraints of a query
caused some model elements to appear in the results. Slicing techniques
in traditional programming environments are used to calculate similar
dependencies between program statements. Here, we introduce a slicing
approach for model queries based on Rete networks, a cache structure
applied for the incremental evaluation of model queries. The proposed
method reuses the structural information encoded in the Rete networks
to calculate and present a trace of operations resulting in some model
elements to appear in the result set. The approach is illustrated on a
running example of validating well-formedness over UML state machine
models using graph patterns as a model query formalism.

Keywords: Program slicing · Model queries · Graph patterns

1 Introduction

Modern industrial and open source modeling tools frequently rely upon various
services built on top of incremental query evaluation techniques [1,2] for
efficient revalidation of well-formedness constraints, recalculation of view models, re-execution of code generators or maintenance of traceability links [3,4].
EMF-INCQUERY [3] is an open source Eclipse project which offers a declarative
graph query language [5] for capturing queries and a scalable query engine for
incremental query evaluation using the Rete algorithm [6].

Industrial domain-specific languages and tools (e.g. in the automotive, avionics or telecommunications domain) necessitate the development of large number
of complex, interrelated queries, which turns out to be an error prone task in
industrial practice. Some constraints may accidentally be omitted, other constraints may be added to a query unintentionally, while patterns may be composed using a wrong order of parameters. While the EMF-IncQuery framework

© Springer International Publishing Switzerland 2016
R. Echahed and M. Minas (Eds.): ICGT 2016, LNCS 9761, pp. 137–152, 2016.
DOI: 10.1007/978-3-319-40530-8_9

contains a type checker and various well-formedness constraints are also checked, such static checks still do not guarantee that query specifications are free of flaws, thus runtime debugging of queries need to be carried out in practice.

Unfortunately, the declarative nature of query languages makes debugging of query specifications a challenging task. The order of clauses in a query specification does not coincide with the actual evaluation order in case of local search based query evaluation [7,8]. Furthermore, incremental evaluation techniques further complicate the issue as all matches of all queries (and subqueries) are readily available immediately at any time.

Model transformation slicing was introduced in [9] as a technique to assist debugging of model transformations. As a conceptual difference with respect to traditional program slicing, a transformation slice includes not only the relevant instructions of the transformation program, but also those model elements that can influence the slicing criterion. A dynamic backward slicing approach was proposed for the transformation languages of VIATRA [9] and static backward slicing approach for ATL [10,11].

In the current paper, we propose a slicing technique for incremental graph patterns evaluated on top of Rete networks. Based upon an observed change in the match set of a graph pattern (e.g. an extra match or a missing match) we traverse the nodes of the Rete network in a bottom-up way to identify those tuples in other Rete nodes which may contribute to the observed aggregate change. We illustrate how this slicing information can be computed in the context of statechart models. Our slicing approach may assist the debugging of model queries by localizing suspicious spots in queries.

The rest of the paper is structured as follows. Section 2 gives a brief overview of graph patterns, and presents why slicing can help to debug incorrect pattern definitions. A formalization of incremental evaluation of model queries using Rete networks is provided in Section 3. The slicing approach itself is presented in Section 4 and is illustrated in the context of our running example. Related work is discussed in Section 5 while Section 6 concludes our paper.

2   Motivating Example and Overview

Graph patterns are a declarative, graph-like formalism representing a condition (or constraint) to be matched against instance model graphs. Graph patterns are used for various purposes in model-driven development, such as defining model transformation rules or defining general purpose model queries including model validation constraints in various advanced tools (such as eMOFLON, Henshin, EMF-INCQUERY or VIATRA).

Informally, a graph pattern can be described as a set of *structural constraints* prescribing the interconnection between nodes and edges of given types. Furthermore, further constraint types, such as *pattern composition constraints* for the reuse of subpatterns, help the description of complex constraints. Finally, a *match* in a model M of a pattern is the binding of all variables to elements of M that satisfies all constraints expressed by the pattern. Efficient caching

```
1   //'tr' represents a transition connecting two statemachines.
2    pattern DifferentStateMachine(tr : Transition) {
3      State.out(src, tr);
4      Transition.target(tr, trg);
5      find StateofMachine(sm, src);
6      neg find StateOfMachine(sm, trg);
7      //find StateOfMachine(sm, trg);
8    }
9
10  //Variable 'st' represents a State of the StateMachine 'sm'.
11   pattern StateOfMachine(sm : StateMachine, st : State) {
12     StateMachine.states(sm, st);
13   }
```

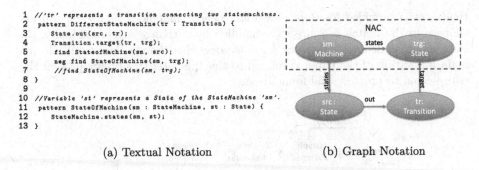

(a) Textual Notation   (b) Graph Notation

Fig. 1. Example graph patterns

techniques based on Rete networks [6] enumerate all matches of a pattern and incrementally update the caches upon model changes.

Example 1. Figure 1 describes a graph pattern using the textual syntax of EMF-INCQUERY [3] that identifies transitions whose source and target states are in different states machines. It uses a subpattern called StateOfMachine (Line 11), connecting two variables of type StateMachine and State with the edge type of StateMachine.states. The main pattern DifferentStateMachine (Line 2) uses four variables to represent a Transition, a source and a target State and a StateMachine. The Transition and the two States are connected with two edge constraints, while the states fr and to are connected to the statemachine by a positive (Line 5) meaning that variable fr has to be connected via the called pattern, and a negative pattern call (Line 6) which prevents to to be connected.

Figure 1b depicts the same pattern using a graphical notation where nodes are entity constraints, edges are relational constraints, positive pattern calls are inlined (copied), and negative pattern call are marked by NAC areas.

During pattern development, engineers may accidentally make faults. For instance, imagine that the neg keyword is omitted from Line 6, and thus the definition of pattern DifferentStateMachine erroneously includes (the commented) Line 7 instead of Line 6. It results in a positive pattern call instead of a negative pattern call making the pattern to represent transitions where both source and target states are in the same state machine, thus completely replacing the correct match set of the pattern with incorrect matches.

During debugging of queries and transformations, when the developer identifies that the match set of a pattern is different from what was expected, he or she frequently wishes to learn what elementary model changes would result in the appearance of new match or the disappearance of an existing match of a pattern. The current paper will present a formal slicing technique for Rete network based caches of graph patterns to answer such questions.

For that purpose, we present a chain of semantic mappings (see Fig. 2) by (1) starting from a Σ-term algebra to formalize graphs and then (2) (a subset of) the graph pattern language of EMF-INCQUERY. (3) A relational algebraic treatment is provided for Rete networks to cache matches of patterns and (4) changes in the match sets are then handled by a change algebra. Finally, (5) Rete slicing is defined as specific formulae over this change algebra. While the main innovation of the paper is related to this final step, we briefly present the entire chain to provide solid foundations.

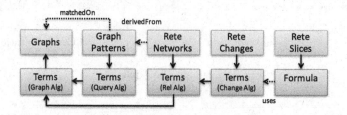

Fig. 2. Overview of formalization

3   Graph Patterns and Rete Networks

We present an algebraic formalization of incremental graph pattern matching with Rete networks following the definitions of [12].

3.1   Graphs and Graph Patterns

Since Rete networks can be adapted to various graph formalisms, we omit the handling of types and use directed labeled and attributed graphs to represent models for the sake of generality and simplicity.

Definition 1 (Directed labeled attributed graph). A *directed labeled and attributed graph* $M = \langle V_M, E_M, L_M, D_M, \mathrm{src}_M, \mathrm{trg}_M, \mathrm{lbl}_M, \mathrm{attr}_M \rangle$ is a tuple, where V_M and E_M denote nodes and edges of the graph, respectively. L_M is a set of labels, while D_M represents a set of data nodes. The nodes, edges and data nodes represent the universe of the graph model $\mathbb{U}_M = V_M \cup E_M \cup D_M$.

Functions src_M and trg_M map edges to their source and target nodes, formally $\mathrm{src}_M : E_M \mapsto V_M$ and $\mathrm{trg}_M : E_M \mapsto V_M$. The labeling function lbl assigns labels to edges, formally $\mathrm{lbl}_M : (V_M \cup E_M) \mapsto L_M$, and the attribute function maps nodes to corresponding attribute values, formally $\mathrm{attr}_M : V_M \mapsto D_M$. We may omit subscript M when graph M is unambiguous.

Graphs will serve as the core underlying semantic domain to evaluate graph patterns but we define an algebraic term representation (in the style of abstract state machines [13]) for a unified treatment of formalization. For that purpose, we rely on some core definitions of terms, substitution, interpretation and formulas.

Definition 2 (Vocabulary and terms). A *vocabulary* Σ is a finite collection of function names. Each function name f has an *arity*, a non-negative integer, which is the number of arguments the function takes. Nullary function names are often called *constants*.

The **terms** of Σ are syntactic expressions generated inductively was follows: (1) Variables $v_0, v_1, v_2, \ldots$ are terms; (2) constants c of Σ are terms; (3) if function f is an n-ary function name and $t_1, \ldots, t_n$ are terms, $f\langle t_1, \ldots t_n \rangle$ is a term.

Since terms are syntactic objects, they do not have a meaning. A term can be evaluated, if elements of the model are assigned to the variables of the term.

Definition 3 (Substitution). Given a directed attribute graph model M, a **substitution** for M is a function s which assigns an element $s(v_i) \in \mathbb{U}_M$ to each variable v_i. A **partial substitution** assigns a value to only certain variables v_i.

Definition 4 (Interpretation of terms). By induction on the length of term t, given a substitution s, a value $[\![t]\!]_s^M \in \mathbb{U}_M$ (the **interpretation of term** t in model M) is defined as follows:

1. $[\![v_i]\!]_s^M := s(v_i)$ (interpretation of variables);
2. $[\![c]\!]_s^M := c^M$ (interpretation of constants);
3. $[\![f\langle t_1, \ldots, t_n \rangle]\!]_s^M := f^M \langle [\![t_1]\!]_s^M, \ldots, [\![t_n]\!]_s^M \rangle$ (interpretation of functions).

A **ground term** is a term with a (full) substitution of variables.

Definition 5 (Formulas). **Formulas** of Σ are generated inductively as follows:

1. Equality (and inequality) of two terms $t_1 = t_2$ is a formula f.
2. If f_1 and f_2 are formulas then $f_1 \wedge f_2$, $f_1 \vee f_2$ are formulas.

A simplified notation is used for predicates (i.e. boolean terms) which may omit $= \top$ and $= \bot$ from equality formulas.

We first define algebraic terms to represent graph patterns which are evaluated over directed labeled attributed graphs as semantic models. A *match* of a pattern is a binding of all variables to model elements or attribute values that fulfill all the constraints of the graph pattern.

Definition 6 (Graph pattern and match set). A *graph pattern* P is a term over a special vocabulary Σ with function symbols for constraints including *structural constraints* (entity, relation), *equality checks*, *pattern definitions* with a disjunction of *pattern bodies* containing conjunction of constraints and *positive and negative pattern calls* and constants (representing model element identifiers and data values). The semantics of P is defined as an interpretation of the term over a graph M and along a substitution s as detailed in Table 1 for the key elements of the EMF-INCQUERY language [3]. For easier formulation, we use $\overline{V^k}$ as a shorthand to represent a vector of variables, formally $f\langle \overline{V^k} \rangle = f\langle t_1, \ldots t_k \rangle$.

A *match* of P in M is a substitution s which satisfies all constraints. The *match set* is the set of all matches of a pattern in a graph model:

$$MS_M^P = \{s \mid [\![P\langle \overline{V^k} \rangle \leftarrow PB_1 \vee \ldots PB_n]\!]_s^M = \top\}$$

Table 1. Algebraic definition of graph patterns

Name	Interpretation
Entity	$[\![ent\langle l, v\rangle]\!]_s^M = \top$, if $\begin{cases} \mathrm{lbl}([\![v]\!]_{s'}^M) = l, \text{where} \\ s' \supseteq s \wedge v \in \mathrm{dom}(s') \end{cases}$
Relation	$[\![rel\langle l, v, v_s, v_t\rangle]\!]_s^M = \top$, if $\begin{cases} \mathrm{lbl}([\![v]\!]_{s'}^M) = l \wedge \\ \mathrm{src}([\![v]\!]_{s'}^M) = [\![v_s]\!]_{s'}^M \wedge \\ \mathrm{trg}([\![v]\!]_{s'}^M) = [\![v_t]\!]_{s'}^M, \text{where} \\ s' \supseteq s \wedge \{v, v_s, v_t\} \subseteq \mathrm{dom}(s') \end{cases}$
Equality check	$[\![eq\langle v_1, v_2\rangle]\!]_s^M = \top$, if $\begin{cases} [\![v_1]\!]_{s'}^M = [\![v_2]\!]_{s'}^M, \text{where} \\ s' \supseteq s \wedge \{v_1, v_2\} \subseteq \mathrm{dom}(s') \end{cases}$
Inequality check	$[\![neq\langle v_1, v_2\rangle]\!]_s^M = \top$, if $\begin{cases} [\![v_1]\!]_{s'}^M \neq [\![v_2]\!]_{s'}^M, \text{where} \\ s' \supseteq s \wedge \{v_1, v_2\} \subseteq \mathrm{dom}(s') \end{cases}$
Pattern Body	$[\![PB\langle v_1, \ldots v_k\rangle \leftarrow c_1 \wedge \ldots c_n]\!]_s^M = \top$, if $\begin{cases} \bigwedge_{i \in 1..n} [\![c_i]\!]_{s'}^M = \top, \text{where} \\ s' \supseteq s \wedge \{v_1, \ldots v_k\} \subseteq \mathrm{dom}(s') \end{cases}$
Graph Pattern	$[\![P\langle v_1, \ldots v_k\rangle \leftarrow PB_1 \vee \ldots PB_n]\!]_s^M = \top$, if $\begin{cases} \bigvee_{i \in 1..n} [\![PB_i]\!]_{s'}^M = \top, \text{where} \\ s' \supseteq s \wedge \{v_1, \ldots v_k\} \subseteq \mathrm{dom}(s') \end{cases}$
Positive Call	$[\![call(P^c\langle v_1, \ldots v_n\rangle)]\!]_s^M = \top$, if $\begin{cases} [\![P^c\langle v_1^c, \ldots v_n^c\rangle]\!]_{s'}^M = \top, \text{where} \\ \forall_{i \in 1..n} : s'(v_i^c) = s(v_i) \end{cases}$
Negative Call	$[\![neg(P^c\langle v_1, \ldots v_n\rangle)]\!]_s^M = \top$, if $\begin{cases} [\![P^c\langle v_1^c, \ldots v_n^c\rangle]\!]_{s'}^M = \bot, \text{where} \\ \forall_{i \in 1..n} : s'(v_i^c) = s(v_i) \end{cases}$

Table 2. The Different State Machines pattern

	Different State Machines	State of Machine (SoM)
Variables	$src, trg, tr, sm, r_1, r_2$	$st, sm, r \in V^{rel}$
Constraints	$ent_1\langle State, src\rangle, ent_2\langle Transition, tr\rangle$ $ent_3\langle State, trg\rangle, ent_4\langle Machine, sm\rangle$ $rel_5\langle State.out, r_1, src, tr\rangle$ $rel_6\langle Transition.target, r_2, tr, trg\rangle$ $call_7(SoM\langle src, sm\rangle), neg_8(SoM\langle trg, sm\rangle)$	$ent_1\langle State, source\rangle$ $ent_2\langle Machine, sm\rangle$ $rel_3\langle Machine.States, r, sm, st\rangle$

Example 2. Table 2 provides the algebraic representation of graph pattern *Different State Machines* of Fig. 1 that identifies transitions which connect elements between different state machines.

3.2 Graph Pattern Matching with Rete Networks

The Rete algorithm [6] is a well-known and efficient technique of rule-based systems which has been adapted to several incremental pattern matchers [12,14, . 15]. The algorithm uses an incremental caching approach that indexes the basic model elements as well as *partial matches* of a graph pattern that enumerate all model element tuples which satisfy a subset of the graph pattern constraints. These caches are organized in a graph structure called Rete network supporting incremental updates upon model changes.

Table 3. Relational algebraic operations of Rete networks

Name	Interpretation
Entity/0	$[\![n^E\langle v\rangle]\!]^M = \{\langle v\rangle \mid [\![ent\langle l, v\rangle]\!]^M = \top\}$
Relation/0	$[\![n^R\langle v, v_s, v_t\rangle]\!]^M = \{\langle v, v_s, v_t\rangle \mid [\![rel\langle l, v, v_s, v_t\rangle]\!]^M = \top\}$
Projection/1	$[\![n^\pi\langle\overline{V^k}\rangle]\!]^M = \pi_{\overline{V^k}}[\![n^1\langle\overline{V^n}\rangle]\!]^M$, where $\overline{V^n} \supseteq \overline{V^k}$
Filter/1	$[\![n^\sigma\langle\overline{V^k}\rangle]\!]^M = \sigma[\![n^1\langle\overline{V^k}\rangle]\!]^M$
Join/2	$[\![n^\bowtie\langle\overline{V^k}\rangle]\!]^M = \begin{cases} [\![n^1\langle\overline{V^i}\rangle]\!]^M \bowtie [\![n^2\langle\overline{V^i}\rangle]\!]^M\text{, where} \\ \overline{V^k} = \overline{V^i} \cup \overline{V^j} \end{cases}$
Anti-join/2	$[\![n^\triangleright\langle\overline{V^k}\rangle]\!]^M = [\![n^1\langle\overline{V^k}\rangle]\!]^M \triangleright [\![n^2\langle\overline{V^j}\rangle]\!]^M$
Disjunction/2	$[\![n^\cup\langle\overline{V^k}\rangle]\!]^M = [\![n^1\langle\overline{V^k}\rangle]\!]^M \cup [\![n^2\langle\overline{V^k}\rangle]\!]^M$

Definition 7 (Rete network). A *Rete network* is a directed acyclic graph $R \equiv \langle N, E, L, Term, \text{src}, \text{trg}, \text{lbl}, \text{attr}\rangle$, where N is a set of *Rete nodes* connected by edges E (along src and trg), $L = Kind \cup Index$ defines node kinds (entity E and relation input R, natural join $\bowtie$, filter σ, projection π, disjunction $\cup$ and anti-join $\triangleright$) as node labels and indices as edge labels (along lbl), while data associated to nodes are specific $Terms$ of type $n^{op}\langle\overline{V^k}\rangle$.

Definition 8 (Memory of a Rete node). Each Rete node $n \in N$ of the Rete network R_P stores all matches of an instance model M which satisfy certain constraints which is denoted as $[\![n\langle\overline{V^k}\rangle]\!]^M$. Each Rete node n relies upon the memory of its parents $[\![n^i\langle\overline{V^k}\rangle]\!]^M$ to calculate its own content inductively by relational algebraic operators which are specified in details in Table 3.

The *memory of an input node* n^I lists entities and relations of the model with a specific label. Positive pattern calls are always mapped to join node, while negative pattern calls are expressed via anti-join nodes. A *production (output) node* in a Rete network contains all matches of a graph pattern P by expressing the complex constraints with a relational algebraic operations (e.g. projection, filter, join, anti-join, disjunction). The compilation of the graph pattern language of EMF-INCQUERY into a corresponding Rete network is out of scope for the current paper and it is studied in [12] in details. We only rely on the correctness of a compilation $comp : P \mapsto N$ to guarantee that a match set of a graph pattern P (see Table 1) equals to the memory of the corresponding Rete node (as defined in Table 3), i.e. $MS_M^P = [\![n\langle\overline{V^k}\rangle]\!]^M$ where $n = comp(P)$.

Example 3. Figure 3a depicts a Rete network for the **Different State Machines** pattern. Its input nodes cache three references: **states** of Machines; **out** references of **States** and **target** references of **Transitions**. The first join node of the network connects the **source** and **target** states, while the second join node adds the container machines of the source patterns by joining the production node of the called pattern *State of Machine*. Finally, the anti-join node (depicted by the red triangle) ensures that the target state is not connected to the same state machine as the source node by filtering matches of its join parent node which also correspond to matches of the **states** node along the called pattern *State of Machine* (which is inlined during compilation).

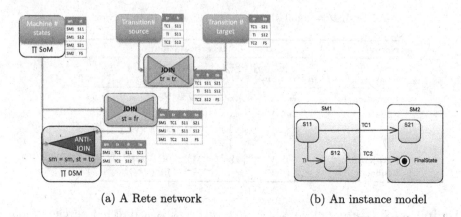

(a) A Rete network   (b) An instance model

Fig. 3. A Rete network for the Different State Machines pattern

We display the cached model elements of the instance model Fig. 3b in a table for each Rete node, describing two state machines with a few states and transitions (some of which cross the boundary of a statemachine).

3.3   Incremental Change-Driven Behavior of Rete Networks

If the memory of a Rete nodes changes, the memory of all its children Rete nodes needs to be updated in accordance with the relational algebraic operation of the Rete node. For that purpose, we define a change algebra over terms n_+ and n_- (jointly denoted as n_Δ) which represent tuples added and removed from a Rete node n. We briefly revisit the semantics of such change terms in Table 4 while the reader is referred to [12] for a detailed discussion.

Definition 9 (Change algebra for Rete nodes). Let M be a graph model cached in a Rete network R and let Δ be a set of elementary model changes (represented by terms for creation and deletion of entities n_+^E, n_-^E or references n_+^E and n_-^E) over this graph. We define a term n_Δ^{op} for each node n^{op} of the Rete network to represent matches that are changed by Δ with respect to M.

The semantics of such terms are inductively defined by using (i) match information n^{op} cached in R for M (i.e. the previous state of the model) and (ii) change already computed at parent nodes n_Δ^1 and n_Δ^2 of n_Δ^{op} split along operations op as detailed in Table 4.

A brief informal explanation of these cases is as follows:

Entity and relation change. A model entity appears in the change set n_Δ^E if (1) it exists in M and it is removed by Δ or (2) it does not exist in M and it is created by Δ (and same holds for model references).

Change in projection and filter nodes. The change set of a projection node is defined as the difference of the new $n^1\langle \overline{V^n} \rangle \cup n_\Delta^1 \langle \overline{V^n} \rangle$ and old $n^1\langle \overline{V^n} \rangle$ memory of the parent nodes. In case of a filter node the change set is the change set of its single parent $n_\Delta^1 \langle V^k \rangle$ filtered using the σ filter operator.

Table 4. Change algebra for Rete nodes

Name	Interpretation
Entity	$[\![n_\Delta^E \langle v \rangle]\!]^{M,\Delta} = \begin{cases} ([\![n^E \langle v \rangle]\!]^M = \top \wedge [\![n_-^E \langle v \rangle]\!]^\Delta = \top) \vee \\ ([\![n^E \langle v \rangle]\!]^M = \bot \wedge [\![n_+^E \langle v \rangle]\!]^\Delta = \top) \end{cases}$
Relation	$[\![n_\Delta^R \langle v, v_s, v_t \rangle]\!]^{M,\Delta} = \begin{cases} ([\![n^R \langle v, v_s, v_t \rangle]\!]^M = \top \wedge [\![n_-^R \langle v, v_s, v_t \rangle]\!]^\Delta = \top) \vee \\ ([\![n^R \langle v, v_s, v_t \rangle]\!]^M = \bot \wedge [\![n_+^R \langle v, v_s, v_t \rangle]\!]^\Delta = \top) \end{cases}$
Projection	$[\![n_\Delta^\pi \langle \overline{V^k} \rangle]\!]^{M,\Delta} = \pi([\![n^1 \langle \overline{V^n} \rangle]\!]^M \cup [\![n_\Delta^1 \langle \overline{V^n} \rangle]\!]^{M,\Delta}) \setminus \pi[\![n^1 \langle \overline{V^n} \rangle]\!]^M$
Filter	$[\![n_\Delta^\sigma \langle \overline{V^k} \rangle]\!]^{M,\Delta} = \sigma[\![n_\Delta^1 \langle \overline{V^k} \rangle]\!]^{M,\Delta}$
Join	$[\![n_\Delta^\bowtie \langle \overline{V^k} \rangle]\!]^{M,\Delta} = \begin{cases} ([\![n_1 \langle \overline{V^i} \rangle]\!]^M \bowtie [\![n_\Delta^2 \langle \overline{V^j} \rangle]\!]_n^{M,\Delta}) \cup \\ ([\![n_\Delta^1 \langle \overline{V^i} \rangle]\!]^{M,\Delta} \bowtie [\![n_2 \langle \overline{V^j} \rangle]\!]^M) \cup \\ ([\![n_\Delta^1 \langle \overline{V^i} \rangle]\!]^{M,\Delta} \bowtie [\![n_\Delta^2 \langle \overline{V^j} \rangle]\!]^{M,\Delta}) \end{cases}$
Anti-join	$[\![n_\Delta^\triangleright \langle \overline{V^k} \rangle]\!]^{M,\Delta} = \begin{cases} [\![n^1 \langle \overline{V^k} \rangle]\!]^M \bowtie \\ (\pi[\![n^2 \langle \overline{V^j} \rangle]\!]^M \setminus \pi([\![n^2 \langle \overline{V^j} \rangle]\!]^{M,\Delta} \cup [\![n_\Delta^2 \langle \overline{V^j} \rangle]\!]^{M,\Delta})) \cup \\ [\![n_\Delta^1 \langle \overline{V^k} \rangle]\!]^{M,\Delta} \triangleright ([\![n^2 \langle \overline{V^j} \rangle]\!]^M \cup [\![n_\Delta^2 \langle \overline{V^j} \rangle]\!]^{M,\Delta}) \end{cases}$
Disjunction	$[\![n_\Delta^\cup \langle \overline{V^k} \rangle]\!]^{M,\Delta} = \begin{cases} \{[\![n_\Delta^1 \langle \overline{V^k} \rangle]\!]^{M,\Delta} \mid ([\![n^2 \langle \overline{V^k} \rangle]\!]^M = \emptyset) \wedge ([\![n_\Delta^2 \langle \overline{V^k} \rangle]\!]^{M,\Delta} = \emptyset)\} \cup \\ \{[\![n_\Delta^2 \langle \overline{V^k} \rangle]\!]^{M,\Delta} \mid ([\![n^1 \langle \overline{V^k} \rangle]\!]^M = \emptyset) \wedge ([\![n_\Delta^1 \langle \overline{V^k} \rangle]\!]^{M,\Delta} = \emptyset)\} \cup \\ \{[\![n_\Delta^1 \langle \overline{V^k} \rangle]\!]^{M,\Delta} \mid [\![n_\Delta^2 \langle \overline{V^k} \rangle]\!]^{M,\Delta}\} \end{cases}$

Change in join nodes. The change set of a join node consists of the union of three change sets: (1) the join of the memory of the first parent node $n^1 \langle \overline{V^i} \rangle$ with the delta coming from the second parent $n_\Delta^2 \langle \overline{V^j} \rangle$; (2) the join of the second parent $n^2 \langle V_i \rangle$ with the first parent delta $n_\Delta^1 \langle \overline{V^j} \rangle$; and (3) the join of the two parent deltas $n_\Delta^1 \langle \overline{V^i} \rangle$ and $n_\Delta^2 \langle \overline{V^j} \rangle$.

Change in anti-join nodes. The change set of an anti-join node is the union of two sets: (1) the elements in the second parent delta $n_\Delta^2 \langle \overline{V^j} \rangle$ that are filtering out pre-existing tuples from the first parent $n^1 \langle \overline{V^k} \rangle$; and (2) the changed elements of the first parent $n_\Delta^1 \langle \overline{V^k} \rangle$ that are not filtered out by the second parent or its changes.

Change in disjunction nodes. The change set of a disjunction node is the union of three sets: (1) the delta of the first parent $n_\Delta^1 \langle \overline{V^k} \rangle$ that was not present in the second parent $n^2 \langle \overline{V^k} \rangle$; (2) the delta of the second parent $n_\Delta^2 \langle \overline{V^k} \rangle$ that was not present in the first parent $n^1 \langle \overline{V^k} \rangle$ and (3) elements that were added or removed by both parent changes.

4 Slicing Rete Networks of Graph Patterns

The change algebra of Table 4 precisely specifies how to propagate changes in Rete networks from input nodes to production nodes corresponding to graph patterns. However, an inverse direction of change propagation needs to be defined for debugging purposes.

Slicing of Rete networks will systematically collect dependencies from (an aggregate) change at a production (pattern) node towards elementary changes at input nodes. More specifically, based on an observed change of a match of a pattern, we need to calculate how to change the caches of each parent node in

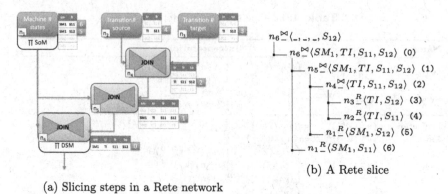

(a) Slicing steps in a Rete network

$n_6^{\bowtie}\langle \_,\_,\_,S_{12}\rangle$
$\quad\lfloor n_6^{\bowtie}\langle SM_1,TI,S_{11},S_{12}\rangle\ (0)$
$\qquad\lfloor n_5^{\bowtie}\langle SM_1,TI,S_{11},S_{12}\rangle\ (1)$
$\qquad\quad\lfloor n_4^{\bowtie}\langle TI,S_{11},S_{12}\rangle\ (2)$
$\qquad\qquad\lfloor n_3{}_{-}^{R}\langle TI,S_{12}\rangle\ (3)$
$\qquad\qquad\lfloor n_2{}_{-}^{R}\langle TI,S_{11}\rangle\ (4)$
$\qquad\quad\lfloor n_1{}_{-}^{R}\langle SM_1,S_{12}\rangle\ (5)$
$\qquad\lfloor n_1{}^{R}\langle SM_1,S_{11}\rangle\ (6)$

(b) A Rete slice

Fig. 4. Sample Rete slice for faulty pattern

the Rete network so that those changes consistently imply the specific changes of the match set of a production node. For instance, if a match is included in n_+^{op} (n_-^{op}, respectively) then it needs to be added to (removed from) the cache of the corresponding Rete node n^{op} to observe a specific change n_{Δ}^{P} of a production node. In a debugging context, if a specific match of pattern P is missed by the engineer then he or she can ask the slicer to calculate possible model changes that would add the corresponding match n_+^{P}.

As a slice, we present complete dependency information from aggregate changes to elementary changes by a logic formula over change terms which is calculated by appending new clauses in the form of (ground) change terms along specific matches s while traversing the Rete network from production nodes to input nodes. This slice is informally calculated as follows:

- The input of slicing is the appearance of a new match s in M or the disappearance of an existing match s in M at a production node P, which is a ground term $[\![n_+^{P}\langle\overline{V^k}\rangle]\!]_s^{M,\Delta}$ or $[\![n_-^{P}\langle\overline{V^k}\rangle]\!]_s^{M,\Delta}$ appended to the slice.
- For each ground term appended to the slice, we calculate what changes are necessitated at their parent Rete nodes, and append those potential changes to the slices one by one. Formulas are calculated in correspondence with Table 5 for the Rete nodes.
  - For instance, when *a match of a join node disappears* (see Join in Table 5(b)) then at least one of the corresponding partial matches of its parent nodes need to be removed, captured in the slice by the change terms $[\![n_-^{1}\langle\overline{V^i}\rangle]\!]_s^{M,\Delta}$ and $[\![n_-^{2}\langle\overline{V^j}\rangle]\!]_s^{M,\Delta}$ as disjunctive branches.
  - When a *new match of a join node appears* (see Join in Table 5(a)) then we add new matches to one or both parent nodes n^1, n^2 which is compliant with the match of the join node.
- Special care needs to be taken for *projection* and *anti-join* nodes which may need to fabricate new entities (identifiers) to create ground terms for unbound variables.
- As a base case of this recursive definition, we stop when

- elementary changes of input nodes are reached (first two lines in Table 5a and b), or
- a match already existing in the cache of a Rete node is to be added by a change (see Table 5(c)), or
- when the deletion of a match is prescribed by a change which does not exist in M (see Table 5(c)).

Definition 10 (Rete Slice). The slice of a change predicate $n_+(t)$ or $n_-(t)$ starting from a node n in a Rete network R over model M and along substitution s is a formula (derived in disjunctive normal form in our case) calculated in accordance with Table 5. □

Example 4. Figure 4 depicts the sliced Rete network of the faulty version of the Different State Machines pattern. The only difference in its network (as opposed to the Rete network of the correct pattern in Fig. 3a) uses a join node instead of an anti-join node as a production node.

The slicing starts with noticing an undesired tuple where the variable *to* equals to the state S_{12}. At this point, we can ask the slicer how to remove this undesired tuple by calculating the slice of the change predicate $n_6 \overset{\bowtie}{-} \langle -, -, -, S_{12} \rangle$.

1. The memory of node n_6 is checked for tuples matching the input predicate; a single tuple $n_6 \overset{\bowtie}{-} \langle SM_1, TI, S_{11}, S_{12} \rangle$ is found and added to the slice formula.
2. To remove the element from the output of the join node, following Table 5b the corresponding input tuples are to be removed from one of its parents. In this case, the node $n_5 \overset{\bowtie}{-} \langle SM_{11}, TI, S_{11}, S_{12} \rangle$ is added to the formula.
3. The first parent node $n_4 \overset{\bowtie}{-} \langle TI, S_{11}, S_{12} \rangle$ is selected and added to the formula.
4. $n_3 \overset{R}{-} \langle TI, S_{12} \rangle$ is selected as the dependency to remove, and added to the formula. At this point, an input node is reached so the recursion terminates.
5. However, we have to backtrack to node n_4, and evaluate the second case for the join node by adding $n_2 \overset{R}{-} \langle TI, S_{11} \rangle$ to a second branch of the formula.
6. Similarly, $n_1 \overset{R}{-} \langle SM_1, S_{12} \rangle$ and $n_1 \overset{R}{-} \langle SM_1, S_{11} \rangle$ are added to new branches.

The final formula looks as follows:

$$[n_6 \overset{\bowtie}{-} \langle v_1, v_2, v_3, v_4 \rangle]_{\{v_4 \mapsto S_{12}\}}^M =$$
$$\left(n_3 \overset{R}{-} \langle TI, S_{12} \rangle \wedge n_4 \overset{\bowtie}{-} \langle TI, S_{11}, S_{12} \rangle \wedge n_5 \overset{\bowtie}{-} \langle SM_{11}, TI, S_{11}, S_{12} \rangle \wedge n_6 \overset{\bowtie}{-} \langle SM_1, TI, S_{11}, S_{12} \rangle \right) \vee$$
$$\left(n_2 \overset{R}{-} \langle TI, S_{11} \rangle \wedge n_4 \overset{\bowtie}{-} \langle TI, S_{11}, S_{12} \rangle \wedge n_5 \overset{\bowtie}{-} \langle SM_{11}, TI, S_{11}, S_{12} \rangle \wedge n_6 \overset{\bowtie}{-} \langle SM_1, TI, S_{11}, S_{12} \rangle \right) \vee$$
$$\left(n_1 \overset{R}{-} \langle SM_1, S_{12} \rangle \wedge n_5 \overset{\bowtie}{-} \langle SM_{11}, TI, S_{11}, S_{12} \rangle \wedge n_6 \overset{\bowtie}{-} \langle SM_1, TI, S_{11}, S_{12} \rangle \right) \vee$$
$$\left(n_1 \overset{R}{-} \langle SM1, S11 \rangle \wedge n_6 \overset{\bowtie}{-} \langle SM_1, TI, S_{11}, S_{12} \rangle \right).$$

□

Although the formula refers to all nodes of the Rete network, the slice describes a reduced model: (1) the model element tuples unrelated to the criteria are not included, and (2) the tuples in a single disjunctive branch describe a possible series of operations that would result in a tuple matching the input predicate to appear or disappear.

Table 5. Definition of slices for Rete networks of graph patterns

Different cases for the same node are handled as disjunctions in the formula.

Node	Change	Append to formula
Entity/0	$[\![n^E_+\langle v\rangle]\!]^{M,\Delta}_s$: $\top$	
Relation/0	$[\![n^R_+\langle v,v_s,v_t\rangle]\!]^{M,\Delta}_s$: $\top$	
Projection/1	$[\![n^\pi_+\langle\overline{V^k}\rangle]\!]^{M,\Delta}_s$: $[\![n^1_+\langle\overline{V^j}\rangle]\!]^{M,\Delta}_s$	
Filter/1	$[\![n^\sigma_+\langle\overline{V^k}\rangle]\!]^{M,\Delta}_s$: $[\![n^1_+\langle\overline{V^k}\rangle]\!]^{M,\Delta}_s \wedge \sigma\langle\overline{V^k}\rangle$	
Join/2	$[\![n^\bowtie_+\langle\overline{V^k}\rangle]\!]^{M,\Delta}_s$: $[\![n^1_+\langle\overline{V^i}\rangle]\!]^{M,\Delta}_s \wedge [\![n^2_+\langle\overline{V^j}\rangle]\!]^{M,\Delta}_s$	
	$[\![n^1\langle\overline{V^i}\rangle]\!]^M_s \wedge [\![n^2_+\langle\overline{V^j}\rangle]\!]^{M,\Delta}_s$	
	$[\![n^2\langle\overline{V^j}\rangle]\!]^M_s \wedge [\![n^1_+\langle\overline{V^i}\rangle]\!]^{M,\Delta}_s$	
Anti-join/2	$[\![n^\rhd_+\langle\overline{V^k}\rangle]\!]^{M,\Delta}_s$: $[\![n^1\langle\overline{V^k}\rangle]\!]^M_s \wedge [\![n^c_-\langle\overline{V^j}\rangle]\!]^{M,\Delta}_s$	
	$[\![n^1\langle\overline{V^k}\rangle]\!]^{M,\Delta}_s \wedge [\![n^2\langle\overline{V^j}\rangle]\!]^M_s = \emptyset$	
Disjunction/2	$[\![n^\cup_+\langle\overline{V^k}\rangle]\!]^{M,\Delta}_s$: $[\![n^1_+\langle\overline{V^k}\rangle]\!]^{M,\Delta}_s$	
	$[\![n^2_+\langle\overline{V^k}\rangle]\!]^{M,\Delta}_s$	

(a) How to update inputs to add match m to output?

Node	Change	Append to formula
Entity/0	$[\![n^E_-\langle v\rangle]\!]^{M,\Delta}_s$: $\top$	
Relation/0	$[\![n^R_-\langle v,v_s,v_t\rangle]\!]^{M,\Delta}_s$: $\top$	
Projection/1	$[\![n^\pi_-\langle\overline{V^k}\rangle]\!]^{M,\Delta}_s$: $[\![n^1_-\langle\overline{V^k},\overline{V^n}\rangle]\!]^{M,\Delta}_s$	
Filter/1	$[\![n^\sigma_-\langle\overline{V^k}\rangle]\!]^{M,\Delta}_s$: $[\![n^1_-\langle\overline{V^k}\rangle]\!]^{M,\Delta}_s$	
Join/2	$[\![n^\bowtie_-\langle\overline{V^k}\rangle]\!]^{M,\Delta}_s$: $[\![n^1_-\langle\overline{V^i}\rangle]\!]^{M,\Delta}_s$	
	$[\![n^2_-\langle\overline{V^j}\rangle]\!]^{M,\Delta}_s$	
Anti-join/2	$[\![n^\rhd_-\langle\overline{V^k}\rangle]\!]^{M,\Delta}_s$: $[\![n^1_-\langle\overline{V^k}\rangle]\!]^{M,\Delta}_s$	
	$[\![n^1\langle\overline{V^k}\rangle]\!]^M_s \wedge [\![n^2_-\langle\overline{V^j}\rangle]\!]^{M,\Delta}_s$	
Disjunction/2	$[\![n^\cup_-\langle\overline{V^k}\rangle]\!]^{M,\Delta}_s$: $[\![n^1\langle\overline{V^k}\rangle]\!]^M_s \neq \emptyset \wedge [\![n^1_-\langle\overline{V^k}\rangle]\!]^{M,\Delta}_s \wedge$ $[\![n^2\langle\overline{V^k}\rangle]\!]^M_s = \emptyset$	
	$[\![n^1\langle\overline{V^k}\rangle]\!]^M_s = \emptyset$ $[\![n^2\langle\overline{V^k}\rangle]\!]^M_s \neq \emptyset \wedge [\![n^2_-\langle\overline{V^k}\rangle]\!]^{M,\Delta}_s$	
	$[\![n^1\langle\overline{V^k}\rangle]\!]^M_s \neq \emptyset \wedge [\![n^1_-\langle\overline{V^k}\rangle]\!]^{M,\Delta}_s \wedge$ $[\![n^2\langle\overline{V^k}\rangle]\!]^M_s \neq \emptyset \wedge [\![n^2_-\langle\overline{V^k}\rangle]\!]^{M,\Delta}_s$	

(b) How to update inputs to remove match m from output?

Node	Change	Append to formula
Add existing tuple	$([\![n\langle\overline{V^k}\rangle]\!]^M_s = \bot) \wedge ([\![n_-\langle\overline{V^k}\rangle]\!]^{M,\Delta}_s)$: $\top$	
Remove missing tuple	$([\![n\langle\overline{V^k}\rangle]\!]^M_s = \top) \wedge ([\![n_+\langle\overline{V^k}\rangle]\!]^{M,\Delta}_s)$: $\top$	

(c) Handling trivial cases

5   Related Work

Traditional program slicing techniques have been regularly and exhaustively surveyed in the past in papers like [16,17]. The current paper focuses on model transformation slicing [9–11], more specifically on incremental model queries. The main difference with respect to traditional approaches is that query slicing has to consider the specification and the model simultaneously.

Slicing of Declarative Programs. The closest related work addresses the slicing of logic programs as declarative graph patterns [5,18] share certain similarities with logic programs. Forward slicing of Prolog programs are discussed in [19] based on partial evaluation, while [20] executes static and dynamic slicing of logic programs based on the procedural behaviour of the programs. [21] augments the data flow analysis with control-flow dependencies in order to identify the source of a bug included in a logic program and was extended in [22] to the slicing of constraint logic programs (with fixed domains). Program slicing for the Alloy language was proposed in [23] as a novel optimization strategy to improve the verification of Alloy specifications. Our conceptual extension to these existing slicing techniques is the incorporation of model elements into the slices.

Slicing Queries over Databases. In the context of databases and data warehousing, related approaches called *data lineage tracing* [24] or *data provenance problem* [25] aim to explain why a selected record exists in a materialized view. These approaches focus on identifying the records of the original tables that contribute to a selected record, and expect the queries be correct. A further difference to our contribution is that storing partial results in a data warehousing context can be impractical due to high (memory) costs while in case of the Rete algorithm, these partial results are already cached to be available for slicing.

Model Slicing. Model slicing [26] techniques have already been successfully applied in the context of MDD. Slicing was proposed for model reduction purposes in [27,28] to make the following automated verification phase more efficient.

Lano et. al. [29] exploits both declarative elements (like pre- and postconditions of methods) and imperative elements (state machines) to construct UML model slices by using model transformations. The slicing of finite state machines in a UML context was studied by Tratt [30], especially, to identify control dependence. A similar study was also executed for extended finite state machines in [31]. A dynamic slicing technique for UML architectural models is introduced in [32] using model dependence graphs to compute dynamic slices based on the structural and behavioral (interactions only) UML models.

Metamodel pruning [33] can also be interpreted as a slicing problem where the effective metamodel is automatically derived as a view. Moreover, model slicing is used in [34] to modularize the UML metamodel into a set of small metamodels for each UML diagram type. Various model slicing techniques are merged by Blouin et al. [35] into a single, generative framework, using different approaches for different models. Still, none of the existing model slicing approaches address the slicing of model queries, the main focus of our work.

Model Transformation Debugging. Slicing can be beneficial for debugging model transformations. The authors of [36] propose a dynamic tainting technique for debugging failures of model transformations, and propose automated techniques to repair input model faults [37]. Colored Petri nets are used for underlying formal support for debugging transformations in [38]. The debugging of triple graph grammar transformations is discussed in [39], which envisions the future use of slicing techniques in the context of model transformations.

6   Conclusion and Future Work

In this paper, we defined a dynamic slicing technique for Rete networks derived from graph patterns. As a slicing criterion, the appearance of a new match or the disappearance of an existing match is selected in a production node of the Rete network. Since a Rete network also caches partial matches, it is possible to follow match dependencies step by step back to the input nodes storing elementary graph nodes and edges. Such dependencies constitute the slice is captured as formulas over terms of a change algebra. As the main contribution, we provided a formal slicing technique for Rete networks of graph patterns constituted from the most frequently used language elements of the EMF-INCQUERY framework. Our slicing technique was illustrated on a running example of UML state machines.

In the future, we plan to integrate this slicing approach into EMF-INCQUERY [3] in order to use it for various tasks, such as presenting this slice together with the Rete networks graphically, easing the debugging of erroneous model queries. Furthermore, the approach seems promising for declarative bidirectional view model synchronization well, as it enables calculating possible source model changes for view model changes automatically.

Acknowledgements. The authors would like to thank István Ráth for the valuable discussions during the preparation of this paper.

References

1. Reder, A., Egyed, A.: Incremental consistency checking for complex design rules and larger model changes. In: France, R.B., Kazmeier, J., Breu, R., Atkinson, C. (eds.) MODELS 2012. LNCS, vol. 7590, pp. 202–218. Springer, Heidelberg (2012)
2. Bergmann, G., Horváth, A., Ráth, I., Varró, D., Balogh, A., Balogh, Z., Ökrös, A.: Incremental evaluation of model queries over EMF models. In: Rouquette, N., Haugen, Ø., Petriu, D.C. (eds.) MODELS 2010, Part I. LNCS, vol. 6394, pp. 76–90. Springer, Heidelberg (2010)
3. Ujhelyi, Z., Hegedüs, A., Bergmann, G., Horváth, A., Ráth, I., Varró, D.: EMF-IncQuery: an integrated development environment for live model queries. Sci. Comput. Program. **98**, 80–99 (2015)
4. Hegedüs, A., Horváth, A., Ráth, I., Starr, R., Varró, D.: Query-driven soft traceability links for models. Softw. Syst. Model. 1–24 (2014). http://link.springer.com/article/10.1007/s10270-014-0436-y

5. Bergmann, G., Ujhelyi, Z., Ráth, I., Varró, D.: A graph query language for EMF models. In: Cabot, J., Visser, E. (eds.) ICMT 2011. LNCS, vol. 6707, pp. 167–182. Springer, Heidelberg (2011)
6. Forgy, C.L.: Rete: A fast algorithm for the many pattern/many object pattern match problem. Artif. Intell. **19**(1), 17–37 (1982)
7. Varró, G., Friedl, K., Varró, D.: Adaptive graph pattern matching for model transformations using model-sensitive search plans. Electron. Notes Theoret. Comput. Sci. **152**, 191–205 (2006). Proceedings of the International Workshop on Graph and Model Transformation (GraMoT 2005)
8. Búr, M., Ujhelyi, Z., Horváth, Á., Varró, D.: Local search-based pattern matching features in EMF-IncQuery. In: Parisi-Presicce, F., Westfechtel, B. (eds.) ICGT 2015. LNCS, vol. 9151, pp. 275–282. Springer, Heidelberg (2015)
9. Ujhelyi, Z., Horváth, A., Varró, D.: Dynamic backward slicing of model transformations. In: Proceedings of the 2012 IEEE Fifth International Conference on Software Testing, Verification and Validation. ICST 2012, Computer Society , pp. 1–10. IEEE, Washington, DC (2012)
10. Clarisó, R., Cabot, J., Guerra, E., de Lara, J.: Backwards reasoning for model transformations: method and applications. J. Syst. Softw. **116**, 113–132 (2016). http://www.sciencedirect.com/science/article/pii/S0164121215001788
11. Burgueno, L., Troya, J., Wimmer, M., Vallecillo, A.: Static fault localization in model transformations. IEEE Trans. Softw. Eng. **41**(5), 490–506 (2015)
12. Bergmann, G.: Incremental model queries in model-driven design. Ph.D. dissertation, Budapest University of Technology and Economics, Budapest (2013)
13. Gurevich, Y.: Sequential abstract-state machines capture sequential algorithms. ACM Trans. Comput. Logic **1**(1), 77–111 (2000)
14. The JBoss Project: Drools - The Business Logic integration Platform (2014). http://www.jboss.org/drools
15. Ghamarian, A., Jalali, A., Rensink, A.: Incremental pattern matching in graph-based state space exploration. Electron. Commun. EASST 32 (2011)
16. Tip, F.: A survey of program slicing techniques. J. Program. Lang. **3**(3), 121–189 (1995)
17. Xu, B., Qian, J., Zhang, X., Wu, Z., Chen, L.: A brief survey of program slicing. ACM SIGSOFT Softw. Eng. Notes **30**(2), 1–36 (2005)
18. Varró, D., Balogh, A.: The model transformation language of the VIATRA2 framework. Sci. Comput. Program. **68**(3), 214–234 (2007)
19. Leuschel, M., Vidal, G.: Forward slicing by conjunctive partial deduction and argument filtering. In: Sagiv, M. (ed.) ESOP 2005. LNCS, vol. 3444, pp. 61–76. Springer, Heidelberg (2005)
20. Weber Vasconcelos, W.: A flexible framework for dynamic and static slicing of logic programs. In: Gupta, G. (ed.) PADL 1999. LNCS, vol. 1551, p. 259. Springer, Heidelberg (1999)
21. Szilágyi, G., Harmath, L., Gyimóthy, T.: The debug slicing of logic programs. Acta Cybernetica **15**(2), 257–278 (2001)
22. Szilágyi, G., Gyimóthy, T., Małuszyński, J.: Static and dynamic slicing of constraint logic programs. Autom. Softw. Eng. **9**, 41–65 (2002)
23. Uzuncaova, E., Khurshid, S.: Kato: A program slicing tool for declarative specifications. In: 29th International Conference on Software Engineering, pp. 767–770. IEEE (2007)
24. Cui, Y., Widom, J., Wiener, J.L.: Tracing the lineage of view data in a warehousing environment. ACM Trans. Database Syst. **25**(2), 179–227 (2000)

25. Freire, J., Koop, D., Santos, E., Silva, C.T.: Provenance for computational tasks: a survey. Comput. Sci. Eng. **10**(3), 11–21 (2008)
26. Kagdi, H., Maletic, J.I., Sutton, A.: Context-free slicing of UML class models. In: 21st International Conference on Software Maintenance ICSM 2005, pp. 635–638. IEEE (2005)
27. Schaefer, I., Poetzsch-Heffter, A.: Slicing for model reduction in adaptive embedded systems development. In: International Workshop on Software Engineering for Adaptive and Self-managing Systems, pp. 25–32. ACM, New York (2008)
28. Shaikh, A., Clarisó, R., Wiil, U.K., Memon, N.: Verification-driven slicing of UML/OCL models. In: 25th IEEE/ACM International Conference on Automated Software Engineering, pp. 185–194. ACM (2010)
29. Lano, K., Kolahdouz-Rahimi, S.: Slicing of UML models using model transformations. In: Petriu, D.C., Rouquette, N., Haugen, Ø. (eds.) MODELS 2010, Part II. LNCS, vol. 6395, pp. 228–242. Springer, Heidelberg (2010)
30. Androutsopoulos, K., Clark, D., Harman, M., Li, Z., Tratt, L.: Control dependence for extended finite state machines. In: Chechik, M., Wirsing, M. (eds.) FASE 2009. LNCS, vol. 5503, pp. 216–230. Springer, Heidelberg (2009)
31. Korel, B., Singh, I., Tahat, L., Vaysburg, B.: Slicing of state-based models. In: IEEE International Conference on Software Maintenance, pp. 34–43 (2003)
32. Lallchandani, J.T., Mall, R.: A dynamic slicing technique for UML architectural models. IEEE Trans. Softw. Eng. **37**(6), 737–771 (2011)
33. Sen, S., Moha, N., Baudry, B., Jézéquel, J.-M.: Meta-model Pruning. In: Schürr, A., Selic, B. (eds.) MODELS 2009. LNCS, vol. 5795, pp. 32–46. Springer, Heidelberg (2009)
34. Bae, J.H., Lee, K., Chae, H.S.: Modularization of the UML metamodel using model slicing. In: 3rd International Conference on Information Technology: New Generations, IEEE, pp. 1253–1254 (2008)
35. Blouin, A., Combemale, B., Baudry, B., Beaudoux, O.: Modeling model slicers. In: Whittle, J., Clark, T., Kühne, T. (eds.) MODELS 2011. LNCS, vol. 6981, pp. 62–76. Springer, Heidelberg (2011)
36. Dhoolia, P., Mani, S., Sinha, V.S., Sinha, S.: Debugging model-transformation failures using dynamic tainting. In: D'Hondt, T. (ed.) ECOOP 2010. LNCS, vol. 6183, pp. 26–51. Springer, Heidelberg (2010)
37. Mani, S., Sinha, V.S., Dhoolia, P., Sinha, S.: Automated support for repairing input-model faults. In: 25th IEEE/ACM International Conference on Automated Software Engineering, pp. 195–204. ACM (2010)
38. Schoenboeck, J., Kappel, G., Kusel, A., Retschitzegger, W., Schwinger, W., Wimmer, M.: Catch me if you can – debugging support for model transformations. In: Ghosh, S. (ed.) MODELS 2009. LNCS, vol. 6002, pp. 5–20. Springer, Heidelberg (2010)
39. Seifert, M., Katscher, S.: Debugging triple graph grammar-based model transformations. In: Fujaba Days, pp. 19–25 (2008)

An SQL-Based Query Language and Engine for Graph Pattern Matching

Christian Krause[1(✉)], Daniel Johannsen[1], Radwan Deeb[1], Kai-Uwe Sattler[2], David Knacker[1], and Anton Niadzelka[1]

[1] SAP SE, Potsdam, Germany
{christian.krause01,daniel.johannsen,radwan.deeb,
david.knacker,anton.niadzelka}@sap.com
[2] Technische Universität Ilmenau, Ilmenau, Germany
kus@tu-ilmenau.de

Abstract. The interest for graph databases has increased in the recent years. Several variants of graph query languages exist – from low-level programming interfaces to high-level, declarative languages. In this paper, we describe a novel SQL-based language for modeling high-level graph queries. Our approach is based on graph pattern matching concepts, specifically nested graph conditions with distance constraints, as well as graph algorithms for calculating nested projections, shortest paths and connected components. Extending SQL with graph concepts enables the reuse of syntax elements for arithmetic expressions, aggregates, sorting and limits, and the combination of graph and relational queries. We evaluate the language concepts and our experimental SAP HANA Graph Scale-Out Extension (GSE) prototype (This paper is not official SAP communication material. It discusses a research-only prototype, not an existing or future SAP product. Any business decisions made concerning SAP products should be based on official SAP communication material.) using the LDBC Social Network Benchmark. In this work we consider only complex read-only queries, but the presented language paves the way for a SQL-based graph manipulation language formally based on graph transformations.

1 Introduction

In contrast to relational database management systems, graph databases employ dedicated data structures and algorithms tailored for analytical and transactional graph processing. Current applications in the domains of social network analysis (e.g., Facebook, LinkedIn), business network analysis (e.g., Ebay, SAP Ariba) and knowledge graphs (e.g., Google Search, Microsoft Office) show that there is a high demand for efficient reasoning on large-scale graph data. There exist a number of low-level graph programming models, the most prominent being Bulk Synchronous Parallel [16]. However, in the context of enterprise applications there is a need for high-level, declarative graph query languages that enable complex analysis scenarios. While SQL is the accepted standard query

© Springer International Publishing Switzerland 2016
R. Echahed and M. Minas (Eds.): ICGT 2016, LNCS 9761, pp. 153–169, 2016.
DOI: 10.1007/978-3-319-40530-8_10

language in the world of relational databases, there currently is no consensus on a standard for a general-purpose graph query language.

OpenCypher [10] is an initiative by the inventors of the Neo4j graph database to define a common graph query language based on their Cypher language. For historical reasons, many companies today use Cypher. However, the language in its current form is ad hoc and lacks tool support by other vendors. SPARQL [18] is the standard query language in the semantic web domain. While it supports graph query concepts, it is geared into the triple store (subject-predicate-object) concept of the Resource Description Framework (RDF). General graph analysis applications may be encoded in SPARQL/RDF, but due the dedicated focus on the semantic web domain, there is a limit for applications with a different scope.

SQL is widely accepted as the standard query language for relational database systems. While extensions for hierarchical [2], geospatial [15] and time series data exist, graph queries have not been considered in the past. Reasons may be the complexity of graph queries (graph pattern matching, path expressions etc.) and too much focus on methods for encoding graph data in relational database tables. Moreover, specifying graph queries directly in SQL is cumbersome and often leads to inefficient query executions. Particularly graph pattern matching and transitive closures usually require dedicated graph query languages and engines.

In this paper, we propose a novel, high-level graph query language that is based on SQL. We build on the syntax and semantics of SQL, transfer its query concepts into the realm of graph databases, and extend them with dedicated graph features. In particular, our language supports graph pattern matching with nested graph conditions [3,7], expressions for traversing paths of fixed length, calculation of transitive closures and definition of distance constraints between matched nodes. The pattern matching is in general non-injective, but it can be customized by adding injectivity constraints for pairs of node variables. Moreover, dedicated functions for computing shortest paths and connected components are included. The general structure of queries follows the one of SQL. The syntax for arithmetic expressions, aggregates, sorting, limits etc. can be reused entirely. Since the result of graph queries are tables, they can be embedded as subqueries in standard relational SQL queries, thereby enabling a smooth integration with relational and other types of engines in heterogeneous database management systems. For instance, when agreeing on SQL as common base language, graph queries could be combined with relational, geospatial or even time series queries. Although the engine implementations are typically separate, a common base language enables the usage of a common query processing infrastructure including parsing, plan generation and query optimization.

We provide an execution engine for the proposed language, referred to as the SAP HANA Graph Scale-Out Extension (GSE) prototype in the rest of this paper. To achieve a high query performance, our engine uses optimized graph data structures instead of relational tables. We evaluate the expressive power of our query language and the performance of the GSE implementation using the LDBC Social Network Benchmark [4]. We focus in this paper on complex, read-only queries. However, our query language lays the foundation for an SQL-based graph manipulation language based on the theory of graph transformations. Therefore, it paves the way for transferring formal methods, such as critical pair analysis for confluence checking [8], into the graph database realm.

Organization: Sect. 2 first gives an overview of the graph model we use. It then discusses graph pattern matching with nested formulas and subsequent relational evaluation. Section 3 introduces our SQL-based graph query language. Section 4 provides an evaluation of the GSE implementation based on an LDBC benchmark. Section 5 gives an overview of related work. Section 6 contains conclusions and future work.

2   Graph Pattern Matching with Relational Evaluation

In this section, we present the models and concepts that form the foundation of our query language and engine.

2.1   Graph Model

We consider directed and undirected graphs with typed nodes, typed edges and typed node properties. Each of these types has a fixed value range which is part of the graph definition. Figure 1 shows the corresponding graph model.

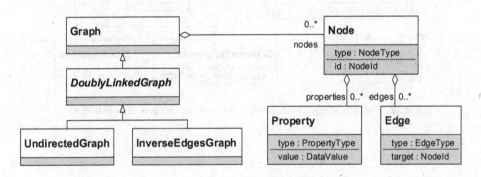

Fig. 1. Typed graph model with node properties and inverse edges

Every node has a unique numeric ID, a list of properties, and a list of edges. A property has a primitive data value of a data type derived from the property type (e.g., string, integer, float). The value of a property can be NULL independently of its data type. We consider the ID and the type of a node as special properties with the respective property types NODEID and NODETYPE.

An edge identifies its target node by its ID. Every node can have at most one edge of the same type and with the same target, i.e., parallel edges of the same type are not allowed. However, edges may target their source node (*loops*).

In a *doubly linked* graph, every edge has a corresponding edge with swapped source and target nodes. This corresponding edge either has the same type as the original edge or an implicitly defined inverse edge type. In the first case, the graph is *undirected*. In the second case, it is an *inverse edge* graph where all edges of regular type are outgoing and all edges of inverse type are incoming.

Our graph model deliberately omits edge properties (e.g., edge weights). This design choice enables an efficient implementation, particularly in a distributed setting. The resulting restriction can be overcome by modeling edges with properties by auxiliary edge nodes with one incoming and one outgoing edge.

2.2   Graph Pattern Matching

In graph pattern matching, the task is to find all matches between a set of pattern variables and the nodes of a target graph that satisfy a set of conditions. Figure 2 shows our graph pattern model and Fig. 3 a basic example pattern.

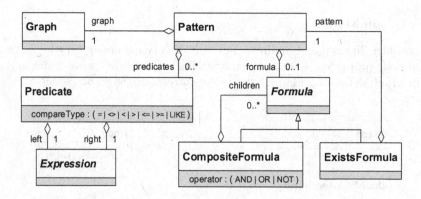

Fig. 2. Pattern model with predicates and nested graph constraints

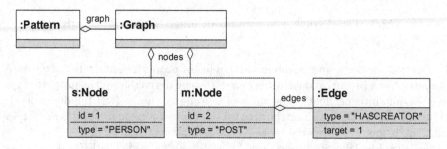

Fig. 3. Basic graph pattern consisting only of a pattern graph with two typed nodes and a typed edge between these nodes

A *pattern* consists of a pattern graph, a set of predicates, and an optional nested formula. The predicates and the formula are defined over a set of *pattern variables* that represent the *pattern nodes*, i.e., the nodes of the pattern graph.

A *match* of a pattern with respect to a given target graph is a map of the pattern variables to nodes of the target graph (called the *target nodes* of the match), such that (i) each pattern node is matched to a target node of the same type, (ii) each pattern edge is matched to a target edge of the same type, (iii) all predicates are satisfied, and (iv) the logical formula is satisfied.

Formally, conditions (i) and (ii) describe a typed graph homomorphism from the pattern graph into the target graph. In general, this graph homomorphism does not have to be injective. Instead, predicates comparing the IDs of the pattern nodes can be used to guarantee that pattern nodes are matched to different target nodes. The same approach is used to assure that a certain set of target nodes is matched only once if the pattern graph has symmetries.

In order to be able to omit the type constraint from conditions (i) and (ii), we add the implicitly defined type ANY to the lists of node types and edge types of the pattern graph. Thus, if a pattern node or edge has the type ANY, it can be matched to any target node or edge, respectively.

A predicate is a binary comparison between two expressions. Figure 4 depicts how we model expressions. Expressions are defined over the pattern variables and, given a potential match, evaluate to primitive data values. Literals simply evaluate to their constant value. Arithmetic expressions and functions are evaluated recursively, i.e., after evaluating their arguments they are evaluated as expressions over primitive data values.

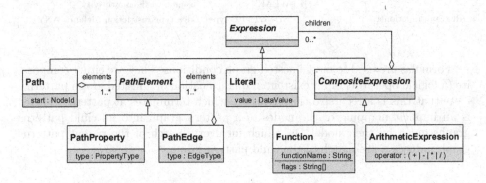

Fig. 4. Expression model

Path expressions form the link between the pattern variables and the properties of the target nodes. In their most basic form, paths consist of a pattern variable (given by the start node ID) and a property type (given by a path property). Such a path is called a *node property path* and evaluates to the property value with the given type of the target node matched to the given variable. All path expressions that occur in a graph pattern have to be node property paths.

This does not restrict the expressive power of the model, since complex path conditions can be expressed in the graph part of the pattern and by nested formulas. We discuss the evaluation of more complex paths in the next section where expressions containing such paths are introduced.

Note that we may access the ID of a target node and use it in other expressions, most notably functions. Such functions are called *graph functions* and are parametrized by optional function flags (see Table 1). They have access to the complete target graph, allowing them to explore the neighborhood of a node (e.g., to calculate degrees) as well as to traverse the global graph structure (e.g., to compute distances, shortest paths, and connected components).

Table 1. Summary of currently supported graph functions

Function	Args	Flag	Result
DEGREE	1		Degree of argument node
		IN	Compute in-degree
		OUT	Compute out-degree (default)
		INOUT	Compute in-degree + out-degree
DISTANCE	2		Node distance (NULL if unconnected)
SHORTEST_PATH	2		Shortest paths sub-graph (as JSON)
		DIRECTED	Traverse edges regularly (default)
		UNDIRECTED	Traverse edges in both directions
		INVERSE	Traverse edges in inverse direction
CONNECTED_COMPONENT	1		Id of node's connected component
		STRONG	Assume strong connectivity (default)
		WEAK	Assume weak connectivity
All graph functions:		EDGETYPE type	Edge type restriction (default: ANY)

Formulas are used to model nested graph conditions. They are either composite (a logical operator) or an existential quantification of a nested graph pattern. Note that this is a recursive tree structure which terminates at patterns. There is an implicit mapping of the nodes of a pattern graph into its child pattern graphs given by their node IDs, which for every match of the parent pattern graph induces a pre-match of the child pattern graph.

2.3   Graph-Relational Evaluation and Nested Paths

The result of the graph pattern matching described in the previous section is a list of matches from the pattern variables to nodes in the target graph. Based on a feature list, *graph-relational evaluation* computes a table of primitive data values from these matches. A *feature* is an expression as defined in the previous section. Thus, given a list of features and a list of matches, we can compute a table where each column corresponds to a feature. For each match, we create a row in the table by evaluating the feature expressions over the match.

In the following, we extend this concept by introducing expressions that do not evaluate to single values but rather to lists of data value tuples.

In the previous section, we introduced the concept of node property paths which evaluate to a single value. Now we extend this notion to nested projection paths which we specify in Fig. 4. Such a nested projection path describes a nested traversal of the target graph and evaluates to a list of property value tuples. For this, path elements are recursively evaluated. Unlike for composite expressions, this evaluation is carried out top down. This means that the parent path element is evaluated first and then passes a target node as an argument to its children.

For example, consider the nested projection path depicted in Fig. 5. It consists of an edge followed by at nested projection to a property and a subpath formed by a second edge and another property. If the pattern variable (start) is matched to a node representing a person, the path evaluation traverses all outgoing WORKSAT edges to find all companies the person worked for. For every company and each location of the company, the evaluation creates a pair of the company name and the location name. Each company generates at least one pair, even if it has no ISLOCATEDIN edge. Likewise, the whole traversal generates at least one pair, even if the person has no outgoing WORKSAT edge (the NULL pair).

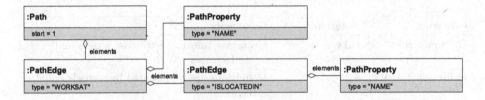

Fig. 5. Example of a nested projection path.

In the following, we discuss nested path expressions in detail, beginning with basic *traversal paths*. A traversal path expression consists of a pattern variable (start) and a sequence of path edges followed by a path property. When evaluated, the path expression first determines the target node matched to its variable and passes it as argument to the first path edge. Every subsequent path edge is evaluated by traversing all edges of its argument node that match its type. For each of these traversed edges, the end-node is passed as argument to the next path edge in the sequence.

The recursion terminates at a path property or if the argument target node of a path edge does not have any edges of the respective type. In the first case, the result is the respective property of the target node that was passed as argument. In the second case, the path expression returns a set containing a single NULL value. From a relational perspective, a traversal path expresses a left outer join over the edge relation. From a graph-theoretic perspective, it expresses a path traversal of the target graph starting at the start node of the path expression and using the edge types defined by the path edges. The result is the list of property values of the traversal path end-nodes specified by the property type of the path property.

If one or more of the expressions forming a feature list contains a traversal path, the one-to-one correspondence between matches and rows of the result table becomes a one-to-many correspondence. When evaluating the feature list for a particular match, we first evaluate all traversal path expressions. Next, we create the cross product of the data value lists that result from the path traversals. Finally, we create a row in the result table for every tuple in the cross product. For this, we first replace all traversal paths in the feature expressions by the corresponding cross product tuple element. Then we evaluate this modified feature list for the current match.

In addition to traversal paths, we also define *nested projection paths*. A nested projection path expression and its path edges may have more than one child path elements. In this case, the result is a list of tuples which is derived by evaluating all child elements and creating the cross product of their result lists. Note that the result lists of the child elements may already be tuple lists. In this case we stay in line with relational algebra and treat the tuples as *shallow*, i.e., concatenate the tuples when creating the cross product (in contrast to cre-

Table 2. Notation used to specify the syntax of our query language

Construct	Notation	Comments
Grammar rule	`rule`	Grammar rules use lowercase letters and underscores
Definition	`=`	Definitions are represented by a single equal signs
Alternation	`...|...`	Alternatives are separated using vertical bars
Grouping	`(...)`	Grouping is represented by enclosing parentheses
Option	`[...]`	Optional parts are represented by enclosing square brackets
Repetition	`...*`	Zero or more repetitions are indicated by the suffix *
Terminal symbol	`KEYWORD`	Language keywords are written in uppercase letters
Terminal character	`"."`	Single-character language symbols are set in double quotes
Terminal literal	`literal`	Literals represent typed string and numerical constants
Terminal identifier	`identifier::id`	Identifiers are indicated by the suffix `::id`
List abbreviation	`rule::list`	Comma-separated lists are abbreviated by the suffix `::list.list_rule = rule (","rule )*`

ating tuples of tuples). From a relational perspective, a nested path expresses a cross join over its child elements. From a graph-theoretic perspective, it expresses a tree traversal of the target graph.

During evaluation, nested projection paths are treated like traversal paths, i.e., the result lists of the nested projection paths become part of the cross product created over the result lists of the traversal paths. However, a feature represented by a nested projection path expression corresponds to several table columns, one for every element in the result tuples of the nested projection path. A nested projection paths may appear as a subexpression of a compound expression (e.g., a function). The compound expression is evaluated for each tuple generated by the nested path expression.

3 Graph Query Language

The syntax and semantics of our graph query language is described in this section. It is closely aligned with the SQL standard. It uses, where possible, SQL syntax and extends it with graph-specific features. Some of these extensions can be also found in a similar form in the query language of SAP HANA Core Data Services (CDS) [14]. We discuss three complex example queries in Sect. 4.

To specify the syntax of our query language, we use the notation defined in Table 2 which is inspired by the Extended Backus-Naur Form (EBNF). For the sake of brevity, we consider identifiers and literals as additional syntax terminals beside the traditional syntax terminals.

Listing 1.1. Overview of graph query syntax

```
query       = SELECT ( "*" | feature::list ) FROM variable::list
              [ USING GRAPH graph::id ] [ WHERE condition ]
              [ GROUP BY expression::list ] [ ORDER BY expression::list ]
              [ LIMIT literal ]

feature     = expression [ AS expression_alias::id ]
variable    = node_type::id [ [ AS ] variable_alias::id ]

condition   = formula | predicate | ( path IN path )

formula     = ( "(" condition ( AND | OR ) condition ")" )
              | ( "(" NOT condition ")" )
              | ( EXISTS variable::list WHERE "(" condition ")" )
predicate   = ( expression comparator expression ) | ( path IS [ NOT ] NULL )
comparator  = "=" | "<>" | "<" | ">" | "<=" | ">=" | LIKE

expression  = literal | arithmetic | function | path
arithmetic  = "(" expression ( "+" | "-" | "*" | "/" ) expression ")"
              | ( "-" expression ")"
function    = function_name "(" flag::id* expression::list ")"

path        = variable::id [ projection ]
projection  = "." element | ( "." "(" element::list "}" )
element     = property::id | ( edge_type::id [ projection ] )
```

3.1 Query Structure

A graph query is a read-only operation that performs graph pattern matching on a given target graph followed by relational evaluation to generate a primitive-typed result table. This table can be subsequently consumed by other relational operators. Therefore, graph queries can in principle be embedded in and combined with standard SQL queries. Syntactically, graph queries follow closely the structure of SELECT statements in SQL. Listing 1.1 summarizes the syntax definition of our graph query language.

The query language syntax provides a clear separation between the graph pattern matching and the graph-relational evaluation. This is achieved by defining each operation in different clauses (with the only overlap of the variable definition in the FROM clause). The pattern matching is defined by the FROM, USING GRAPH, and WHERE clauses, whereas the relational evaluation is defined by the SELECT, FROM, GROUP BY, ORDER BY, and LIMIT clauses.

The following query finds all matches to the graph pattern shown in Fig. 3:

```
1   SELECT s.ID, m.CONTENT, s.WORKSAT.{NAME,ISLOCATEDIN.NAME} AS COMPANY
2   FROM PERSON s, POST m
3   WHERE s IN m.HASCREATOR
```

It then evaluates for each matched person and post the nested projection path depicted in Fig. 5. Each matched person and post together with a traversed company and location produce a row in the result table, containing the respective properties.

3.2 Pattern Matching

As discussed in Sect. 2.2, in graph pattern matching we compute all matches from the node variables of a graph pattern to the nodes of a target graph that satisfy the conditions of a given graph pattern.

The USING GRAPH clause of a graph query identifies the target graph by its name. Thus, multiple graphs stored in a graph database can be distinguished.

The variables in the FROM clause consist of a node type and a variable alias. Although syntactically similar to standard SQL, the semantics of these variables differs from relational queries. In a graph query, the variables represent the pattern variables introduced in Sect. 2.2. During the graph pattern matching, these variables are matched to nodes in the target graph rather than to relational tuples. Moreover, the node type of a variable declaration is already part of the graph pattern and defines the type of the corresponding pattern node.

The main part of the graph pattern is defined in the condition part of the WHERE clause. The syntax of formulas, predicates, expressions, and paths directly translates to the corresponding models discussed in Sect. 2.2.

Semantically, only paths differ from standard SQL, since they are defined over pattern variables and not over relational variables. Note that not all syntactically correct paths are admitted at every position in a graph query. Depending on whether the position allows for one result or a result list and for a single value or a tuple, traversals and nested projections may be forbidden. Moreover, the end-property of a path is optional and defaults to NODEID.

The main difference between standard SQL conditions and graph query conditions is the syntax and semantics of the IN keyword. In an IN condition, the path before the IN keyword must be a node property path. The path after the IN keyword has to be a traversal path. Semantically, the IN conditions are mainly used to define the edges and the edge types of the pattern graph.

3.3 Relational Evaluation

The relational part of a graph query is defined by the SELECT, FROM, GROUP BY, ORDER BY, and LIMIT clauses. The optional GROUP BY, ORDER BY, and LIMIT clauses of a graph-relational evaluation follow the standard SQL syntax and semantics (see Listing 1.1), generating tables as result of graph queries.

The graph-specific part of the relational evaluation is defined in the SELECT and FROM clauses. In the previous section, we already discussed the FROM clause and established that semantically it defines the pattern variables. The feature list syntax directly reflects and maps to the concepts introduced in Sect. 2.3. There, we already discussed in detail the evaluation of feature lists, including the implicit cross-join introduced by traversal paths and nested paths.

We also allow the SELECT * syntax known from standard SQL. In a graph query, the star symbol is expanded to a feature list which contains a nested projection path x.{NODEID, NODTYPE, P1,..., Pk} for every variable x in the FROM clause, where P1,..., Pk are all property types for which at least one node of the variable's type has a non-NULL value.

Besides the graph functions defined in Table 1, function expressions can also be classical SQL aggregates such as COUNT or AVG and arithmetic functions such as ABS or MOD. Since the result of the graph-relational evaluation is a table, the semantics of aggregates (in particular in conjunction with a GROUP BY clause) carries over from standard relational queries.

4   Evaluation

In this section, we evaluate our graph query language and the GSE prototype engine implementation using the LDBC Social Network Benchmark.

4.1   Implementation

In the following we give an overview of the GSE prototype implementation of a graph query engine that supports major parts of our query language.

The high-level architecture of the GSE is shown in Technical Architecture Modeling notation in Fig. 6. Application development is supported via a high-level programming API (C/C++, Java) and a Neo4j-compatible REST-API. An additional connector provides an integration with SAP HANA Vora [5] – a scale-out extension of SAP HANA for massively parallel data processing integrating with the Hadoop framework.

The core of the graph engine uses a distributed in-memory graph store which implements the graph model shown in Fig. 1. Data adapters enable the loading and saving of graph data from (distributed) file systems such as HDFS, and relational tables in SAP HANA/SAP HANA Vora.

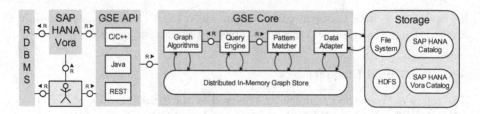

Fig. 6. High-level GSE architecture

The query engine parses textual queries as described in Sect. 3, translates them to pattern models (Fig. 2) and uses the pattern matcher and graph algorithm implementations to execute queries (see Table 1 for supported graph functions). The query execution and pattern matching are then parallelized and distributed across a cluster. The pattern matching engine is based on an encoding into a constraint satisfaction problem [13].

4.2 LDBC Social Network Benchmark

The Linked Data Benchmark Council (LDBC) is a non-profit organization defining benchmarks for graph data management software. We use here the Social Network Benchmark (SNB), specifically the Interactive Workload [4], consisting of 29 queries which are split into three categories: complex read, short read and update queries. This benchmark includes different "choke points" for query engines, such as aggregation performance and data access locality. Graphs to run this workload against can be generated with the help of a given data generator, which produces a social network graphs of a given scale factor.

4.3 Complex Read Queries

There are 14 complex read queries in the LDBC Social Network Benchmark. Out of these, 13 can be represented in our graph query language. Figure 7 shows three selected queries including informal descriptions and graphical representations of their respective pattern graphs.

In Query 9, the implicitly defined node type ANY (line 4) is used in combination with NODETYPE (lines 7–8) enabling a simple kind of node type inheritance. The predicate involving the DISTANCE-function (line 9) ensures that the two PERSON nodes are connected by a KNOWS-path of length at most 2. Note that the :KNOWS[1..2]-path in the graphical notation refers to the length of a shortest path between the two matched nodes. The predicate s<>p is used to selectively ensure injective matching of the two PERSON nodes.

Query 5 consists of a cyclic pattern graph. This query is also an example of a conversion of edges with attributes to nodes and node properties (see Sect. 2.1). The original edge HASMEMBER from PERSON to FORUM is modeled as a node of type HASMEMBER, which contains all properties of the original

```
1  SELECT p.ID, p.FIRSTNAME, p.LASTNAME,
2    m.ID, m.CONTENT, m.IMAGEFILE,
3    m.CREATIONDATE
4  FROM PERSON s, PERSON p, ANY m
5  WHERE s.ID={1}
6  AND s<>p
7  AND (m.NODETYPE='POST' OR
8       m.NODETYPE='COMMENT')
9  AND DISTANCE(EDGETYPE KNOWS s,p)<=2
10 AND p IN m.HASCREATOR
11 AND m.CREATIONDATE<{2}
12 ORDER BY m.CREATIONDATE DESC, m.ID ASC
13 LIMIT 20
```

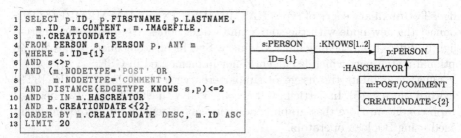

Query 9: Recent posts and comments by friends or friends of friends: Given a start Person, find the (most recent) Posts/Comments created by that Persons friends or friends of friends (excluding start Person). Only consider the Posts/Comments created before a given date (excluding that date). Return the top 20 Posts/Comments, and the Person that created each of those Posts/Comments. Sort results descending by creation date of Post/Comment, and then ascending by Post/Comment identifier.

```
1  SELECT f.TITLE, COUNT(o)
2  FROM PERSON s, PERSON p, POST o,
3        HASMEMBER m, FORUM f
4  WHERE s.ID={1}
5  AND s<>p
6  AND DISTANCE(EDGETYPE KNOWS s,p)
7                              <= 2
8  AND m IN f.HASMEMBER_IN
9  AND p IN m.HASMEMBER_OUT
10 AND m.JOINDATE>{2}
11 AND o IN f.CONTAINEROF
12 AND p IN o.HASCREATOR
13 GROUP BY f.ID
14 ORDER BY COUNT(o) DESC, f.ID ASC
15 LIMIT 20
```

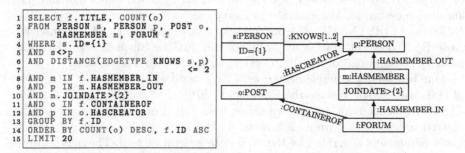

Query 5: New Groups: Given a start Person, find the Forums which that Persons friends and friends of friends (excluding start Person) became Members of after a given date. Return top 20 Forums, and the number of Posts in each Forum that was Created by any of these Persons. For each Forum consider only those Persons which joined that particular Forum after the given date. Sort results descending by the count of Posts, and then ascending by Forum identifier

```
1  SELECT t.NAME AS TAGNAME, COUNT(t) AS tc
2  FROM PERSON s, PERSON p, TAG t, POST m
3  WHERE s.ID={1}
4  AND s IN p.KNOWS
5  AND t IN m.HASTAG
6  AND p IN m.HASCREATOR
7  AND m.CREATIONDATE>={2}
8  AND m.CREATIONDATE<{3}
9  AND NOT EXISTS PERSON p2, POST m2
10 WHERE (p2 IN s.KNOWS
11       AND p2 IN m2.HASCREATOR
12       AND m2.CREATIONDATE<{2}
13       AND t IN m2.HASTAG
14       AND m<>m2
15 )
16 GROUP BY t.NAME
17 ORDER BY COUNT(t) DESC, t.NAME ASC
18 LIMIT 10
```

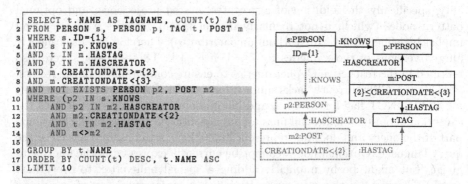

Query 4: New Topics: Given a start Person, find Tags that are attached to Posts that were created by that Persons friends. Only include Tags that were attached to friends Posts created within a given time interval, and that were never attached to friends Posts created before this interval. Return top 10 Tags, and the count of Posts, which were created within the given time interval, that this Tag was attached to. Sort results descending by Post count, and then ascending by Tag name.

Fig. 7. Selected queries of the LDBC Social Network Benchmark

edge. Two auxiliary edges of types HASMEMBER_IN and HASMEMBER_OUT connect the new node with the source and target nodes. Note that our Query 5 slightly differs from the original SNB query. Specifically, our query is missing an optional match part which is currently not implemented in GSE.

Query 4 exhibits the usage of subpatterns with the help of the EXISTS-keyword (lines 9–15). In particular, it is combined with NOT to specify a negated graph condition. Note that graph conditions can in general be nested and combined using Boolean operators.

4.4   Performance

For performance evaluation, we use generated LDBC data of scale-factor 3, which corresponds to a graph with 23.7M nodes and 84.8M edges. Minor adaptations of the generated data include a conversion of string-based dates into integer-encoded timestamps. We generate property indices for the property types ID, NAME and FIRSTNAME. The benchmarks were executed on a 24-core Intel(R) Xeon(R) X5650 workstation at 2.67 GHz and 96 GB main memory. The total memory consumption of the LDBC graph at scale-factor 3 is 12.5 GB.

Our benchmark is implemented as a Java application using the Java APIs of GSE and Neo4j. We use the Community Edition of Neo4j in version 2.3.0. The Neo4j benchmark queries are taken from [11]. Both systems were benchmarked independently from each other on the same machine, using the same query parameters generated by the SNB data generation tool. The queries were executed between 20 and 50 times, depending on query complexity. To warm up both engines, all queries were executed with randomly chosen parameters beforehand.

Out of the 14 SNB queries, 8 queries were implemented completely. For the queries 1, 3 and 5, the GSE implementation is currently missing some functionality, specifically the addition of aggregates, nested projections, and optional pattern nodes (which, if not regularly matched, still generate a match to an implicitly defined null-vertex without any attributes). Therefore, we used slightly altered versions of those queries for our benchmark. Theses changes were also reflected in the respective Cypher queries to ensure comparability of the results. For Query 7, GSE currently lacks support for the SQL CASE-keyword, for Query 10, a DISTINCT flag for COUNT and modulo-operation. Notice, that with the exception of optional nodes, all missing implementation belongs to the relational part of the query and does not represent limitations of the pattern model or the query language. Optional nodes cannot be currently expressed in the query language (but might be by implicitly adding a special null-vertex to every host graph). Query 14 is the only query for which a major extension of our query language is required, since it includes shortest paths weighted by sub-patterns.

Table 3 shows the query run-times of Neo4j and GSE in milliseconds and the respective speed-up factors. GSE is faster for all implemented queries, staying under the one-second mark for most of them. The long running queries, particularly Query 9, yield large match sets and result tables which need to be sorted.

Our analysis shows that the pattern matching is efficient in these queries, but the generation of the result table and its sorting requires most of the time.

Table 3. LDBC complex read benchmark: mean of runtimes in milliseconds and speed-up factors. Queries marked with * include minor modifications.

Query	1*	2	3*	4	5*	6	8	9	11	12	13
Neo4j	7,041	2,122	5,496	12,262	10,074	44,625	161	405,457	197	5186	5
GSE	40	195	83	105	2,468	1,325	30	13,616	12	42	2
Speed-up	174	10	66	116	4	33	5	29	16	123	2

5   Related Work

SPARQL [18] supports graph pattern matching using an RDF-triple syntax, e.g.

```
PREFIX LDBC: <http://ldbcouncil.org/developer/snb>
SELECT ?m ?n WHERE { ?x LDBC:NAME ?m . ?x LDBC:KNOWS ?y . ?y LDBC:NAME ?n }
```

Note that both edges and primitive-valued properties are described using triples. Conditions on properties are defined using FILTER, existential quantification using EXISTS, and alternative patterns using the UNION keyword. Furthermore, optional patterns parts can be specified using the OPTIONAL keyword.

The openCypher [10] query language uses a different query structure, e.g.

```
MATCH (x:PERSON)-[:KNOWS]-(y:PERSON) RETURN x.NAME, y.NAME
```

Alternative pattern parts are defined using UNION and optional patterns using OPTIONAL MATCH. There is no existential quantification of patterns but negation and collections can be used to define more complex graph conditions. Both the SPARQL and openCypher syntax make use of SQL-keywords. However, the general query structure is too specific to be integrated with SQL. In contrast, our query language is closely aligned to the SQL standard and therefore enables a smooth integration with relational database systems.

The Gremlin [12] language which is part of the Apache TinkerPop project can be characterized as a functional embedded DSL for graph traversals.

A recent performance comparison of graph and relational databases is given in [6]. The paper shows that state-of-the-art relational databases can compete or even exceed the performance of graph databases in certain graph pattern matching scenarios. However, the employed benchmark uses rather simple graph patterns without additional (nested) graph conditions.

An encoding of graph transformation rules in SQL is presented in [17]. A graphical syntax is used for modeling the graph transformation rules. In general, except for the limitation of being read-only, our query language provides a number of features that are usually not found in graph transformations, such as distance constraints, shortest paths, and relational operations including sorting, limits and aggregations. Note also that graph transformations usually operate on *one* match, whereas our approach always considers *all* matches.

6   Conclusions and Future Work

We presented a novel SQL-based graph query language supporting graph pattern matching with nested graph conditions [7] and distance constraints, as well as calculation of nested projections, shortest paths and connected components. Since it is based on SQL, the syntax for arithmetic expressions, aggregations, sorting etc. can be reused entirely, and graph queries can be embedded as subqueries in relational queries. We evaluated the language features and the GSE prototype implementation using the LDBC Social Network Benchmark.

As future work, we plan to incorporate optional matching (see [10,18]) and concepts for shortest paths weighted by sub-patterns as required in LDBC-SNB Query 14. We further plan to define a SQL-based graph manipulation language formally based on amalgamated graph transformations [1]. The planned graph manipulation language will build on the pattern matching syntax proposed in this paper and extend it to provide the same expressive power as graph transformation rules. The syntax of FROM and WHERE clauses will be reused and extended by an UPDATE clause for specifying created and deleted graph parts, and updates of property values. Semantically, the relational evaluation will be replaced by an amalgamated graph transformation step. The use of graph transformations is potentially enabling formal analysis techniques, such as critical pair analysis for confluence checking [8]. For the implementation, we plan to add nested projections and optional matching to the GSE query engine, to incorporate Bulk Synchronous Parallel [9,16] for distributed graph algorithms, and to compare the performance of our (distributed) engine with other graph database engines such as Sparksee and Virtuoso.

References

1. Boehm, P., Fonio, H., Habel, A.: Amalgamation of graph transformations: a synchronization mechanism. J. Comput. Syst. Sci. **34**(2/3), 377–408 (1987)
2. Brunel, R., Finis, J., Franz, G., May, N., Kemper, A., Neumann, T., Färber, F.: Supporting hierarchical data in SAP HANA. In: Proceedings of ICDE 2015, pp. 1280–1291. IEEE (2015)
3. Ehrig, H., Ehrig, K., Habel, A., Pennemann, K.-H.: Constraints and application conditions: from graphs to high-level structures. In: Ehrig, H., Engels, G., Parisi-Presicce, F., Rozenberg, G. (eds.) ICGT 2004. LNCS, vol. 3256, pp. 287–303. Springer, Heidelberg (2004)
4. Erling, O., Averbuch, A., Larriba-Pey, J., Chafi, H., Gubichev, A., Prat-Pérez, A., Pham, M., Boncz, P.A.: The LDBC social network benchmark: interactive workload. In: Proceedings of 2015 ACM SIGMOD, pp. 619–630. ACM (2015)

5. Goel, A.K., Pound, J., Auch, N., Bumbulis, P., MacLean, S., Färber, F., Gropengießer, F., Mathis, C., Bodner, T., Lehner, W.: Towards scalable real-time analytics: an architecture for scale-out of OLxP workloads. PVLDB 8(12), 1716–1727 (2015)
6. Gubichev, A., Then, M.: Graph pattern matching - do we have to reinvent the wheel? In: Proceedings of GRADES 2014, pp. 8:1–8:7. ACM (2014)
7. Habel, A., Pennemann, K.: Correctness of high-level transformation systems relative to nested conditions. Math. Struct. Comput. Sci. 19(2), 245–296 (2009)
8. Heckel, R., Küster, J.M., Taentzer, G.: Confluence of typed attributed graph transformation systems. In: Corradini, A., Ehrig, H., Kreowski, H.-J., Rozenberg, G. (eds.) ICGT 2002. LNCS, vol. 2505, pp. 161–176. Springer, Heidelberg (2002)
9. Krause, C., Tichy, M., Giese, H.: Implementing graph transformations in the bulk synchronous parallel model. In: Gnesi, S., Rensink, A. (eds.) FASE 2014 (ETAPS). LNCS, vol. 8411, pp. 325–339. Springer, Heidelberg (2014)
10. Neo Technology Inc., OpenCypher (2015). http://www.opencypher.org
11. Prat, A., Boncz, P., Larriba, J.L., Angles, R., Averbuch, A., Erling, O., Gubichev, A., Spasić, M., Pham, M.D., Martínez, N.: LDBC Social Network Benchmark (SNB) - v0.2.2 first public draft (2015). http://github.com/ldbc/ldbc_snb_docs/blob/master/LDBC_SNB_v0.2.2.pdf
12. Rodriguez, M.A.: The Gremlin graph traversal machine and language (invited talk). In: Proceedings of DBPL 2015 (DBPL 2015), pp. 1–10. ACM (2015)
13. Rudolf, M.: Utilizing constraint satisfaction techniques for efficient graph pattern matching. In: Ehrig, H., Engels, G., Kreowski, H.-J., Rozenberg, G. (eds.) TAGT 1998. LNCS, vol. 1764, pp. 238–252. Springer, Heidelberg (2000)
14. SAP SE: SAP HANA core data services (CDS) reference (2015). http://help.sap.com/hana/SAP_HANA_Core_Data_Services_CDS_Reference_en.pdf
15. SAP SE: SAP HANA spatial reference (2015). http://help.sap.com/hana/sap_hana_spatial_reference_en.pdf
16. Valiant, L.G.: A bridging model for parallel computation. Commun. ACM 33(8), 103–111 (1990)
17. Varró, G., Friedl, K., Varró, D.: Implementing a graph transformation engine in relational databases. Softw. Syst. Model. 5(3), 313–341 (2006)
18. World Wide Web Consortium (W3C): SPARQL 1.1 query language (2013). http://www.w3.org/TR/2013/REC-sparql11-query-20130321

On the Operationalization of Graph Queries with Generalized Discrimination Networks

Thomas Beyhl, Dominique Blouin, Holger Giese$^{(\boxtimes)}$, and Leen Lambers$^{(\boxtimes)}$

Hasso-Plattner Institute at the University of Potsdam,
Prof.-Dr.-Helmert-Str. 2-3, 14482 Potsdam, Germany
{thomas.beyhl,dominique.blouin,
holger.giese,leen.lambers}@hpi.uni-potsdam.de

Abstract. Graph queries have lately gained increased interest due to application areas such as social networks, biological networks, or model queries. For the relational database case the relational algebra and generalized discrimination networks have been studied to find appropriate decompositions into subqueries and ordering of these subqueries for query evaluation or incremental updates of queries. For graph database queries however there is no formal underpinning yet that allows us to find such suitable operationalizations. Consequently, we suggest a simple operational concept for the decomposition of arbitrary complex queries into simpler subqueries and the ordering of these subqueries in form of *generalized discrimination networks* for graph queries inspired by the relational case. The approach employs graph transformation rules for the nodes of the network and thus we can employ the underlying theory. We further show that the proposed generalized discrimination networks have the same expressive power as nested graph conditions.

1 Introduction

The model of typed graphs and related graph queries to explore existing graphs and their properties has lately gained increased importance due to application areas of increasing relevance such as social networks, biological networks, and model queries [14] and technologies like graph databases [2] or model-driven development [4] where graphs rather than relations are the main characteristics of the employed models and queries.

While the definition of typed graphs by means of schemas, metamodels, or grammars is a formally well studied topic, there is yet no clear formal underpinning for graph queries concerning their specification as well as their operationalization (cf. [2,16]). For the *operationalization* of the query evaluation and incremental query updates of relational queries the *relational calculus* [1] and *generalized discrimination networks* (GDN) have been suggested (cf. [13]) as a formal framework to study which decomposition into subqueries and ordering of

---

This work was partially developed in the course of the project Correct Model Transformations II (GI 765/1-2), which is funded by the Deutsche Forschungsgemeinschaft.

R. Echahed and M. Minas (Eds.): ICGT 2016, LNCS 9761, pp. 170–186, 2016.
DOI: 10.1007/978-3-319-40530-8_11

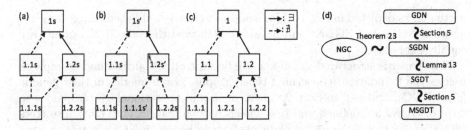

Fig. 1. GDNs in form of a SGDN (a) and SGDTs (b)(c) for a social network query

these subqueries is most appropriate. As depicted in Fig. 1(a), in such a network each node (numbered block) is responsible for evaluating a subquery and for this purpose it may compose subquery evaluations of nodes it depends on. The overall result is then the query evaluation of the terminal node. However, such a formal framework does not exist for graph queries so far.

Consequently, inspired by the relational case we suggest motivated by our practical work on view maintenance for graph databases [6] a simple operational concept for the decomposition of arbitrary complex *graph queries* into a suitable ordering of simpler subqueries in form of GDNs. Rather than considering one particular kind of GDN with particular computation nodes, we suggest employing *graph transformation* (GT) rules for these computation nodes such that we are also able to employ the well understood GT theory [9] as a basis. The basic idea to define our notion of GDN related to GT systems is to employ extra marking nodes and edges to encode the results of subqueries and specific graph transformation rules to describe the propagation behavior of the network nodes via creating and reading markings.

We study in this paper what are the core ingredients required to approach graph query evaluation based on an operational specification using the above-described GDNs while having the same expressiveness as *declarative graph queries* based on *nested graph conditions* (NGC) [12]. The latter have the expressive power of first order logic on graphs and constitute as such a natural formal foundation for pattern-based graph queries.

We assume in the following that a *graph query* is characterized by a *request graph L* delivering its answers in form of a set of matches for L into the queried graph G fulfilling some additional properties as described in the graph query.[1,2] Based on the answer set semantics we were able to establish equivalence of NGCs with GDNs including different specific subsets such as so-called simple GDNs (SGDNs), simple tree-like GDNs (SGDT), and minimal SGDTs (MSGDT). In

---

[1] It is to be noted that a simple record as provided by an SQL-statement is also a special form of graph where no links are included.

[2] While in practice the requested number of answers is often limited to a fixed upper bound of answers, for our more theoretical considerations in this paper, we can assume w.l.o.g. that all matches of L for G that fulfill the additional properties that must hold are building the correct set of answers.

particular as depicted in Fig. 1(d), as a main result we established the equivalence between NGCs and SGDNs and in addition showed that all GDN variants are equally expressive.

The paper is structured as follows: We first introduce our running example as well as the foundations concerning typed graphs, graph queries in their generic form, NGCs, and GT in Sect. 2. Then, in Sect. 3 operational graph queries in form of GDNs are defined and it is shown how to transform SGDNs into trees (SGDTs). That SGDNs and declarative queries based on NGCs have the same expressive power follows in Sect. 4 and we discuss the different variants of GDNs concerning their expressiveness and applicability w.r.t. optimization and incremental updates for graph queries in Sect. 5. Finally, we conclude the paper and provide an outlook on planned future work.

2   Prerequisites

After outlining our running example, we will introduce typed graphs, based on that a generic notion of graph query (language) together with the concept of equivalence, the notion of graph conditions with arbitrary nesting level (NGCs), and GT systems. Moreover, we introduce in particular the answer set of graph queries based on NGCs.

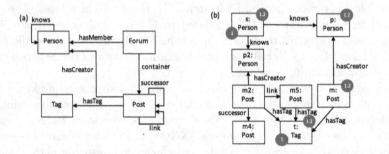

Fig. 2. Excerpt of social network type graph and an example graph G

Example 1 (social network query). As running example we use a social network model and a slightly adjusted graph query employed by the LDBC benchmark [8]. A class diagram outlining the possible graph models as well as an example graph to apply the query are depicted in Fig. 2(a) resp. (b). The considered complex graph query looks for pairs of Tags and Persons (1) such that the Tag is new in the Posts by a friend of this Person. To be a Post of a friend, the Post must be from a second Person the Person knows (1.2). In order to be new, the Tag must be linked in the latest Post of the second Person (and thus in a Post that has no successor Post) (1.2.1) and there has to be no former Post by any other or the same friend that is not her last one and where the same Tag has been already used (1.1). In both cases only Tags that are not simply inherited from a linked Post should be considered (1.1.1 and 1.2.2). Note that the employed numbering of the conditions relates to the tree-like network depicted in Fig. 1(c). Occurrences for the positive sentences (1) and (1.2) in the example graph are depicted accordingly as markers in form of blue circles with the respective number in Fig. 2(b). The circular blue markers (1) on the graph denote the occurrence of the request graph consisting of the person s and tag t. Marker (1.2) denotes the extra condition that the searched tag t must be attached (hasTag) to a post created by person p that is known by person s. Note that the markers (1) denote the only correct answer for the query. Thereby the required match for the positive subquery (1.2) depicted by the markers (1.2) is such that indeed no match exists for the negative subsubqueries (1.2.1) and (1.2.2). Furthermore, as required no match for the negative subquery (1.1) consistent with (1) exists such that no match for the negative subsubquery (1.1.1) of (1.1) can be found. Consequently, no match for (1.1) is visualized.

We briefly reintroduce the notion of typed graphs and graph morphisms [9]. A *graph* $G = (G^V, G^E, s^G, t^G)$ consists of a set G^V of nodes, a set G^E of edges, a source function $s^G : G^E \rightarrow G^V$, and a target function $t^G : G^E \rightarrow G^V$. Given the graphs $G = (G^V, G^E, s^G, t^G)$ and $H = (H^V, H^E, s^H, t^H)$, a *graph morphism* $f : G \rightarrow H$ is a pair of mappings, $f^V : G^V \rightarrow H^V, f^E : G^E \rightarrow H^E$ such that $f^V \circ s^G = s^H \circ f^E$ and $f^V \circ t^G = t^H \circ f^E$. A graph morphism $f : G \rightarrow H$ is a *monomorphism* if f^V and f^E are injective mappings. Finally, two graph morphisms $m : H \rightarrow G$ and $m' : H' \rightarrow G$ are *jointly epimorphic* if $m^V(H^V) \cup m'^V(H'^V) = G^V$ and $m^E(H^E) \cup m'^E(H'^E) = G^E$. A *type graph* is a distinguished graph $TG = (TG^V, TG^E, s^{TG}, t^{TG})$. TG^V and TG^E are called the vertex and the edge type alphabets, respectively. A tuple $(G, type)$ of a graph G together with a graph morphism $type : G \rightarrow TG$ is then called a *typed graph*. Given typed graphs $G_1^T = (G_1, type_1)$ and $G_2^T = (G_2, type_2)$, a *typed graph morphism* $f : G_1^T \rightarrow G_2^T$ is a graph morphism $f : G_1 \rightarrow G_2$ such that $type_2 \circ f = type_1$. We further denote the set of all graphs typed over some type graph TG by $\mathcal{L}(TG)$.

An example for a *typed graph* G and the type graph TG related to the social network query Example 1 are depicted in Fig. 2.

In the rest of the paper we will compare the answer sets of graph queries to analyze them for equivalence. Since we will compare queries stemming from different query languages, we introduce here a generic notion of query (language) equivalence that we will refine in the rest of the paper to particular queries and query languages. As the most generic form of a graph query language we just assume that it consists of a set of graph queries, where each graph query is characterized by a request graph L typed over some type graph TG. The query then expresses some extra properties that need to hold for the request graph L that is searched for in the queried graph G. The answer set for this query then describes all matches of L in the queried graph that fulfill these extra properties.

Definition 2 (graph query (language)). *Given a type graph TG, then a graph query is characterized by a so-called request graph L, which is a finite graph typed over TG. A graph query language is a set of graph queries.*

Definition 3 (answer set mapping, equivalence). *Given some graph query language $\mathcal{L}$, an answer set mapping ans for $\mathcal{L}$ maps each pair (q_L, G) with q_L a graph query in $\mathcal{L}$ with request graph L typed over TG and G a graph from $\mathcal{L}(TG)$ to a set of graph morphisms typed over TG with domain L and co-domain G.*

Given queries q_L and q'_L for some request graph L typed over TG belonging to the graph query languages $\mathcal{L}$ and $\mathcal{L}'$ with answer set mappings ans and ans', resp., then q_L and q'_L are equivalent if for every graph G in $\mathcal{L}(TG)$ it holds that $ans(q_L, G) = ans'(q'_L, G)$. Two graph query languages $\mathcal{L}$ and $\mathcal{L}'$ are equivalent if for any query $q_L \in \mathcal{L}$ for some request graph L there exists some query $q'_L \in \mathcal{L}'$ for L such that $q_L \sim q'_L$ and vice versa. We denote equivalence also with $\sim$.

We reintroduce the notion of *nested graph conditions* (NGC) from [12], since they represent the declarative kind of graph queries that we will consider in this paper. Given a finite graph L, a *nested graph condition* (NGC) over L is defined inductively as follows: (1) *true* is a NGC over L. We say that *true* has nesting level 0. (2) For every morphism $a : L \to L'$ and NGC $c_{L'}$ over a finite graph L' with nesting level n such that $n \geq 0$, $\exists(a, c_{L'})$ is a NGC over L with nesting level $n + 1$. (3) Given NGCs over L, c_L and c'_L, with nesting level n and n', respectively, $\neg c_L$ and $c_L \wedge c'_L$ are NGCs over L with nesting level n and $max(n, n')$, respectively. We restrict ourselves to finite NGCs, i.e. each conjunction of NGCs is finite. We define when a morphism $q : L \to G$ *satisfies* a NGC c_L over L inductively: (1) Every morphism q satisfies *true*. (2) A morphism q satisfies $\exists(a, c_{L'})$, denoted $q \models \exists(a, c_{L'})$, if there exists a monomorphism $q' : L' \to G$ such that $q' \circ a = q$ and $q' \models c_{L'}$. (3) A morphism q satisfies $\neg c_L$ if it does not satisfy c_L and satisfies $\wedge_{i \in I} c_{L,i}$ if it satisfies each $c_{L,i}$ $(i \in I)$. Note that *false*, $\vee$, and $\Rightarrow$ can be mapped as usual to the introduced logical connectives. Moreover we abbreviate $\exists(\emptyset \to L', c_{L'})$ with $\exists(L', c_{L'})$, $\exists(a, true)$ with $\exists a$ and $\forall(a, c_{L'})$ with $\neg\exists(a, \neg c_{L'})$. NGCs can be equipped with *typing* over a given type graph TG as usual [9] by adding typing morphisms from each graph to TG and by requiring type-compatibility with respect to TG for each graph morphism.[3]

Definition 4 ($\mathcal{L}_{NGC}$, ans_{NGC}). *The graph query language $\mathcal{L}_{NGC}$ is the set of all NGCs. Given some NGC c_L over L, L represents the so-called request graph. The answer set mapping ans_{NGC} for $\mathcal{L}_{NGC}$ is given by*

$$ans_{NGC}(c_L, G) = \{q : L \to G | q \text{ is a monomorphism and } q \models c_L\}$$

with $c_L \in \mathcal{L}_{NGC}$ a NGC with L typed over some type graph TG and G in $\mathcal{L}(TG)$.

---

[3] W.l.o.g. we restrict our notion of condition satisfaction to the existence of monomorphisms. In particular, in [12] it is shown how to translate conditions relying on general morphism matching/satisfaction into equivalent conditions relying on monomorphism matching/satisfaction and the other way round.

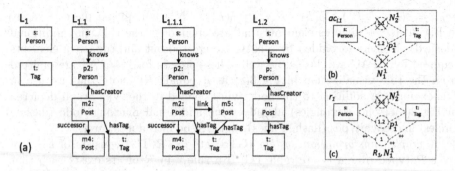

Fig. 3. Graphs for the NGC c_1 and its subconditions (a) and the application condition $ac_{L_1} = \exists(L_1 \rightarrow P_1^1) \wedge \nexists(L_1 \rightarrow N_1^1) \wedge \nexists(L_1 \rightarrow N_2^1)$ (b) and simple marking rule $r_1 = (L_1 \rightarrow R_1, ac_{L_1})$ (c)

An example NGC for the social network query of Example 1, where the subconditions refer to the introduced numbering, is the following: $c_1 = c_{1.1} \wedge c_{1.2}$ with $c_{1.1} = \neg\exists(n_{1.1} : L_1 \rightarrow L_{1.1}, c_{1.1.1})$, $c_{1.2} = \exists(p_{1.2} : L_1 \rightarrow L_{1.2}, c_{1.2.1} \wedge c_{1.2.2})$, $c_{1.1.1} = \neg\exists(n_{1.1.1} : L_{1.1} \rightarrow L_{1.1.1}, true)$, $c_{1.2.1} = \neg\exists(n_{1.2.1} : L_{1.2} \rightarrow L_{1.2.1}, true)$, and $c_{1.2.2} = \neg\exists(n_{1.2.2} : L_{1.2} \rightarrow L_{1.2.2}, true)$. The graphs L_1, $L_{1.1}$, $L_{1.1.1}$, and $L_{1.2}$ are depicted exemplarily (see [5] for the complete example) in Fig. 3(a). Morphisms are implied by equally named objects.

As foundation for an operational graph query evaluation we will employ typed GT systems with priorities. We start with reintroducing GT and thereby assume the double-pushout approach (DPO) with injective matching and non-deleting rules [9] with application conditions of arbitrary nesting level (AC) [12]. A plain GT rule $p : L \rightarrow R$ is a graph monomorphism. We say that the graphs L and R are the left-hand side (LHS) and right-hand side (RHS) of the rule, respectively. A *GT rule* $r = \langle p, ac_L \rangle$ consists of a plain rule $p : L \rightarrow R$ and a so-called application condition ac_L being a graph condition over L. If the application condition $ac_L = \wedge_{i \in I} \exists p_i \wedge \wedge_{j \in J} \nexists n_j$, then we say that $\exists p_i$ or $\neg\exists n_j$ is a positive application condition (PACs) or negative application condition (NAC) over L, respectively. A rule r is *applicable* to a graph G via a graph monomorphism $m : L \rightarrow G$ if $m \models ac_L$. A *direct GT* via rule $r = \langle p, ac_L \rangle$ consists of a pushout over p and m such that $m \models ac_L$. If there exists a direct transformation from G to G' via rule r and match m, we write $G \Rightarrow_{m,r} G'$. If we are only interested in the rule r, we write $G \Rightarrow_r G'$. If a rule r in a set of rules $\mathcal{R}$ exists such that there exists a direct transformation via rule r from G to G', we write $G \Rightarrow_{\mathcal{R}} G'$. A *GT*, denoted as $G_0 \Rightarrow^* G_n$, is a sequence $G_0 \Rightarrow G_1 \Rightarrow \cdots \Rightarrow G_n$ of $n \geq 0$ direct GT. GT rules and GTs can be equipped with *typing* over a given type graph TG as usual [9] by adding typing morphisms from each graph to TG and by requiring type-compatibility with respect to TG for each graph morphism.

An example for a GT rule with AC in the context of the social network query of Example 1 is $r_1 = (L_1 \rightarrow R_1, ac_{L_1})$ as depicted in Fig. 3(c) following the compact notation where all graphs are embedded into a single one. In particular,

$ac_{L_1} = \exists(L_1 \to P_1^1) \land \nexists(L_1 \to N_1^1) \land \nexists(L_1 \to N_2^1)$ is depicted more precisely in Fig. 3(b). ++ denotes elements that are created by the rule, the additional (dashed) elements forbidden by a NAC are crossed out and the extra elements required by a PAC are dashed as well. These crosses for NAC N_1^1 are omitted from the rule visualization in Fig. 3(c) as it equals R_1^1. Note that we use in this example in addition to the node types defined in the type graph depicted in Fig. 2(a) (solid rectangles) already some additional marking node (dashed circles) and edge types (dashed lines) that will be introduced later.

A *graph transformation system* (GTS) gts $= (\mathcal{R}, \text{TG})$ consists of a set of rules $\mathcal{R}$ typed over a type graph TG. If a rule r in $\mathcal{R}$ of gts exists such that a direct transformation $G \Rightarrow_r G'$ via r exists, we also write $G \Rightarrow_{\text{gts}} G'$. If for some graph G it holds that r is not applicable to G, then we write $G \nRightarrow_r$. Moreover, if no rule in gts exists that is applicable to G, then we write $G \nRightarrow_{\text{gts}}$. A GTS *with priorities* $\text{gts}_p = ((\mathcal{R}, TG), p)$ consists of a GTS $(\mathcal{R}, TG)$ and a transitive and asymmetric relation $p \subset \mathcal{R} \times \mathcal{R}$. We write $G \Rightarrow_{\text{gts}_p} G'$ if a rule r in $\mathcal{R}$ of gts_p exists with a direct transformation $G \Rightarrow_r G'$ such that $\nexists r' \in \mathcal{R}$: $(r, r') \in p \land G \Rightarrow_{r'} G''$. For a GTS with priorities gts_p and an initial graph G_0 the *set of reachable graphs* $\text{REACH}(\text{gts}_p, G_0)$ is defined as $\{G \mid G_0 \Rightarrow_{\text{gts}_p}^* G\}$ and the *set of terminal reachable graphs* $\text{TERM}(gts_p, G_0)$ is defined as $\{G | G \in \text{REACH}(\text{gts}_p, G_0) : G \nRightarrow_{gts_p}\}$.

3   Generalized Discrimination Networks

In the following we introduce our suggestion for the operationalization of graph queries employing generalized discrimination networks with computation nodes based on GT rules.

Example 5 (GDN (informal)). A possible GDN for the social network query Example 1 is depicted in Fig. 1(a). Node 1.1.1s and 1.2.2s produce their output independently. Then, node 1.1s and 1.2s can compute the output depending on the output of these two other nodes. Finally, the terminal node 1s can compute its output based on the output of the nodes 1.1s and 1.2s. We further distinguish in Fig. 1(a) positive and negated dependencies accordingly visualized by arrows with a single solid line when representing a PAC ($\exists$) and by arrows with a single dashed line when representing a NAC ($\nexists$).

Our queried graph G typed over TG will be marked with so-called marking nodes and edges to keep track of (sub-)query answer sets. In particular, so-called marking rules in a GDN will take care of that. A (simple) marking rule r_i is a restricted form of GT rule typed over a marking type graph TG'. The latter is equal to TG but for each marking rule r_i it is extended with a so-called marking node type t_i as well as an marking edge type t_v per node v present in r_i's LHS L_i. This allows r_i to mark each node v from L_i by adding a marking node i uniquely corresponding to r_i via its marking node type t_i, called the defined type, and by adding a marking edge e_v from this special marking node i to each node v in L_i. These marking edges encode again via their type t_v which node v in L_i they mark. Finally the application conditions in each marking rule allow for referring to the marking elements (and therefore indirectly to already matched elements) created by other rules.

The required extension for the type graph TG for the social network query Example 5 for rule r_1, which captures that a s:Person and t:Tag exist for which additional conditions must hold, are depicted in Fig. 3(c). Additional nodes visualized as circles with number 1, 1.1, and 1.2, where 1 denotes the created marking node of the rule r_1 and 1.1 and 1.2 are marking nodes of the other rules $r_{1.1}$ and $r_{1.2}$ all use types in TG' but not TG. The edges between the circles and the rectangles also belong to TG' but not TG. We do not visualize their direction, since they always point to nodes of a type from TG.

Definition 6 (marking type graph). *Given a set of graphs* $(L_i)_{i \in I}$ *typed over* TG *via* $type_i : L_i \to TG$, *the* marking type graph TG' *for* $(L_i)_{i \in I}$ *has node set* $TG'^V = TG^V \uplus \{t_i | i \in I\}$ *and edge set* $TG'^E = TG^E \uplus \{t_v | v \in L_i^V, i \in I\}$ *s.t.* $s^{TG'}(e) = s^{TG}(e)$ *and* $t^{TG'}(e) = t^{TG}(e)$ *for* $e \in TG^E$ *and* $s^{TG'}(t_v) = t_i$ *and* $t^{TG'}(t_v) = type_i^V(v)$ *for each* $v \in L_i^V$ *and* $i \in I$ *otherwise. We say that the nodes in* $\{t_i | i \in I\}$ *are* marking node types *and edges in* $\{t_v | v \in L_i^V, i \in I\}$ *are* marking edge types, *respectively. Given a graph* G *typed over* TG', *then we say that a node or edge in* G *such that its type equals a marking node or edge type in* TG' *is a* marking node or edge *in* G, *resp..*

Definition 7 ((simple) marking rule, defined type). *Given a set of graphs* $(L_i)_{i \in I}$ *typed over* TG *via* $type_i : L_i \to TG$, *a* marking rule *(MR) is a GT rule* $r_i = \langle p_i : L_i \to R_i, \nexists p_i \wedge c_{L_i} \rangle$ *typed over the marking type graph* TG' *for* $(L_i)_{i \in I}$ *such that (1)* L_i *inherits its typing from* $type_{L_i}$, *(2)* $R_i^V = L_i^V \uplus \{i\}$ *with* i *of type* t_i *the so-called* marking node *and* t_i *the so-called* defined type *of rule* r_i, *and (3)* $R_i^E = L_i^E \uplus \{e_v | v \in L_i^V\}$ *such that each* e_v *has type* t_v *and* $s^{R_i}(e_v) = i$ *and* $t^{R_i}(e_v) = v$.

A simple marking rule *(SMR) is a marking rule where the application condition* $c_{L_i} = \bigwedge_{j \in J}(\exists p_j : L_i \to P_j) \wedge \bigwedge_{k \in K}(\nexists n_k : L_i \to N_k)$ *such that for each* $j \in J$ *and* $k \in K$ *it holds that* $P_j^V \setminus (p_j(L_i))^V$ *and* $N_k^V \setminus (n_k(L_i))^V$, *resp., consist of exactly one marking node.*

In addition to the defined type of its created marking node each marking rule induces so-called referred types in the marking type graph. Based on these referred and defined types of MRs we define a dependency relation between MRs.

Definition 8 (referred types, dependency relation). *Given a set of graphs* $(L_i)_{i \in I}$ *typed over* TG *and a (simple) marking rule* $r_i = \langle p_i : L_i \to R_i, \nexists p_i \wedge c_{L_i} \rangle$ *typed over the marking type graph* TG' *for* $(L_i)_{i \in I}$ *the set of* referred types $rt(r_i)$ *is the set of all node types in* TG'^V *for nodes occurring in some (co-)domain graph of a morphism employed in* c_{L_i}.

Given a GTS $(\mathcal{R} = (r_i)_{i \in I}, TG')$ *with each rule* r_i *a (simple) marking rule, a* dependency relation $\rightsquigarrow_d \subseteq \mathcal{R} \times \mathcal{R}$ *consists of all rule pairs* (r_i, r_j) *such that the defined type* t_j *of rule* r_j *belongs to the set of referred types* $rt(r_i)$.

Note that by definition a MR r_i can only depend on itself if its defined type t_i is employed for typing elements in the application condition c_{L_i}.

The SMRs for the SGDN for the social network query of Example 5 are depicted in Fig. 4. We use here and in the following the more compact notation for SMRs where all graphs including the PACs and NACs are embedded into a single one as presented in Fig. 3(c), moreover the RHS as well as the NAC equal to p_i are omitted since they can be reconstructed from the rule's LHS uniquely.

Based on the previously introduced MRs or SMRs to encode the behavior of the computation nodes of a GDN, we can now introduce our form of GDN or SGDN, respectively.

Definition 9 (GDN, SGDN, $\mathcal{L}_{GDN}$, $\mathcal{L}_{SGDN}$). *Given a finite graph L typed over TG and a GTS $(\mathcal{R} = (r_i)_{i \in I}, TG')$ of (simple) marking rules typed over the marking type graph TG' for $(L_i)_{i \in I}$, then $gdn_L = ((\mathcal{R}, TG'), \leadsto_d^+)$ is a (simple) generalized discrimination network over L if the following conditions hold: (1) the transitive closure $\leadsto_d^+$ is acyclic, (2) there is a unique so-called terminal rule r_t with LHS $L_t = L$ for some $t \in I$, and (3) $\forall i \in I$ s.t. $i \neq t$ it holds that (r_t, r_i) is in $\leadsto_d^+$. The graph query language $\mathcal{L}_{GDN}$ ($\mathcal{L}_{SGDN}$) is the set of all GDNs (SGDNs). Given some GDN gdn_L (SGDN $sgdn_L$) over L, L represents the so-called request graph.*

Note that it follows directly from this definition that no rule of the GDN transitively depends on the terminal rule otherwise the transitive closure of the dependency relation would contain a cycle.

An example for a SGDN is depicted in Figs. 1(a) and 4, where Fig. 1(a) shows the dependencies between the nodes and Fig. 4 shows the rules for the nodes r_{1s}, $r_{1.1s}$, $r_{1.2s}$, $r_{1.1.1s}$, and $r_{1.2.2s}$.

In the following definitions we assume an operational query in the form of a GDN. In particular, each GDN represents a GTS with priorities. We consider each graph reachable via the GDN to encode an intermediate query result and the terminal graph then encodes the final query result. As shown in the subsequent lemma this terminal graph is indeed unique.

Lemma 10 (unique terminal graph). *Given a GDN $gdn_L = ((\mathcal{R}, TG'), \leadsto_d^+)$ for L typed over TG, then $TERM(gdn_L, G)$ consists of exactly one graph.*

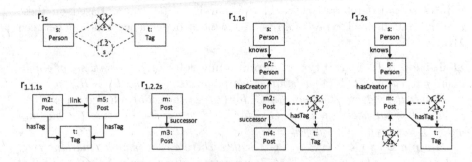

Fig. 4. SMRs for the SGDN of the social network example

Proof. (sketch; more details see [5]) As there is an upper bound on matches that can be marked and rule applications always add exactly one such marking, gdn_L terminates. As the priorities expressed by $\leadsto_d^+$ exclude conflicting applications of different rules and acyclicity of $\leadsto_d^+$ excludes conflicting applications of a rule with itself, gdn_L is also confluent.

Definition 11 (*ans_{GDN}*). *Given the graph query language $\mathcal{L}_{GDN}$, the answer set mapping ans_{GDN} for $\mathcal{L}_{GDN}$ is given by*

$$ans_{GDN}(gdn_L, G) := \{o\colon L \to G \mid G_i \Rightarrow_{o',r_t} G_i' \text{ is a direct GT in } t \wedge o(L) = o'(L)\}$$

with $gdn_L = ((\mathcal{R}, TG'), \leadsto_d^+)$ some GDN such that L is typed over TG, G a graph in $\mathcal{L}(TG)$, r_t the terminal rule of gdn_L and $t : G \Rightarrow_{gdn_L}^* G'$ some transformation with $\{G'\} = TERM(gdn_L, G)$.

The above definition is well-defined, since matches are never destroyed because of dealing only with non-deleting rules and no conflicting direct transformations arise because of the priorities encoded with $\leadsto_d^+$ and acyclicity of $\leadsto_d^+$ (as mentioned also w.r.t. terminal graph uniqueness). Moreover, for $o' : L \to G_i$ it holds that $o'(L)$ is a subgraph of G.

In practice, it is important for efficiency reasons that we can reconstruct the answer set $ans_{GDN}(gdn_L, G)$ from the markings in the terminal graph G' without having to consider the transformation t leading to G'. Under the condition that we only query graphs without parallel edges of the same type this can be done uniquely (see [5]).

The following result shows that for each SGDN an equivalent tree-like SGDN exists in which no two rules exist that directly depend on the same rule and each dependency is caused by exactly one PAC/NAC. As the considerations in the following section are considerably simpler when operating on tree-like SGDNs, we will w.l.o.g (cf. Lemma 13) in the following restrict to tree-like networks.

Definition 12 (**SGDT**, **$\mathcal{L}_{SGDT}$**). *A simple generalized discrimination tree (SDGT) is a SGDN $sgdn_L = ((\mathcal{R} = (r_i)_{i \in I}, TG'), \leadsto_d^+)$ such that (1) for each $(r_i, r_j) \in \leadsto_d$ no $k \in I$ with $k \neq i$ exists s.t. $(r_k, r_j) \in \leadsto_d$ and (2) for each $i \in I$ it holds that for each PAC or NAC of r_i no other PAC or NAC in r_i exists referring to the same marking node type. The graph query language $\mathcal{L}_{SGDT}$ is the set of all SGDTs.*

Lemma 13 (**$\mathcal{L}_{SGDN} \sim \mathcal{L}_{SGDT}$**). *Given a SGDN $sgdn_L$ for a graph L typed over TG, then it holds that a SGDT $sgdt_L$ exists such that $sgdn_L \sim sgdt_L$. Moreover, $\mathcal{L}_{SGDN} \sim \mathcal{L}_{SGDT}$.*

Proof. (sketch, details see [5]) We can show by induction over the depth of $\leadsto_d^+$ that we can construct an equivalent tree by employing copied rules with disjoint markings. Since each SGDT is in particular also a SGDN, it directly follows that $\mathcal{L}_{SGDN} \sim \mathcal{L}_{SGDT}$.

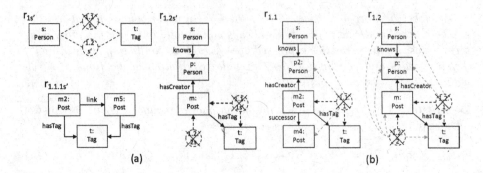

Fig. 5. SMRs for the SGDT for the social network example (a) and with maximal context (b) as denoted by the orange dashed lines.

The SMRs of the SGDT related to the SGDN of Fig. 1(a) depicted in Fig. 1(b) where multiple referenced SMRs are simply replicated are presented in Fig. 5(a). The rules $r_{1.1s}$, $r_{1.1.1s}$, and $r_{1.2.2s}$ of Fig. 4 are not shown in Fig. 5 since they remain the same. Rules $r_{1s'}$ and $r_{1.2s'}$, which differ from the rules r_{1s} and $r_{1.2s}$ of Fig. 4 only concerning the referenced other rules are shown, along with rule $r_{1.1.1s'}$, which is a replication of rule $r_{1.1.1s}$ that differs only w.r.t. created elements (omitted from the visualization).

4   Equivalence to Nested Graph Conditions

In order to prove that each NGC can be represented by some equivalent SGDT, we first show in the following Lemmas that the standard operators in NGCs (true, existential quantification, negation and binary conjunction) (Def. see Sect. 2) can be simulated by equivalent constructions in a SGDT.

Lemma 14 (*true*). *Given the NGC true over L, there exists some SGDT $sgdt_L$ such that $sgdt_L \sim true$.*

Proof. Let $sgdt_L = (\{r_{L,true}\}, TG'), \leadsto^+_d)$ for L typed over TG with marking rule $r_{L,true} = \langle p : L \to R, \nexists p\rangle$, then for each graph G typed over TG, $ans_{GDN}(sgdt_L, G)$ consists of all morphisms $p : L \to G$. This means that $sgdt_L \sim true$.

Lemma 15 ($\exists(a : L \to L', c_{L'})$). *Given some NGC $\exists(a : L \to L', c_{L'})$ and SGDT $sgdt'_{L'}$ such that $sgdt'_{L'} \sim c_{L'}$, there exists some SGDT $sgdt_L$ such that $sgdt_L \sim \exists(a : L \to L', c_{L'})$.*

Proof. Suppose that $sgdt'_{L'}$ has the terminal rule $r'_t = \langle p'_t : L' \to R', \nexists p'_t \wedge c'_{L'}\rangle$. We construct the SGDT $sgdt_L$ by having an additional rule $r_{L,\exists a} = \langle p : L \to R, \nexists p \wedge \exists(p'_t \circ a, true)\rangle$ w.r.t. $sgdt'_{L'}$ as terminal rule. Consider $ans_{GDN}(sgdt_L, G)$ consisting of all morphisms $o : L \to G$ s.t. $r_{L,\exists a}$ created a marking to $o(L)$. Because of the PAC $\exists(p'_t \circ a, true)$ in the terminal rule $r_{L,\exists a}$ this can only be the case if r'_t created a marking for some $o'(L')$ with $o' : L' \to G$ a morphism in $ans_{GDN}(sgdt'_{L'}, G)$. Since $sgdt'_{L'} \sim c_{L'}$ we know that r'_t created a marking to $o'(L')$ iff $o' \models c_{L'}$. Therefore we conclude that $o \models \exists(a : L \to L', c_{L'})$ and thus $sgdt_L \sim \exists(a : L \to L', c_{L'})$.

Lemma 16 ($\neg c_L$). *Given some NGC $\neg c_L$ and SGDT $sgdt'_L$ such that $sgdt'_L \sim c_L$, there exists some SGDT $sgdt_L$ such that $sgdt_L \sim \neg c_L$.*

Proof. Suppose that $sgdt'_L$ has the terminal rule $r = \langle p' : L \to R', \nexists p' \wedge c'_L \rangle$. Then consider the SGDT $sgdt_L$ having an additional rule $r_{L,\neg} = \langle p : L \to R, \nexists p \wedge \nexists p' \rangle$ w.r.t. $sgdt'_L$ as terminal rule. Consider $ans_{GDN}(sgdt_L, G)$ consisting of all morphisms $o : L \to G$ s.t. $r_{L,\neg}$ created a marking to $o(L)$. Because of the NAC $\nexists p'$ in the terminal rule $r_{L,\neg}$ this can only be the case if r did not create a marking to $o(L)$. Since $sgdt'_L \sim c_L$ we know that r created a marking to $o(L)$ iff $o \models c_L$. Therefore we conclude that $o \models \neg c_L$ and thus $sgdt_L \sim \neg c_L$.

Lemma 17 ($c_{1,L} \wedge c_{2,L}$). *Given some NGC $c_{1,L} \wedge c_{2,L}$ and SGDTs $sgdt^1_L$ and $sgdt^2_L$ such that $sgdt^1_L \sim c_{1,L}$ and $sgdt^2_L \sim c_{2,L}$, there exists some SGDT $sgdt_L$ such that $sgdt_L \sim c_{1,L} \wedge c_{2,L}$.*

Proof. Let $r_1 = \langle p_1 : L \to R_1, \nexists p_1 \wedge c_L \rangle$ and $r_2 = \langle p_2 : L \to R_2, \nexists p_2 \wedge c'_L \rangle$ be the terminal rules for $sgdt^1_L$ and $sgdt^2_L$, respectively. Consider the SGDT $sgdt_L$ consisting of the subtrees $sgdt^1_L$ and $sgdt^2_L$ with the additional rule $r_{L,\wedge} = \langle p : L \to R, \nexists p \wedge \exists p_1 \wedge \exists p_2 \rangle$ as terminal rule. Consider $ans_{GDN}(sgdt_L, G)$ consisting of all morphisms $o : L \to G$ s.t. $r_{L,\wedge}$ created a marking to $o(L)$. Because of the PACs $\exists p_1$ and $\exists p_2$ in the terminal rule $r_{L,\wedge}$ this can only be the case if r_1 as well as r_2 created a marking to $o(L)$. Since $sgdt^1_L \sim c_{1,L}$ resp. $sgdt^2_L \sim c_{2,L}$ we know that r_1 resp. r_2 created a marking to $o(L)$ iff $o \models c_{1,L}$ resp. $o \models c_{2,L}$. Therefore we conclude that $o \models c_{1,L} \wedge c_{2,L}$ and thus $sgdt_L \sim c_{1,L} \wedge c_{2,L}$.

Now we can prove that each NGC can be emulated by an equivalent SGDT.

Proposition 18. *Given a NGC c_L, there exists a SGDT $sgdt_L$ s.t. $sgdt_L \sim c_L$.*

Proof. We prove this by induction over the nesting level of NGCs and the way they are constructed.
Base case: By Lemma 14 it follows that for $c_L = true$ with nesting level 0 an equivalent SGDT with a single marking rule exists. From Lemmas 16 and 17 it follows that for any combination of conditions of nesting level 0 we can still construct an equivalent SGDT.
Induction step: By Lemmas 15 and the induction hypothesis it follows that for any condition $\exists(a : L \to L', c_{L'})$ of nesting level $n+1$ it follows that an equivalent SGDT exists. From Lemmas 16 and 17 it follows that for any combination of conditions of nesting level $n+1$ we can still construct an equivalent SGDT.

We still need to show that also each SGDT can be emulated by an equivalent NGC. An important first step thereby is the construction of a transformation of some SGDT into a SGDT with so-called maximal context. Marking rules in

GDNs are able to pass merely the context necessary for the next subquery, which is a practical property for efficiency reasons, but not for showing equivalence with NGCs based on maximal context passing. With context propagation we therefore introduce a mechanism transforming marking rules passing only partial context into rules passing maximal context. We moreover show that this context propagation does not alter the answer set semantics of the corresponding SGDT.

Definition 19 (maximal context). *Given a SGDT $sgdt_L$ for a graph L typed over TG then $sgdt_L$ has maximal context if for each two SMRs $r_i = \langle p_i : L_i \to R_i, \nexists p_i \wedge \bigwedge_{j \in J_i} (\exists p_j^i : L_i \to P_j^i) \wedge \bigwedge_{k \in K_i} (\nexists n_k^i : L_i \to N_k^i) \rangle$ and $r_l = \langle p_l : L_l \to R_l, \nexists p_l \wedge \bigwedge_{j \in J_l} (\exists p_j^l : L_l \to P_j^l) \wedge \bigwedge_{k \in K_l} (\nexists n_k^l : L_l \to N_k^l) \rangle$ with marking node l s.t. $(r_i, r_l) \in \rightsquigarrow_d$ because for some $j \in J_i$ (or $k \in K_i$) p_j^i (or n_k^i, resp.) uses a type equal to the type t_l of l, the sets V_j^i (or V_k^i, resp.) constructed as follows are empty:*

$$V_j^i = \{n | n \in L_i^V \ s.t. \ \nexists e \in (P_j^i)^E \ with \ type \ of \ s^{P_j^i}(e) = t_l \wedge t^{P_j^i}(e) = p_j^i(n)\}$$

$$V_k^i = \{n | n \in L_i^V \ s.t. \ \nexists e \in (N_k^i)^E \ with \ type \ of \ s^{N_k^i}(e) = t_l \wedge t^{N_k^i}(e) = n_k^i(n)\}$$

Lemma 20 (context propagation). *Given a SGDT $sgdt_L$ for a graph L typed over TG with two rules r_i and r_l such that $(r_i, r_l) \in \rightsquigarrow_d$ with non-empty V_j^i (or V_k^i) (as given in Definition 19), then there exists some $sgdt_L^c$ in which (r_i, r_l) has been replaced by a SGDT with maximal context such that $sgdt_L^c \sim sgdt_L$.*

Proof. (sketch; details see Lemma 20) We construct a $sgdt_L^c$ in which marking rules with propagated context check in contrast to r_l the presence of additional nodes and edges in the queried graph G that would otherwise have been searched for anyway by rule r_i after all matches for r_l had been found. Marking these elements earlier does not change the overall answer set.

Lemma 21 (maximal context). *For a SGDT $sgdt_L$ for a graph L typed over TG their exists a SGDT $sgdt'_L$ with maximal context such that $sgdt'_L \sim sgdt_L$.*

Proof. We proof this lemma by induction on the height of the tree.
Base case: Suppose that we have $sgdt_L$ with height 0, then it trivially holds that $sgdt_L$ has maximal context already.
Induction step: Suppose that we have $sgdt_L$ with height $n + 1$. Then apply subsequently for each $(r_t, r_i) \in \rightsquigarrow_d$ context propagation to $sgdt_L$ obtaining according to Lemma 20 an equivalent $sgdt_L^c$ of height $n + 1$. Now consider for each r_i the subtree $sgdt_{L_i^c}^{r_i}$ in $sgdt_L^c$ of height n. Then for each $sgdt_{L_i^c}^{r_i}$ by induction hypothesis an equivalent SGDT $sgdt'_{L_i^c}$ with maximal context exists. Replacing in $sgdt_L^c$ each $sgdt_{L_i^c}^{r_i}$ with $sgdt'_{L_i^c}$ we obtain a SGDT $sgdt'_L$ with maximal context s.t. $sgdt'_L \sim sgdt_L$.

Two of the modified SMRs of the SGDT depicted in Fig. 1(c) with maximal context related to the SGDN of Fig. 1(a) are presented in Fig. 5(b). While the rules $r_{1.1}$ and $r_{1.2}$ already have maximal context and therefore differ from the

$r_{1.1s}$ and $r_{1.2s'}$ only concerning the referenced other rules and additional links to bind the propagated context as depicted in Fig. 5(b) by the orange edges, the rules $r_{1.1.1}$, $r_{1.2.1}$, and $r_{1.2.2}$ are extended with propagated context concerning the rules $r_{1.1.1s}$, $r_{1.1.1s'}$, and $r_{1.2.2s}$ and in addition have to reference the new rules.

Now we are ready to prove that for each SGDT there exists an equivalent NGC and consequently also that the languages $\mathcal{L}_{SGDT}$ and $\mathcal{L}_{NGC}$ are equivalent.

Proposition 22. *Given, a SGDT $sgdt_L$ for a graph L typed over TG, then there exists a NGC c_L s.t. $sgdt_L \sim c_L$.*

Proof. Because of Lemma 21 we can assume w.l.o.g. that $sgdt_L$ has maximal context. We perform the proof by induction on the height of the tree.

Base case: If $sgdt_L$ has height 0, then it consists merely of some terminal rule without any PACs or NACs. Then $ans_{gdn}(sgdt_L, G)$ consists of all matches of the terminal rule into G. If we choose c_L equal to true over L then it returns exactly the same set of morphisms s.t. $sgdt_L \sim c_L$.

Induction step: Suppose that $sgdt_L$ has height $n+1$ and that it has terminal rule $r = \langle p : L \to R, \not\exists p \wedge \bigwedge_{j\in J}(\exists p_j : L \to P_j) \wedge \bigwedge_{k\in K}(\not\exists n_k : L \to N_k)\rangle$. Then we have a subtree $sgdt_{L_j}$ and $sgdt_{L_k}$ for each p_j and each n_k, respectively. Because of induction hypothesis it holds that for each $sgdt_{L_j}$ and $sgdt_{L_k}$ there exists an equivalent NGC c_{L_j} and c_{L_k}, respectively. Since $sgdt_L$ has maximal context, we moreover know that there exist morphisms $l_j : L \to L_j$ and $l_k : L \to L_k$. Consider the NGCs $c_L^j = \exists(l_j, c_{L_j})$ and $c_L^k = \not\exists(l_k, c_{L_k})$ such that $c_L = \wedge_{j\in J}c_L^j \wedge \wedge_{k\in K}c_L^k$. Now $ans_{GDN}(sgdt_L, G)$ for some G consists of all morphisms $o : L \to G$ such that the terminal rule of each $sgdt_{L_j}$ and $sgdt_{L_k}$ has been applied and not been applied, respectively. The latter is equivalent with the fact that for each $j \in J$ a morphism $o_j : L_j \to G$ exists s.t. $o_j \circ l_j = o$ with $o_j \in ans_{GDN}(sgdt_{L_j}, G) = ans_{NGC}(c_{L_j}, G)$. Analogously for each $k \in K$ there does not exist a morphism $o_k : L_k \to G$ s.t. $o_k \circ l_k = o$ and $o_k \in ans_{GDN}(sgdt_{L_k}, G) = ans_{NGC}(c_{L_k}, G)$. This is exactly what also each morphism $o : L \to G$ in $ans_{NGC}(c_L, G)$ needs to fulfill s.t. we can conclude that $sgdt_L \sim c_L$.

Theorem 23. $\mathcal{L}_{SGDN} \sim \mathcal{L}_{SGDT} \sim \mathcal{L}_{NGC}$

Proof. From Propositions 18 and 22 we can follow directly that $\mathcal{L}_{SGDT} \sim \mathcal{L}_{NGC}$. From Lemma 13 we can conclude that $\mathcal{L}_{SGDN} \sim \mathcal{L}_{SGDT}$.

5   Discussion

In this section, we will discuss a more expressive variant, a minimal variant, as well as some observations and implications for optimization of graph queries and incremental updates concerning GDNs and the proposed SGDNs.

In particular, we can show that for *minimal* SGDT (MSGDT) – SGDT with at most two direct dependencies per SMR, where all rules adhere to one of the four rule schemes introduced in Lemmata 14, 15, 16, and 17, and where in addition all rules for existential quantification are limited to at most one additional element in form of a node or edge – holds that $\mathcal{L}_{MSGDT} \sim \mathcal{L}_{NGC}$ (see [5]) and thus the additional restrictions do not result in any loss of expressive power. As often the tree-like simplification is not wanted, we further name SGDN that are not MSGDT but fulfill all conditions besides the tree nature as MSGDN.

There are several approaches for optimization of graph queries or incremental updates of graph queries based on RETE networks (cf. [10]) such as [7] and VIA-TRA [4] that can be conceptually mapped to MSGDN. In these cases the RETE network structure supports only at most two direct dependencies like MSGDN and the computations of the nodes of the RETE network can be matched to the four permitted cases of MSGDN. Our results also indicate that these approaches have the same expressiveness as NGC.

In our own practical work on graph queries [6], we conceptually employ SGDN with marking rules in form of graph transformation rules for optimization of queries and incremental updates of graph queries. We were able to show that the more powerful capabilities of a single node (marking rule) and advanced dynamic pattern matching strategies [11] can lead to considerable improvements concerning the computation speed and memory consumption for SGDN compared to the restricted case of MSGDN (resp. RETE network). Similar results have been obtained also in the relational case where it has been shown that the more general GATOR networks can outperform RETE networks [13]. Consequently, it seems reasonable to study the broader class of SGDN for optimization of queries and incremental updates of graph queries and not more restricted forms such as MSGDN or MSGDT. In particular the context propagation (see Definition 19) and its inverse context elimination seem useful tools here to minimize the effort for subqueries and the propagation of their results in the network.

As outlined in [5] in more detail, we can also have *more expressive* generalized discrimination networks as given in Definition 9 for which we can show that they will not lead to an increase of expressive power such that the language equivalence $\mathcal{L}_{GDN} \sim \mathcal{L}_{NGC}$ holds. However this result only applies unless we leave the realm of pattern-based property specification concepts such as NGC and consider also path-related properties [15] or we permit cycles in the network in a controlled manner as in our own practical work on graph queries [6] to be able to support path-related properties (analogously to the controlled and repeated rule applications to support path-related properties used in [3]).

6   Conclusion and Future Work

Analog to the relational database case where the relational calculus and generalized discrimination networks have been studied to find appropriate decompositions into subqueries and ordering of these subqueries for query evaluation or incremental updates of queries, we present in this paper GDN for graph queries a simple operational concept where graph transformation describe the node behavior. We further show that the proposed GDNs in different forms all have the same expressive power as NGC.

We plan to study in our future work the complexity of evaluating and updating SGDN, their optimization, and possible extensions of SGDNs towards path-related properties to also formally cover our own practical work on graph queries [6] supporting cycles in the network.

Acknowledgments. We are grateful to Johannes Dyck for his contribution to our discussions and feedback to draft versions of the paper.

References

1. Abiteboul, S., Hull, R., Vianu, V. (eds.): Foundations of Databases: The Logical Level, 1st edn. Addison-Wesley Longman Publishing Co., Inc., Boston (1995)
2. Angles, R.: A comparison of current graph database models. In: Proceedings of the 28th International Conference on Data Engineering, pp. 171–177. IEEE (April 2012)
3. Becker, B., Lambers, L., Dyck, J., Birth, S., Giese, H.: Iterative development of consistency-preserving rule-based refactorings. In: Cabot, J., Visser, E. (eds.) ICMT 2011. LNCS, vol. 6707, pp. 123–137. Springer, Heidelberg (2011)
4. Bergmann, G., Ökrös, A., Ráth, I., Varró, D., Varró, G.: Incremental pattern matching in the viatra model transformation system. In: Proceedings of the 3rd International Workshop on Graph and Model Transformations, GRaMoT 2008, pp. 25–32. ACM (2008)
5. Beyhl, T., Blouin, D., Giese, H., Lambers, L.: On the Operationalization of Graph Queries with Generalized Discrimination Networks. Technical report 106, Hasso Plattner Institute at the University of Potsdam (2016)
6. Beyhl, T., Giese, H.: Incremental view maintenance for deductive graph databases using generalized discrimination networks. In: Electronic Proceedings in Theoretical Computer Science, Graphs as Models 2016 (2016, to appear)
7. Bunke, H., Glauser, T., Tran, T.H.: An efficient implementation of graph grammars based on the RETE matching algorithm. In: Kreowski, H.-J., Ehrig, H., Rozenberg, G. (eds.) Graph Grammars 1990. LNCS, vol. 532, pp. 174–189. Springer, Heidelberg (1991)
8. Council, L.D.B.: LDBC Social Network Benchmark (SNB) - First Public Draft Release v0.2.2 (2015). https://github.com/ldbc/ldbc_snb_docs/blob/master/LDBC_SNB_v0.2.2.pdf
9. Ehrig, H., Ehrig, K., Prange, U., Taentzer, G.: Fundamentals of Algebraic Graph Transformation. Springer, Heidelberg (2006)
10. Forgy, C.L.: Rete: a fast algorithm for the many pattern/many object pattern match problem. Artif. Intell. **19**(1), 17–37 (1982)
11. Giese, H., Hildebrandt, S., Seibel, A.: Improved flexibility and scalability by interpreting story diagrams. In: Magaria, T., Padberg, J., Taentzer, G. (eds.) Proceedings of the 8th International Workshop on Graph Transformation and Visual Modeling Techniques, vol. 18. Electronic Communications of the EASST (2009)
12. Habel, A., Pennemann, K.H.: Correctness of high-level transformation systems relative to nested conditions. Math. Struct. Comput. Sci. **19**, 1–52 (2009)
13. Hanson, E.N., Bodagala, S., Chadaga, U.: Trigger condition testing and view maintenance using optimized discrimination networks. Trans. Knowl. Data Eng. **14**(2), 261–280 (2002)

14. He, H., Singh, A.K.: Graphs-at-a-time: query language and access methods for graph databases. In: Proceedings of the 2008 ACM SIGMOD International Conference on Management of Data, pp. 405–418. ACM (2008)
15. Poskitt, C.M., Plump, D.: Verifying monadic second-order properties of graph programs. In: Giese, H., König, B. (eds.) ICGT 2014. LNCS, vol. 8571, pp. 33–48. Springer, Heidelberg (2014)
16. Wood, P.T.: Query languages for graph databases. SIGMOD Rec. 41(1), 50–60 (2012)

Applications

The Incremental Advantage: Evaluating the Performance of a TGG-based Visualisation Framework

Roland Kluge[1(✉)] and Anthony Anjorin[2]

[1] Technische Universität Darmstadt, Darmstadt, Germany
roland.kluge@es.tu-darmstadt.de
[2] University of Paderborn, Paderborn, Germany
anthony.anjorin@upb.de

Abstract. Triple Graph Grammars (TGGs) are best known as a *bidirectional* model transformation language, which might give the misleading impression that they are wholly unsuitable for unidirectional application scenarios. We believe that it is more useful to regard TGGs as just graph grammars with "batteries included", meaning that TGG-based tools provide simple, default execution strategies, together with algorithms for incremental change propagation. Especially in cases where the provided execution strategies suffice, a TGG-based infrastructure may be advantageous, even for unidirectional transformations.

In this paper, we demonstrate these advantages by presenting a TGG-based, read-only visualisation framework, which is an integral part of the metamodelling and model transformation tool eMoflon. We argue the advantages of using TGGs for this visualisation application scenario, and provide a quantitative analysis of the runtime complexity and scalability of the realised incremental, unidirectional transformation.

Keywords: Graph transformation · Triple graph grammars · Incremental model transformation

1 Introduction

Triple Graph Grammars (TGGs) [23] provide a declarative, rule-based means of specifying how two modelling languages are related. This is done in a direction-agnostic manner using rules that describe how related models can be simultaneously generated. TGGs are best known for their application to bidirectional model transformation, as both forward and backward transformations can be derived automatically from a TGG. When choosing the right model transformation for a certain task, one might thus assume that TGGs, being "bidirectional", are somehow inherently unsuitable for unidirectional tasks. This is perhaps due to the assumption that there must be some "overhead" involved in specifying both directions at once. While this might be conceptually true, we believe it is more helpful to regard TGGs as just graph grammars, but with "batteries

© Springer International Publishing Switzerland 2016
R. Echahed and M. Minas (Eds.): ICGT 2016, LNCS 9761, pp. 189–205, 2016.
DOI: 10.1007/978-3-319-40530-8_12

included". This means that TGG-based tools provide a set of default, out-of-the-box execution strategies including a forward transformation, a backward transformation, simultaneous model generation [22], incremental forward and backward change propagation [2], and consistency checking [15]. We suggest to base the decision to use TGGs less on the question of bidirectionality and more on the following:

Is the transformation task simple enough to be handled by one of the default execution strategies? The forward and backward transformations that can be derived automatically from a TGG are rather simple, performing only a single pass over the input model (each element is visited and marked exactly once). A transformation that is inherently complex, requiring rules with advanced application conditions that create or delete auxiliary elements to trigger the application of other rules, most probably cannot be expressed elegantly (or at all) as a TGG. The same argument applies to deeply nested, recursive control flow structures. Simple, straightforward transformations are a much better fit for TGGs.

Is incremental change propagation required? The true potential of TGGs lies in the formally founded infrastructure for incremental change propagation. Without any additional specification effort, the automatically derived forward and backward transformations can be executed in an incremental mode, updating existing related models appropriately. Incrementality is crucial when the output model cannot be recreated from scratch without losing information [7]. In many situations (large models and small changes), incrementality can also speed-up the transformation process [7]. Even for "simple, straightforward" unidirectional transformations, providing support for incremental updates is non-trivial, especially concerning a choice of sensible semantics.

Contribution. In this paper, we present a case study for using TGGs (i. e., graph transformations) to visualise various models used in the specification and execution of graph transformations. The generated visualisations are rendered in a read-only view, meaning that the transformation is currently unidirectional. We argue that using a TGG-based tool for this task is nonetheless advantageous, as the provided infrastructure can be suitably leveraged to enable declarative, compact specifications that can be executed incrementally.

To address justified concerns of scalability, we perform a detailed quantitative analysis of the transformation, which has been implemented as a general visualisation framework and is currently an integral part of the metamodelling and model transformation tool eMoflon [16]. To enable a realistic evaluation, we make use of a substantial set of real-world models collected over five years of using eMoflon in diverse applications including industrial case studies, bootstrapping eMoflon as much as possible (this includes the TGGs for the TGG-based visualisation framework itself!), and a substantial collection of unit and system tests (see [6] for a repository containing some of these examples).

Structure. Section 2 presents our TGG-based visualisation framework and is complemented by a quantitative evaluation of the realised incremental transformation provided in Sect. 3. Section 4 gives a brief overview of related case studies. Section 5 concludes the paper with a summary and a brief outlook on future work.

2   A TGG-Based Visualisation Framework

A schematic overview of the TGG-based visualisation framework realised in eMoflon is depicted in Fig. 1. Further examples can be found in the appendix.

The top part of the diagram (❶ and ❷) represents what is seen by the *end-user*: ❶ is a tree-view representation of a source model. ❷ is a generated visualisation of the currently selected model element, which is a TGG rule (red box). The visualisation of the TGG rule is in concrete syntax, as defined by an underlying transformation (specified as a TGG). Changes made to the source model inside the tree-view editor are propagated *incrementally* to the visualisation as soon as the editor content is saved.

The bottom part of the diagram (between ❸ and ❹) depicts the chain of transformations used to generate the visualisation from the source model. In the most general case, the end-user makes a change Δ_S (referred to in the following as *delta*) to the source model G_S and triggers an update by saving the editor (represented by ❸). In this context, a "batch" transformation is just a special case with an empty source model G_S. Note that the source model is a typed graph G_S with type graph TG_S. As TGG-based synchronisers operate on *triple graphs*, a correspondence graph G_C and visualisation model G_{vis} are maintained in the background by the tool as a consistent typed triple graph $G_S \leftarrow G_C \rightarrow G_{vis}$ with type triple graph $TG_S \leftarrow TG_C \rightarrow TG_{vis}$. In the context of the visualisation framework, the type graph TG_{vis} is fixed, representing the visualisation capabilities that the framework currently supports.

For every source type graph TG_S that is to be visualised, a *transformation designer* must provide a TGG that specifies how triples $G_S \leftarrow G_C \rightarrow G_{vis}$ are to be constructed. This TGG entails decisions on how source model elements are to be mapped to visualisation elements such as rectangles, arrows, labels, colours, and other available shapes. A forward synchroniser *fwd* is derived automatically from this TGG and is used to forward propagate the applied source model delta Δ_S to yield correspondence and visualisation deltas Δ_C and Δ_{vis}, respectively. As depicted in Fig. 1, applying these computed deltas results in a new consistent triple graph $G'_S \leftarrow G'_C \rightarrow G'_{vis}$, with consistently updated correspondence graph G'_C and visualisation G'_{vis}. Note that this resulting triple graph is also well-typed even though this is not shown explicitly in Fig. 1.

In a final step, G'_{vis} is used to regenerate the visualisation presented to the end-user via a model-to-text transformation *m2t* to produce a file in the `.dot` format, which is then converted to an image (e. g., `.jpeg`) via the Graphviz[1] command line tool *dot*. This final image ❹ is what the end-user can observe in

---

[1] http://www.graphviz.org.

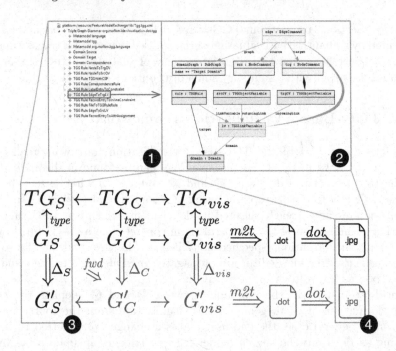

Fig. 1. Overview of TGG-based visualisation framework (Color figure online)

a corresponding view ❷. In contrast to *fwd*, note that *m2t* and *dot* are both currently non-incremental (indicated in Fig. 1 by omitting the deltas).

3 Evaluation

In this section, we present and discuss a quantitative analysis of the forward transformation (*fwd* in Fig. 1) used in our visualisation framework. In the following, we briefly describe the five types of source models that can be visualised, currently. For each transformation, we provide the following statistics to give a rough impression of the complexity of the transformation: (1) the total number of TGG rules n_{tot}, (2) the number of abstract TGG rules n_{abs},[2] and (3) the average number of (object and link) variables per TGG rule $\overline{n_{\text{var}}}$.

TGG [n_{tot}=14, n_{abs}=4, $\overline{n_{\text{var}}}$=22.1]: A *TGG rule*, such as depicted in Fig. 1, is a monotonic triple rules (triples of story patterns without deletion).

SDM [n_{tot}=11, n_{abs}=3, $\overline{n_{\text{var}}}$=15]: *Story diagrams*, a dialect of programmed graph transformations that is similar to simplified UML activity diagrams, are used to specify control flow structures in eMoflon.

---

[2] An *abstract TGG rule* serves to, e. g., extract commonalities of multiple TGG rules, but cannot itself be applied. TGG rules may refine other (abstract or non-abstract) rules to reuse common elements. Refinement is roughly comparable to the purpose of inheritance in object-oriented programming languages. See [3] for more details.

SP $[n_{\text{tot}}=9,\ n_{\text{abs}}=1,\ \overline{n_{\text{var}}}=16.5]$: A *story pattern* represents a regular graph transformation rule that is embedded in an activity node of a story diagram. A story pattern is a graph with annotated nodes and edges, formally representing a graph transformation rule $r : L \to R$ in the SPO approach.

TM $[n_{\text{tot}}=3,\ n_{\text{abs}}=0,\ \overline{n_{\text{var}}}=16]$: A *triple match* represents the match[3] of a TGG rule in an input model and is similar to a story pattern.

PG $[n_{\text{tot}}=3,\ n_{\text{abs}}=0,\ \overline{n_{\text{var}}}=11.7]$: A *precedence graph* is a—predominantly a-cyclic—intermediate data structure representing all possible triple matches of all TGG rules in an input model, together with all resulting dependencies between these triple matches. Precedence graphs are used to control the TGG-based synchronisation process in eMoflon [2].

Our first two research questions to be investigated with this analysis focus on the performance of *fwd* when executed in batch mode, i. e., the first time a user opens a source model in an editor and chooses to visualise it.

RQ 1a: Does *fwd* scale? More precisely does the runtime of *fwd* grow non-exponentially with source model size when executed in batch mode?

RQ 1b: Is the batch runtime of *fwd* acceptable for realistic source models?

The next three research questions concern the incremental execution of *fwd*:

RQ 2a: How large is the speed-up in runtime obtained via incremental change propagation? More precisely, how large is the ratio of runtime of *fwd* in incremental mode compared to batch mode?

RQ 2b: Is this speed-up in runtime perceivable for realistic source models? Would an end-user notice the difference in runtime for re-translating the whole source model as compared to incrementally propagating changes?

RQ 2c: Is it better (wrt. attained speed-up) to synchronise frequently, i. e., after every small change, or to accumulate changes before synchronising?

Finally, the last research question investigates the optimality of *fwd*:

RQ 3: To what extent is incremental change propagation coupled to model size? Optimal would be no coupling at all, i. e., constant time for propagating the same change independent of model size.

Evaluation Setup. The dataset of the evaluation comprises two subsets:

D1: To provide for "realistic" models, required for RQs 1b and 2b, we collected instances of all five metamodels from the current eMoflon developer workspace and all test suite workspaces. These models have been used and collected for over five years from various industrial case studies, the development of eMoflon itself, and numerous examples and tests. To obtain realistic (or even pessimistic) results, all runtime data for D1 was acquired on a typical business notebook (i7-4600U with $2 \times 3.3\,\text{GHz}$, 12 GB RAM) running Windows 8.1 (64bit).

---

[3] A match of a rule $r : L \to R$ in a graph G is an occurrence $m : L \to G$ of the left-hand side L of the rule in G.

Table 1. Characteristics of the evaluation datasets D1 and D2

Property	TGG(D1)	TGG(D2)	SDM(D1)	SP(D1)	TM(D1)	PG(D1)
Model count	3,660	12	4,395	11,191	11,310	293
Mean model size	173	206,945	342	45	69	2,991
Median model size	124	182,436	269	28	56	272

D2: To evaluate scalability for large models, required for RQs 1a, 2a, and 3, we derived a TGG-based model generator [22] from the TGG for visualising TGG rules and used it to synthesise large TGG rules. We chose TGG rules for this complementary synthetic data generation as the corresponding visualisation is currently the most often used one in eMoflon and is thus the richest (uses most visual elements). To be able to run the evaluation in a reasonable amount of runtime, the data for D2 were acquired on a workstation (i7-2600, 4×3.4 GHz, 8 GB RAM) with Windows 7 Professional (64bit).

On both machines, we used Eclipse Mars 4.5 (-Xmx4G), eMoflon 2.12.0[4] and version 1.0.0 of our evaluation application[5]. Table 1 summarises the core characteristics of the datasets. The size of a model is the number of its contained nodes (*EObjects*) and edges (*EReferences*). Both the mean and median of all model sizes are provided to indicate the presence of outliers, e.g., in **PG** (D1).

3.1 RQ1: Scalability of Batch Transformation

Figure 2 depicts the runtime for batch transformation in milliseconds plotted over model size for all models. The caption of each subplot shows the dataset and number n of models. Each data point is the median execution time of 5 runs. When comparing the plots, it is important to note that the x- and y-axes of all plots are of vastly different scale. The characteristic runtime values are additionally summarised in Table 2.

Table 2. Characteristic batch runtime values

Property		TGG(D1)	TGG(D2)	SDM(D1)	SP(D1)	TM(D1)	PG(D1)
Maximum	[ms]	56.7	1,558,500.9	55.4	122.9	38.6	9,019.6
Mean	[ms]	6.0	417,717.3	6.4	3.7	7.0	112.9
Median	[ms]	5.0	145,397.2	6.3	2.1	6.3	3.9

---

[4] http://www.emoflon.org.
[5] https://github.com/eMoflon/paper-icgt2016/releases/tag/icgt2016-v1.0.0.

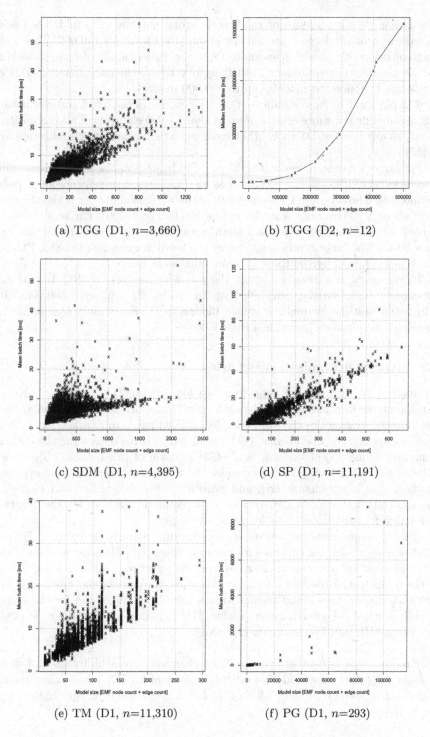

(a) TGG (D1, n=3,660)

(b) TGG (D2, n=12)

(c) SDM (D1, n=4,395)

(d) SP (D1, n=11,191)

(e) TM (D1, n=11,310)

(f) PG (D1, n=293)

Fig. 2. Runtime of batch transformation over model size (n: model count)

Discussion. While the size of the largest story pattern (SP) in D1 is below 700, the size of the largest precedence graph (PG) is above 100,000. This wide range of real-world model sizes shows that the visualisation of large models is *not* an unrealistic requirement, justifying our complementary dataset (D2) of synthetically generated models (up to 500,000 in size).

The plots in Fig. 2a, c–e indicate that the runtime of the batch transformation is generally linear for model sizes of up to about 1,000 elements. This is a positive result, as our dataset D1 shows that most realistic models for visualisation are in this range.

For larger models, however, Fig. 2b and f indicate non-linear behaviour. This is to be expected, as the complexity class for TGG-based transformation is polynomial [14]. The absolute values are still arguably reasonable for a visualisation task: about 8 (25) min for a model of size 300,000 (500,000). For small models, Table 2 shows that the mean and median runtimes for models in D1 are less than 10 ms. The large gap between mean and median execution time for PG can easily be explained by the three extreme outliers in D1.

In summary, our results suggest the following answers to RQ 1: (1a) *fwd* appears to scale satisfactorily even up to model sizes of over 500,000, and (1b) batch runtime for realistic source models in D1 is certainly acceptable for visualisation purposes (being less than 10 ms).

3.2   RQ2: Synchronisation Behaviour

To analyse synchronisation behaviour, we focused on TGG models. For each data point, we first performed a batch forward transformation and then applied the following changes to the source model (a TGG rule), to mimic typical modifications applied by an end-user: (**C1**) addition of three object variables, (**C2**) renaming of one object variable, and (**C3**) removal of two random object variables. To investigate RQ 2c, we consider the following two situations: synchronisation after every change (II), and synchronisation only after performing all changes (III). In both cases, we compare the required time with (I), the duration of a batch forward transformation after all changes.

Table 3 summarises the results of this experiment for both datasets, D1 and D2. For each model, the runtimes for each situation (I)–(III) is the median of five runs. The last two rows show the runtime of the synchronisation as a percentage of the batch transformation (the lower the value, the greater the speed-up). The given maximum, mean, and median values have been calculated for the metric of each row, i.e., the maximum value for (II)/(I) is *not* equal to the ratio of the maximum (II) and the maximum (I) values.

Discussion. The benefit of synchronising changes incrementally instead of retransforming the entire model is particularly evident for D2. The synchronisation only takes between 0.2 % and 1.5 % of the batch transformation time in the mean

Table 3. Comparison of batch transformation **b** in I and synchronisation **s** in II, III.

			D1				D2			
			Max	Mean	Median	Min	Max	Mean	Median	Min
I	C1+C2+C3+b	[ms]	101.0	11.4	7.8	0.4	1,827,980	504,472	186,756	233.6
II	C1+s+C2+s+C3+s	[ms]	17.8	1.8	1.6	0.3	2,387.2	990.8	802.4	29.9
III	C1+C2+C3+s	[ms]	24.6	2.2	1.1	0.1	822.7	330.8	277.4	2.1
II/I	Rem. runtime	[%]	166.7	32.5	22.2	2.1	12.8	1.5	0.5	0.1
III/I	Rem. runtime	[%]	90.9	27.5	20.5	0.8	0.9	0.2	0.2	0.0

and median cases. In the worst case, when the ratio of synchronisation time and batch re-transformation time is maximal, synchronising still takes only 12.8 % to 0.9 % of batch runtime. For D1, i.e., real-world models, the speed-up is less impressive but still remarkable: In the mean and median cases between about 70 % and 80 % of the runtime is saved.

Our results thus suggest the following answers to our research questions: the speed-up enabled by incrementality is substantial and, as can be expected, increases with model size (RQ 2a), even though the speed-up is still substantial for D1, for most realistic models, an end-user probably will not notice a difference of only a few milliseconds in our visualisation scenario (RQ 2b), finally, incremental propagation appears to perform somewhat better if changes are collected (RQ 2c). Although this is not so clear for small- and medium-sized models in D1, the difference is evident for the larger models in D2. This is because (1) every synchronisation run has a certain overhead that increases with model size due to technical reasons, and (2) certain optimisations can be performed by the algorithm, propagating multiple changes at the same time.

3.3 RQ3: Coupling of Incremental Change Propagation to Model Size

The plots in Fig. 3 show the runtime of synchronisation over model size for D1 (Fig. 3a, c and e) and D2 (Fig. 3b, d and f). In each row of Fig. 3, the left figure shows the runtime behaviour for (realistic) model sizes of up to about 1,200, while the right figure shows asymptotic runtime behaviour for large synthetic models. When comparing plots, note that the x- and y-axes of left and right plots have vastly different scales.

Discussion. For an optimal synchroniser, incremental propagation time would be constant, i.e., independent of model size. Figure 3 shows that this is not really the case in practise (for eMoflon). Due to technical reasons and challenges involved with using EMF collections for large models, there is a certain coupling with model size. The results are, nonetheless, reasonably positive: for real-world models (left plots) synchronisation time is almost constant with only a slight linear increase (less than 0.5 ms). For large models, the linear increase is evident but with a very small gradient: for all changes, it only takes about half a second longer to visualise a model with 300,000 more elements.

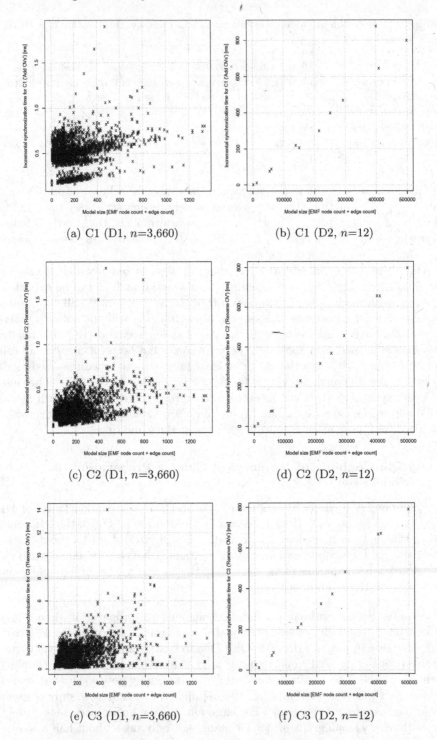

Fig. 3. Synchronisation time over model size (TGG, n: model count)

Our answer to RQ 3 is that a certain coupling of incremental propagation to model size is indeed still present in practise, but it is linear and reasonably small for our application scenario.

3.4 Threats to Validity

Our primary concern is *external validity*, i. e., can our results be generalised beyond our specific case study. This is a justified concern and has two orthogonal dimensions: (1) Do our results hold for other TGG-based tools or are they specific for eMoflon? (2) Do our results hold for other possibly more "complex" application scenarios that require, e. g., hundreds of TGG rules. Concerning (1), TGG comparison papers [12,18] have shown that TGG-based tools are quite diverse, especially concerning their underlying synchronisation algorithms. It is thus difficult to argue that our results hold in any way for "TGGs in general". More evidence should be provided with a new comparison paper, comparing current TGG-based tools using, e. g., our data from this case study. To mitigate (2), we have mined all our workspaces and collected a substantial number of real-world metamodels for the measurements (our dataset D1). This ensures that at least the input data for the transformation is somewhat realistic and not completely synthesised. The task of visualisation is also quite varied, ranging from story diagrams that are deeply nested tree-like structures, to flat, highly connected, more graph-like patterns (TGG rules, story patterns, triple matches). The primary limitation of our case study is more the "complexity" of the TGGs used for the visualisation, the largest TGG having only 14 rules. Although we have 5 TGGs, so in total 40 TGG rules, this is still not comparable to other application scenarios requiring hundreds of TGG rules. We argue, however, that equating complexity with number of rules is naïve: we have encountered cases that are essentially trivial 1–1 bijections, but still require hundreds of rules as the source and target models simply have many types.

Finally, using two different machines to perform the runtime measurements on D1 and D2, respectively, may be considered a threat to *construct validity*. We emphasise here, however, that the objectives of performing the experiments on the two datasets were rather different: While the measurements on D1 focused on applicability in terms of acceptable runtime for realistic models, the measurements on D2 served to observe the behaviour of the TGG-based visualisation for large, synthesised models.

4 Related Work

This paper builds on our previous work in [16,17] and shares the common goal of bootstrapping eMoflon. In [16], Leblebici et al. present the various model transformations used in eMoflon and describe how we have progressed from an initial implementation of our import/export in C#, to a bootstrapped version with story diagrams (a unidirectional programmed graph transformation dialect), to a final bootstrapped TGG-based version that is still being optimised and extended up until today. A runtime comparison of story diagrams and TGGs is provided, showing on the positive side that TGGs are expressive enough to derive both forward (export) and backward (import) transformations from the same specification. Noteworthy is also that both directions perform comparably well, exhibiting almost linear behaviour for up to 10,000 elements, and then polynomial until running out of memory for about 300,000 elements. This indicates that TGGs are inherently symmetric and do not favour any direction. On the negative side, however, the measurements show that TGGs are still 10–15 times slower than story diagrams with lots of room for improvement in this regard. In comparison to this paper, the TGG-based transformations in [16] were *not* executed incrementally and the provided measurements thus give no indication of how feasible or useful this might be. The measurements are also for a single TGG and a single pair of source and target metamodels, while we provide evidence for various research questions using 5 TGGs and a substantial number of diverse source metamodels collected over five years in our test and development workspaces (the visualisation, i. e., target metamodel is fixed in all cases).

In [17], Leblebici et al. compare TGGs implemented in eMoflon with Medini QVT[6], showing that TGGs outperform Medini QVT up to a factor of 20 for model sizes of about 1,000–200,000 elements. In comparison to this paper, the focus of [17] is on showcasing multi-amalgamation, a new language feature of TGGs. Only a single "toy" TGG and a fixed pair of source and target metamodels are used for the comparison. Finally, just as with [16], the transformation is *not* executed in an incremental mode.

There has also been comparable work in the TGG community such as [12,18], which provide a comparison of various TGG tools, including runtime measurements. In contrast to this paper, the focus of [12,18] is on comparing the different tools and not on providing evidence for the performance or advantages of TGGs in general. To ensure that all tools could be used for the exact same TGG, a very simple toy example is used, and only synthetic data is generated for the measurements. The results indicate, however, that there are considerable differences between TGG tools regarding runtime efficiency and expressiveness. This means that our results are primarily valid for eMoflon and cannot be directly generalised to all other TGG tools (see the discussion in Sect. 3.4).

A further source for TGG runtime measurements and comparison with other tools is the annual transformation tool contest (TTC). For example, [11,13, 20] present TGG-based solutions to various contests. Although these results provide evidence for the expressiveness and applicability of TGGs, it is difficult to compare solutions in many cases: for example, [13] and [11] provide solutions using different TGG tools, but the degree of freedom of the contest (the choice of the source metamodel) makes it impossible to compare absolute runtime values. In many cases, it is also impossible to discern the runtime complexity of the

---

[6] http://projects.ikv.de/qvt.

solutions, as the provided test cases are more often used to ensure correctness. Nonetheless, both [20] and [11], for example, provide encouraging evidence that the involved TGG transformation is *not* necessarily the bottleneck in practical model transformation chains. Our experience corroborates this as `dot` dominates our transformation chain for large diagrams, especially as it is not incremental.

Finally, there is some evidence indicating that TGGs can be used successfully for industrial scale applications. In [10], Hermann et al. report on using TGGs for the translation of satellite procedures. Their results show that a pragmatic mix of programmed graph transformation and TGGs can be made to be "more efficient than what is needed for practical use" by applying advanced optimisation techniques. Such powerful domain- and even task-specific optimisations are feasible mainly due to the formal and declarative nature of graph transformations. The application scenario of [10] provides an interesting contrast to this paper as it is also unidirectional but *not* incremental. The main motivation for using TGGs in [10] is the formal guarantee of correctness, while in the case of our visualisation, correctness is important but not crucial; incrementality, conciseness, and readability are arguably more useful for our application scenario.

An inherently incremental industrial application scenario is presented in [4]. Blouin et al. demonstrate how a synchronisation layer between textual and graphical editors can be established using TGGs. As explained in [4], incrementality, expressiveness, and scalability are crucial for the application scenario. Unfortunately, no evaluation and measurement results are provided by Blouin et al., making it hard to conclude more than that the TGG-based solution was "fast enough" for practical usage. Other (industrial) case studies include work from Giese et al., e. g., in [8] for a TGG-based synchronisation between SysML and AUTOSAR models, and from Greenyer et al., e. g., in [9] for a TGG-based transformation of sequence diagram specifications to timed game automata.

5 Conclusion and Future Work

In this paper, we presented a TGG-based visualisation framework, which is currently being used as an integral part of the metamodelling and model transformation tool eMoflon. This is an example of a real-world, unidirectional, and incremental application scenario for TGGs. With a detailed quantitative analysis, we have shown that the realised transformation scales with model size, and that incrementality provides a substantial speed-up. The case study highlights the major advantage of TGGs: due to their declarative nature, multiple default execution strategies for the same TGG can be provided by a TGG tool. Specifically, we investigated the derived forward incremental mode, and made use of the simultaneous mode for generating large models for our scalability analysis.

A current limitation of the visualisation framework is that some steps in the tool chain are not incremental (e. g., *dot* used to render diagrams), and become the bottleneck of the framework. Future work includes, therefore, use cases that further realise the potential of TGGs such as: (1) allowing manual adjustments of the layout in the visualisation (i. e., coping with information loss), and

(2) implementing a completely incremental tool chain such as a TGG-based code generator for EMF. Case studies considering the correspondence between the concrete and abstract syntax (or semantics) of a specification, as discussed, e.g., in [1,5], would also be interesting for further investigating the potential of graph transformation in general and TGGs in particular. Finally, a comparison of our TGG-based model generator to other model generators such as [19,21] would be illuminating: it is, for instance, currently impossible to enforce certain statistical properties using our model generator, but the models are at least guaranteed to be translatable with the underlying TGG. This would not be the case with general purpose (random) model generators and would require a potentially large set of additional constraints to adequately control the generation.

Acknowledgements. This work has been funded by the German Research Foundation (DFG) as part of projects A01 within the Collaborative Research Centre (CRC) 1053 – MAKI.

Appendix: Examples from the eMoflon Handbook

We show concrete examples of visualised source models taken from the eMoflon handbook,[7] whose illustrative example is Leitner's learning box, a system, e.g., for language learning. This system works by creating cards, sorted into sequential partitions, with a front face showing the known word (e.g., "hello" in English) and a back face showing the to-be-learnt word (e.g., "Hallo" in German). While exercising, the learner takes a card from a partition, tries to guess the back-face word based on the front-face word, and, if successful, may move the card to the next partition. A so-called fast card contains easy-to-learn words and may be moved to the last partition upon success, immediately.

The story diagram in Fig. 4a shows the logic of checking a card: If the answer is correct (story pattern `checkCard`) and if the card is a so-called fast card (story pattern `isFastCard`), then this card is promoted to the last partition, as shown in the story pattern depicted in Fig. 4b.

Another task in the eMoflon handbook is to synchronise (using TGGs) a learning box with a dictionary, whose entries can be thought of as simple key-value pairs. Figure 4c shows the precedence graph resulting from translating the sample box in the handbook into a dictionary. The root node `BoxToDictionary-Rule 0` indicates that the box is first of all translated into an empty dictionary, before translating all cards to dictionary entries. Finally, Fig. 4d depicts the triple match that corresponds to `CardToEntryRule 5` in Fig. 4c. This match shows that the card containing "Question One" is mapped to the entry with content "One : Eins".

---

[7] http://www.emoflon.org/.

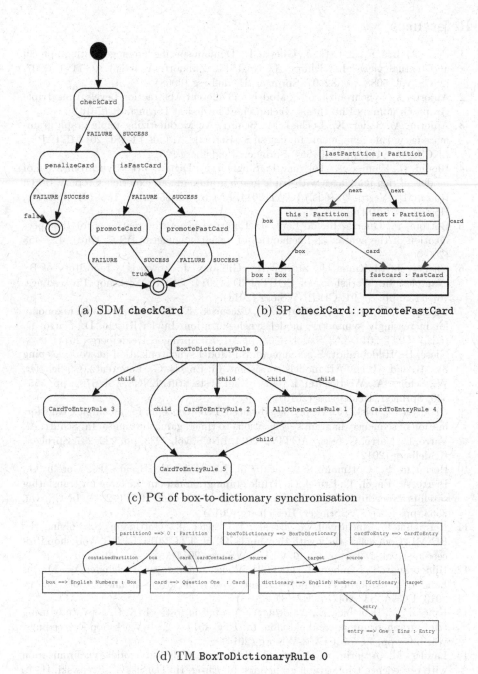

(a) SDM `checkCard`   (b) SP `checkCard::promoteFastCard`

(c) PG of box-to-dictionary synchronisation

(d) TM `BoxToDictionaryRule 0`

Fig. 4. Visualisations of sample models (a) SDM `checkCard` (b) SP `checkCard::promoteFastCard` (c) PG of box-to-dictionary synchronisation, (d) TM `BoxToDictionaryRule 0`

References

1. Pérez Andrés, F., de Lara, J., Guerra, E.: Domain specific languages with graphical and textual views. In: Schürr, A., Nagl, M., Zündorf, A. (eds.) AGTIVE 2007. LNCS, vol. 5088, pp. 82–97. Springer, Heidelberg (2008)
2. Anjorin, A.: Synchronization of Models on Different Abstraction Levels using Triple Graph Grammars Phd thesis, Technische Universität Darmstadt (2014)
3. Anjorin, A., Saller, K., Lochau, M., Schürr, A.: Modularizing triple graph grammars using rule refinement. In: Gnesi, S., Rensink, A. (eds.) FASE 2014 (ETAPS). LNCS, vol. 8411, pp. 340–354. Springer, Heidelberg (2014)
4. Blouin, D., Plantec, A., Dissaux, P., Singhoff, F., Diguet, J.-P.: Synchronization of models of rich languages with triple graph grammars: an experience report. In: Di Ruscio, D., Varró, D. (eds.) ICMT 2014. LNCS, vol. 8568, pp. 106–121. Springer, Heidelberg (2014)
5. Bottoni, P., Guerra, E., de Lara, J.: Enforced generative patterns for the specification of the syntax and semantics of visual languages. JVLC **19**(4), 429–455 (2008)
6. Cheney, J., McKinna, J., Stevens, P., Gibbons, J.: Towards a repository of Bx examples. In: Workshops of EDBT/ICDT 2014. CEUR Workshop Proceedings, vol. 1133, pp. 87–91. CEUR-WS.org (2014)
7. Diskin, Z., Wider, A., Gholizadeh, H., Czarnecki, K.: Towards a rational taxonomy for increasingly symmetric model synchronization. In: Di Ruscio, D., Varró, D. (eds.) ICMT 2014. LNCS, vol. 8568, pp. 57–73. Springer, Heidelberg (2014)
8. Giese, H., Hildebrandt, S., Neumann, S.: Model synchronization at work: keeping SysML and AUTOSAR models consistent. In: Engels, G., Lewerentz, C., Schäfer, W., Schürr, A., Westfechtel, B. (eds.) Nagl Festschrift. LNCS, vol. 5765, pp. 555–579. Springer, Heidelberg (2010)
9. Greenyer, J., Rieke, J.: Applying advanced TGG concepts for a complex transformation of sequence diagram specifications to timed game automata. In: Schürr, A., Varró, D., Varró, G. (eds.) AGTIVE 2011. LNCS, vol. 7233, pp. 222–237. Springer, Heidelberg (2012)
10. Hermann, F., Gottmann, S., Nachtigall, N., Ehrig, H., Braatz, B., Morelli, G., Pierre, A., Engel, T., Ermel, C.: Triple graph grammars in the large for translating satellite procedures. In: Di Ruscio, D., Varró, D. (eds.) ICMT 2014. LNCS, vol. 8568, pp. 122–137. Springer, Heidelberg (2014)
11. Hermann, F., Nachtigall, N., Braatz, B., Engel, T., Gottmann, S.: Solving the FIXML2Code-case study with HenshinTGG. In: TTC 2014. CEUR Workshop Proceedings, vol. 1305, pp. 32–46. CEUR-WS.org (2014)
12. Hildebrandt, S., Lambers, L., Giese, H., Rieke, J., Greenyer, J., Schäfer, W., Marius Lauder, A., Anjorin, A. Schürr : A survey of triple graph grammar tools. In: BX 2013. ECEASST, vol. 57. EASST (2013)
13. Kulcsár, G., Leblebici, E., Anjorin, A.: A solution to the FIXML case study using triple graph grammars and eMoflon. In: TTC 2014. CEUR Workshop Proceedings, vol. 1305, pp. 71–75. CEUR-WS.org (2014)
14. Lauder, M., Anjorin, A., Varró, G., Schürr, A.: Efficient model synchronization with precedence triple graph grammars. In: Ehrig, H., Engels, G., Kreowski, H.-J., Rozenberg, G. (eds.) ICGT 2012. LNCS, vol. 7562, pp. 401–415. Springer, Heidelberg (2012)
15. E. Leblebici: Towards a graph grammar-based approach to inter-model consistency checks with traceability support. In: BX 2016. CEUR Workshop Proceedings, vol. 1571. CEUR-WS.org (2016)

16. Leblebici, E., Anjorin, A., Schürr, A.: Developing eMoflon with eMoflon. In: Di Ruscio, D., Varró, D. (eds.) ICMT 2014. LNCS, vol. 8568, pp. 138–145. Springer, Heidelberg (2014)

17. Leblebici, E., Anjorin, A., Schürr, A., Taentzer, G.: Multi-amalgamated triple graph grammars. In: Parisi-Presicce, F., Westfechtel, B. (eds.) ICGT 2015. LNCS, vol. 9151, pp. 87–103. Springer, Heidelberg (2015)

18. Leblebici, E., Anjorin, A., Schürr, A., Hildebrandt, S., Rieke, J., Greenyer, J.: A comparison of incremental triple graph grammar tools. In: GT-VMT 2014. ECE-ASST, vol. 67. EASST (2014)

19. Mougenot, A., Darrasse, A., Blanc, X., Soria, M.: Uniform random generation of huge metamodel instances. In: Paige, R.F., Hartman, A., Rensink, A. (eds.) ECMDA-FA 2009. LNCS, vol. 5562, pp. 130–145. Springer, Heidelberg (2009)

20. Peldszus, S., Kulcsár, G., Lochau, M.: A Solution to the java refactoring case study using eMoflon. In: TTC 2015. CEUR Workshop Proceedings, vol. 1524, pp. 118–122. CEUR-WS.org (2015)

21. Scheidgen, M.: Generation of large random models for benchmarking. In: BigMDE 2015. CEUR Workshop Proceedings, vol. 1406, pp. 1–10. CEUR-WS.org (2015)

22. Schleich, A.: Skalierbare und effiziente Modellgenerierung mit Tripel-Graph-Grammatiken Master's thesis. TU Darmstadt, Germany (2015)

23. Schürr, A.: Specification of graph translators with triple graph grammars. In: Mayr, E.W., Schmidt, G., Tinhofer, G. (eds.) WG 1994. LNCS, vol. 903, pp. 151–163. Springer, Heidelberg (1995)

Automatic Inference of Graph Transformation Rules Using the Cyclic Nature of Chemical Reactions

Christoph Flamm[2,8], Daniel Merkle[1(✉)],
Peter F. Stadler[2,3,4,5,6,7], and Uffe Thorsen[1]

[1] Department of Mathematics and Computer Science,
University of Southern Denmark, DK-5230 Odense, Denmark
`daniel@imada.sdu.dk`
[2] Institute for Theoretical Chemistry, University of Vienna, 1090 Wien, Austria
[3] Bioinformatics Group, Department of Computer Science,
and Interdisciplinary Center for Bioinformatics,
University of Leipzig, 04107 Leipzig, Germany
[4] Max Planck Institute for Mathematics in the Sciences, 04103 Leipzig, Germany
[5] Fraunhofer Institute for Cell Therapy and Immunology, 04103 Leipzig, Germany
[6] Center for Non-coding RNA in Technology and Health,
University of Copenhagen, DK-1870 Frederiksberg, Denmark
[7] Santa Fe Institute, 1399 Hyde Park Rd, Santa Fe Nm 87501, USA
[8] Research Network Chemistry Meets Microbiology,
University of Vienna, 1090 Wien, Austria

Abstract. Graph transformation systems have the potential to be realistic models of chemistry, provided a comprehensive collection of reaction rules can be extracted from the body of chemical knowledge. A first key step for rule learning is the computation of atom-atom mappings, i.e., the atom-wise correspondence between products and educts of all published chemical reactions. This can be phrased as a maximum common edge subgraph problem with the constraint that transition states must have cyclic structure. We describe a search tree method well suited for small edit distance and an integer linear program best suited for general instances and demonstrate that it is feasible to compute atom-atom maps at large scales using a manually curated database of biochemical reactions as an example. In this context we address the network completion problem.

Keywords: Chemistry · Atom-atom mapping · Maximum common edge subgraph · Integer linear programming · Network completion

1 Introduction

The individual records in databases of chemical reactions typically describe, apart from more or less detailed meta-information, the transformation of a set

© Springer International Publishing Switzerland 2016
R. Echahed and M. Minas (Eds.): ICGT 2016, LNCS 9761, pp. 206–222, 2016.
DOI: 10.1007/978-3-319-40530-8_13

of educts into a set of products [30,31]. Both the product and the educt molecules have representations as labeled graphs, where vertices designate atoms and edges refer to chemical bonds. Chemical reactions therefore may be understood as transformations of not necessarily connected graphs [5,32]. Chemical graph transformations must respect the fundamental conservation principles of matter and charge and therefore imply the existence of a bijection between vertex sets (atoms) of the educts and products which is commonly known as the atom-atom map (AAM).

Chemical graph transformation are by no means arbitrary even when the conservation laws imposed by the underlying physics are respected. Instead, they conform to a large, but presumably finite, set of rules which in chemistry are collectively known as reaction mechanism and "named reactions". A chemical reaction partitions the sets of atoms and bonds of the participating molecules into a *reaction center* comprising the bonds that change during the reactions and their incident atoms, and an remainder that is left unchanged. By virtue of being a bijection of the vertex (atom) sets, the AAM unambiguously determines the bonds that differ between educt and product molecules and thus it identifies the reaction center. The restriction of a chemical transformation to the reaction center, on the other hand, serves as minimal description of the underlying reaction rule.

The task to infer transformation rules from empirical chemical knowledge therefore would be greatly facilitated if each known reactions, i.e., each concrete pair of educt and product molecules would imply a unique graph transformation. Unfortunately, the true AAM is unknown in general, and even where the chemical mechanism, and thus the actual graph transformation, has been reported in the chemical literature, this information is in general not stored together with the educt/product pair in a database. The inference of chemical reaction mechanisms therefore requires that we first solve the problem of inferring AAMs for the known chemical reactions.

Several computational methods for the AAM problem have been devised and tested in the past [9]. The most common formulations are variants of the maximum common subgraph (isomorphism) problem [13]. In the NP-complete Maximum Common Edge Subgraph (MCES) variant an isomorphic subgraphs of both the educt and product graph with a maximal number of edges is identified. An alternative formulation as Maximum Common Induced Subgraph (MCIS) problem [1] is also NP complete. Algorithmic solutions decompose the molecules until only isomorphic sub-graphs remain [1,11]. In the context of graph transformation systems, few methods to infer transformation rules have been published [20], and none applicable in the context of AAMs.

Neither solutions of MCES nor MCIS necessarily describe the true atom map, however. There is no reason why the re-organization of chemical bonds in a chemical reaction should maximize a subgraph problem. Instead, they follow strict rules that are codified, e.g., in the theory of imaginary transition states (ITS) [16,18]. There is only a limited number of ITS "layouts" for single step reactions, corresponding to the cyclic electron redistribution pattern usually involving less than 10 atoms [19]. In a most basic case, an elementary reaction, the broken and newly formed bonds form an alternating cycle of a length rarely exceeding 6 or at most 8 [18]. In [23] we made use of this chemical constraint to devise a Constraint Programming approach for elementary homovalent reactions, i.e., those chemical transformations that do not change the charge and oxidation state of an atom. Here, we use an extended representation of chemical graphs that explicitly represents lone pairs and bond orders; in this manner the graph representation incorporates more detailed chemical information.

Advances in bioinformatics technologies made it possible to infer large-scale metabolic networks automatically from genomic information [6,14,25]. These network models, however, suffer from structural gaps in pathways [7,28], caused by orphan metabolic activities, for which no sequences are known and which cannot be inferred from genomic data. Thus there is an urgent need to infer missing metabolic reactions by other means. We illustrate the potential of AAM for the discovery of novel metabolic reactions. To this end we determine whether chemically plausible AAMs can be founds connecting hypothetical educt/product pairs each consisting of one or two known metabolites.

2   Chemical Reactions Are Cyclic

We model each molecule as a labeled, edge-weighted graph with loops. While the graph model used here is similar to most other formalizations of chemical graphs, it differs in several subtle, but important, details, such as the way charges and lone pairs are modeled:

Definition 1 (Molecule Graph). *A molecule graph* $G = (V, E, l, w)$ *is a labeled, edge-weighted, undirected graph with loops. The label function* $l: V \cup E \to \Sigma_V \cup \Sigma_E$ *denotes vertex and edge labels, and the weight function* $w: E \to \mathbb{Z}$ *denotes the weight of edges. These are assigned so that*

- *Atoms are vertices, with labels denoting which type of atom.*
- *Bonds are edges, with labels denoting the bond type and a weight encodes the number of involved electron pairs. Hence 1, 2, and 3 corresponds to single, double and triple bonds.*
- *Lone pairs, i.e., pairs of non-bonding electrons, are modeled by loops. Again the weight refers to the number of lone pairs.*
- *Charges are modeled using a single special vertex together with edges from this special vertex to the charged atoms. The edge weight equals the atom's charge.*
- *Free radicals, single non-bonding electrons, are modeled using a single special vertex together with edges from this special vertex to the atom with the free radical. The edge weight equals the number of free radicals.*
- *Aromatic complexes are modeled by adding a special vertex for each aromatic complex in the molecules. Each atom participating in the aromatic complex has an edge to the special vertex with weight equal to the number of electrons at the atom taking part in the aromatic complex. The aromatic bonds themselves are edges with weight one, but are distinguished from single bonds by the edge label.*

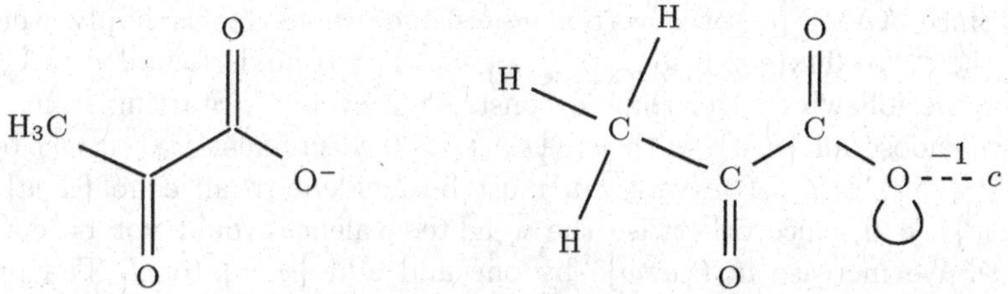

Fig. 1. Usual depiction and molecule graph for pyruvate. Edge labels omitted. Edge weights shown by number of parallel edges (except where negative).

See Fig. 1 for example of molecule graph.

In the following two definitions it will be convenient to consider instead of E the set E^* of all possible edges on V with edges in $e \in E^* \setminus E$ having weight $w(e) = 0$.

Definition 2 (Atom-Atom Mapping). *Given two molecule graphs $G_1 = (V_1, E_1, l_1, w_1)$ and $G_2 = (V_2, E_2, l_2, w_2)$, an atom-atom mapping from G_1 to G_2 is a bijection $\psi: V_1 \to V_2$ that preserves vertex labels, i.e., $l_1(v) = l_2(\psi(v))$ for all $v \in V_1$. With ψ we associate the cost $c[\psi] = \sum_{e \in E_1^*} |w_2(\psi(e)) - w_1(e)|$.*

The cost measures the total number of electron pairs by which G_1 and G_2 differ w.r.t. to a given AAM. Minimizing $c(\psi)$ can be seen as an edit problem [17,21,27] and is equivalent to the NP-hard MCES problem problem [2,4,9,13, 26]. Here we are only interested in MCES instances that correspond to balanced chemical reactions. The complexity results, however, also remains valid also in this case. Next we investigate in some more detail what exactly changes between G_1 and G_2 when an AAM ψ is fixed.

Definition 3 (Transition State). *The transition state of an AAM $\psi: G_1 \to G_2$ is the edge weighted graph $T_\psi = (V_\psi, E_\psi, w_\psi)$ where $E_\psi = \{e \in E_1^* \mid w_1(e) \neq w_2(\psi(e))\}$, $w_\psi(e) = w_2(\psi(e)) - w_1(e)$, and $V_\psi \subseteq V_1$ are all vertices incident to edges in E_ψ.*

By construction of molecule graphs, the weight of each edge is the number of valence electrons. The atom type, i.e., the label of a vertex determines the weighted degree $d_w(v) = \sum_{e \in \delta(v)} w(e)$. Here, loops are counted twice. This reflects that the two electrons per bond order are shared between the incident atoms, while both electrons of a lone pair belong to the same atom. As a consequence, $d_w(v)$ is invariant under all chemically acceptable atom maps. This restriction has important consequences for the structure of transition states:

Proposition 1 (Cyclic Transition States). *The transition state T_ψ of an AAM ψ can be decomposed into a collection of (not necessarily vertex disjoint) cycles $C_1, C_2, \ldots, C_k$ with weights $w_{C_1}, w_{C_2}, \ldots, w_{C_k}$ that are alternating between $+1$ and -1 along the cycles such that $w_\psi(e) = \sum_{i=1}^k w_{C_i}(e)$ for all $e \in E_\psi$.*

Proof. Since AAMs preserve vertex labels and vertex labels imply weighted degree the "zero-flux condition" $\sum_{e\in\delta(v)} w_\psi(e) = 0$ holds for all $v \in V_\psi$. We consider the following algorithm to construct a cycle C. Starting from a vertex v we choose an $\{v, u\}$, with $w_\psi(\{v, u\}) > 0$, decrement $w_\psi(\{v, u\})$ by one and add $\{v, u\}$ to C. The vertex u must be incident to an edge $\{u, w\}$ with $w_\psi(\{u, w\}) < 0$, since otherwise the weighted valence would not be constant under ψ. We increase $w_\psi(\{u, w\})$ by one and add $\{u, w\}$ to C. The process is repeated until we return to v, which is guaranteed by the finiteness of V. Clearly, C is an Eulerian graph, i.e., all its vertex degrees are even. The procedure is repeated until no edges with $w_\psi \neq 0$ is left. If C contains a vertex with degree larger than two, we repeat the procedure recursively on C until we are left with elementary cycles only.   □

The (weighted) degree $\delta_\psi(v) := \sum_{e:v\in e} |w_\psi(e)|$ of a vertex in T_ψ is even because in each step of the proof the value of $\delta_\psi(v)$ is reduced by 2. Thus T_ψ is a generalization of an Eulerian graph, and Proposition 1 is the corresponding variant of Veblen's theorem [29], which states that a graph is Eulerian if and only if it is an edge-disjoint union of cycles.

3   Finding Atom-Atom Mappings

The cyclic nature of the transition states established in Proposition 1 inspires two methods for finding minimum cost AAMs described below. The idea was used in [23] in a much more restrictive chemical setting.

3.1   AltCyc — A Search Tree Approach

The idea of AltCyc is to construct a candidate transition state with a given cost ℓ in a stepwise fashion and to simultaneously map V_1 to V_2. The search for transition states proceeds depth first. The validity of a candidate is then checked by testing whether $G_1 \setminus E_\psi$ and $G_2 \setminus \psi(E_\psi)$ are isomorphic. Finally, the parameter ℓ is increased until a valid mapping is found. A recursive definition of AltCyc is given as Algorithm 1.

Algorithm 1. AltCyc(ψ, P, k, σ)

---
  if $k = 1$ **then**
      if $w_1(P.\text{head}, P.\text{tail}) + \sigma = w_2(\psi(P.\text{head}), \psi(P.\text{tail}))$ **then**
         Complete(ψ, P)
  else
      for $i \in V_1 \wedge i \notin \text{dom}(\psi)$ **do**
         for $p \in V_2 \wedge p \notin \text{range}(\psi)$ **do**
            if $l_1(i) = l_2(p) \wedge w_1(P.\text{head}, i) + \sigma = w_2(\psi(P.\text{head}), p)$ **then**
               $\psi \leftarrow \psi \cup \{i \mapsto p\}$
               AltCyc$(\psi, P.\text{append}(i), k - 1, -1 \cdot \sigma)$

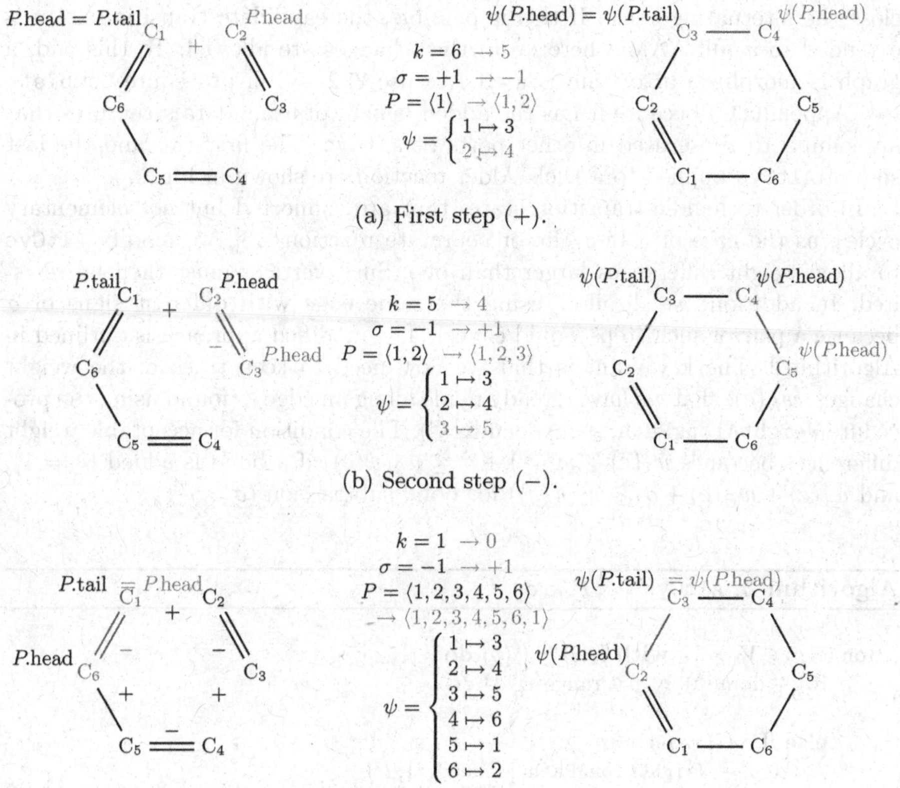

$$P.head = P.tail$$

$$\psi(P.head) = \psi(P.tail)$$

$$k = 6 \;\rightarrow\; 5$$
$$\sigma = +1 \;\rightarrow\; -1$$
$$P = \langle 1 \rangle \;\rightarrow\; \langle 1, 2 \rangle$$
$$\psi = \begin{cases} 1 \mapsto 3 \\ 2 \mapsto 4 \end{cases}$$

(a) First step (+).

$$k = 5 \;\rightarrow\; 4$$
$$\sigma = -1 \;\rightarrow\; +1$$
$$P = \langle 1, 2 \rangle \;\rightarrow\; \langle 1, 2, 3 \rangle$$
$$\psi = \begin{cases} 1 \mapsto 3 \\ 2 \mapsto 4 \\ 3 \mapsto 5 \end{cases}$$

(b) Second step (−).

$$k = 1 \;\rightarrow\; 0$$
$$\sigma = -1 \;\rightarrow\; +1$$
$$P = \langle 1, 2, 3, 4, 5, 6 \rangle$$
$$\quad\rightarrow \langle 1, 2, 3, 4, 5, 6, 1 \rangle$$
$$\psi = \begin{cases} 1 \mapsto 3 \\ 2 \mapsto 4 \\ 3 \mapsto 5 \\ 4 \mapsto 6 \\ 5 \mapsto 1 \\ 6 \mapsto 2 \end{cases}$$

(c) Final step (−, and closing the cycle).

Fig. 2. Stepwise execution of `AltCyc`. Cyan marks the changes within the step. Magenta marks the considered edges and incident vertices.

To explain the algorithm, we first restrict ourself to mappings with transition states consisting of a single elementary cycle. The four parameters are a partial atom-atom mapping candidate ψ, the partial transition state P constructed so far encoded as a list of vertices from V_1, and the number k of edges still to be identified, and the variable $\sigma \in \{-1, 1\}$ that determines whether the current step will add or remove weight.

The search in `AltCyc` starts from all pairs (i, p) with $i \in V_1$ and $p \in V_2$ with $l_2(p) = l_1(i)$; the map ψ is initalized $\psi(i) = p$ and the path starts with $P = \{i\}$. W.l.o.g., the first step is a positive change of weight, i.e., $\sigma = 1$. In each step in the algorithm, a new pair $(i, p) \in V_1 \times V_2$ with matching labels is found and if the $w_1(\{P.head, i\})$ and $w_2(\{\psi(P.head), p\})$ differ by exactly one, i is appended to P, ψ is extended such that $\psi(i) = p$ and the algorithm is called again with k replaced by $k - 1$. If $k = 1$ has been reached, it only remains to close the alternating cycle. If this is possible, the candidate transition state is extended to a full AAM where no further changes are allowed. To this end, a graph isomorphism algorithm is used. We use VF2 [10] in procedure Complete (see Appendix C) because it has the added benefit of using data structures that are similar to those used in other parts of AltCyc. The first two and the last step of AltCyc applied to a Diels-Alder reaction are shown in Fig. 2.

In order to handle transition states that are connected but not elementary cycles, as the case of a bi-cyclic or coarctate reaction [18], we modify AltCyc to allow weight differences larger than one. Such vertices must then be revisited. In addition, we disallow using the same edge with different signs of σ because a pair of such steps would cancel. The modified approach is outlined in Algorithm 2. The key point is that we now need to keep track of the weight changes, $w_P(e)$, that we have already made along an edge e (found using the procedure WeightAlongPath, see Appendix C). The condition for acceptable weight differences becomes $w_1(e) + w_P(e) + \sigma \leq w_2(\psi(e))$ if a bond is added ($\sigma = 1$), and $w_1(e) + w_P(e) + \sigma \geq w_2(\psi(e))$ for bond subtraction ($\sigma = -1$).

Algorithm 2. AltCyc$^*(\psi, P, k, \sigma)$

// As AltCyc...
  for $(i, p) \in V_1 \times V_2$ **with** $l_1(i) = l_2(p)$ **do**
    if $i \notin \mathrm{dom}(\psi) \ \wedge \ p \notin \mathrm{range}(\psi)$ **then**
      // As AltCyc, but using $\leq$ and $\geq$ ...
    else if $\psi(i) = p$ **then**
      $w_P \leftarrow$ WeightAlongPath$(\{P.\mathrm{head}, i\}, P)$
      if $w_P \geq 0 \ \wedge \ \sigma = 1$ **then**
        if $w_1(P.\mathrm{head}, i) + w_P + \sigma \leq w_2(\psi(P.\mathrm{head}), p)$ **then**
          AltCyc$^*(\psi, P.\mathrm{append}(i), k - 1, -1 \cdot \sigma)$
      else if $w_P \leq 0 \ \wedge \ \sigma = -1$ **then**
        // Symmetric case...

There is no guarantee that the transition state is connected. To accommodate disconnected transition states it suffices to replace the path P by a list of paths, where the last path is the current path and all previous paths are kept in order to correctly calculate $w_P(e)$. If a path closes before $k = 0$ is reached, the current cycle is completed and the algorithm restarts to build new path from another initial vertex.

The stepwise approach in AltCyc naturally allows for an elucidation of the mechanism underlying an AAM found by the algorithm. In Fig. 3 the automatic inference of such a mechanism is illustrated. Each step in the figure, the usual way of drawing arrow pushing diagrams, corresponds to two steps in AltCyc.

Taken together, AltCyc uses $O((n^2)^k) = O(n^{2k})$ recursive calls, where $n = |V_1| = |V_2|$. Exploiting the fact that only edges to the special vertex for a charge can be negative, this reduces to $O(n^{k+l})$, where l is the number of components in the transition state candidate, because it suffices to examine only the $O(1)$

```
   C — C         C — C                    C — C          C — C
  /     \       /     \                  /     \        /     \
 C       C   C         C = O            C       C   C          C = O
 \\      //   \\      /                 \\      //  ⌒\\       /
  C   C         C   O                    C   C          C — O
 / |   \       / \                      / |   \       /   \
C  |    \     C — C    H                C  |    \     C — C    H⁺
   C          C — C                        C
```
(a) Initial molecule.   (b) After two bond changes.

```
           C — C         C — C                        C — C          C — C
          /     \       /     \                      /     \        /     \
  H⁺   C         C — C         C = O         H⁺   C         C — C          C = O
      \\    //    \           /                   ⌒\\          \          /
        C   C       C — O                            C   C       C — O
       / |   \     /   \                            / |   \     /   \
      C  |    \   C — C                            C  |    \   C — C
         C                                            C
```
(c) After four bond changes.   (d) After six bond changes.

```
               C — C         C — C
              /     \       /     \
       H — C         C — C         C = O
            \           \         /
              C — C       C — O
             / |   \     /   \
            C  |    \   C — C
               C
```
(e) Resulting molecule.

Fig. 3. An example AAM for Stork's cyclisation of farnesyl acetic acid to ambreinolide [33]. Note that only the single hydrogen in the transition state is shown, and while it is assumed in the model that it is the same hydrogen leaving and later entering, in actual chemistry it is a different hydrogen.

edges incident to P.head or $\psi(P$.head$)$ depending on whether we are making a negative or positive step in the algorithm. In addition, AltCyc incurs the cost of the graph isomorphism check for completing the mapping.

In practice, however, the runtime is much lower since vertex labels must match. The runtime nevertheless still depends heavily on k, and thus the length of the optimal mapping of the instance. However, as discussed k can be assumed to be small for the case of inferring chemical transformation rules. Due to depth first strategy, the memory consumption of AltCyc is $O(n)$.

3.2 ILP2 — An Integer Linear Program

The AAM problem can also be phrased as an ILP. We use binary variables m_{ip} to encode the mapping ψ as $m_{ip} = 1$ iff $\psi(i) = p$ and $m_{ip} = 0$ for all other combinations of i and p. To enforce that ψ is vertex label preserving we set $m_{ip} = 0$ for $l_1(i) \neq l_2(p)$, and to ensure ψ is a bijection we formulate the following linear constraints.

$$\forall i \in V_1: \sum_{p \in V_2} m_{ip} = 1 \quad \text{and} \quad \forall p \in V_2: \sum_{i \in V_1} m_{ip} = 1$$

The most obvious way to proceed would be to keep track of the mapping between the edge sets using either binary variables describing whether a bond is mapped or not as in [15], or integer variables that denote the weight change if a bond is mapped, and zero for unmapped bonds. For such variables we would need $O(|V|^4)$ constraints, however. Empirically we found that ILP-solvers quickly run out of memory and become very slow for such a model.

Though there already exist ILP formulations of similar problems with only $O(|V|^2)$ constraints [22], obtained by exploiting the sparseness of molecule graphs, we propose a new ILP formulation based on the Kaufmann and Broeckx linearization of the quadratic assignment problem [8], which also needs only $O(|V|^2)$ constraints.

We introduce integer variables $c_{ip}^+ \in \mathbb{N}_0$ and $c_{ip}^- \in \mathbb{N}_0$ that model the positive and negative weight changes respectively of all edges incident to vertex $i \in V_1$ if $\psi(i) = p$. Both c_{ip}^+ and c_{ip}^- are zero for all other combinations of i and p. Making use of the fact that weight changes are balanced, i.e. $\sum_{e \in \delta(v)} w_\psi(e) = 0$ for all $v \in V_\psi$, we can use the following constraint for all $i \in V_1$:

$$\sum_{p \in V_2} c_{ip}^+ = \sum_{p \in V_2} c_{ip}^-$$

We also substitute them in the objective function:

$$obj = \sum_{(i,p) \in V_1 \times V_2} c_{ip}^+ + \sum_{(i,p) \in V_1 \times V_2} c_{ip}^-$$

Since the change variables are included in the objective function they will implicitly be constrained from above. In order to constrain them from below we use the following constraints for all $(i,p) \in V_1 \times V_2$:

$$c_{ip}^+ \geq (m_{ip} - 1) \cdot M + \sum_{(j,q) \in V_1 \times V_2} m_{jq} \cdot \max\{0, w_2(\{p,q\}) - w_1(\{i,j\})\}$$

$$c_{ip}^- \geq (m_{ip} - 1) \cdot M + \sum_{(j,q) \in V_1 \times V_2} m_{jq} \cdot \max\{0, w_1(\{i,j\}) - w_2(\{p,q\})\}$$

where M is a suitably large constant. It suffices to set M to the largest weighted degree to void the constraint when $m_{ip} = 0$. The first term voids the constraints if $m_{ip} \neq 1$. The sums correspond to the sum of all positive (negative) changes of edges incident to i and p respectively, if indeed these edges are mapped to each other.

Unlike AltCyc we have little control over intermediate steps in the reaction, but using ILP2 we have much freedom to modify the cost model used. Assuming we have an integer linear programming solver available ILP2 takes very little time to implement.

3.3 Enumeration of All Optimal Atom-Atom Mappings

So far we have focused on the problem of finding a single AAM. The solution of the optimization problem is in general not unique, however. A particular problem in this context are symmetries of the educt or product molecules, because this may bloat the number of AAMs. We are therefore interested only in non-equivalent AAMs.

Definition 4 (Equivalent Atom-Atom Mappings). *For a given AAM define $G_\psi = (V_\psi, E_\psi, l_\psi)$ with vertex set $V_\psi = V_1$, edge set $E_\psi = E_1 \cup \psi^{-1}(E_2)$, and label function $l_\psi(x) = (l_1(x), l_2(\psi(x)))$. If $x \notin \mathrm{dom}(l_i)$ then $l_i(x) = \varepsilon_i$, where ε_i is some label not in $\mathrm{range}(l_i)$, denoting a non-edge. We say two atom-atom mappings, ψ and φ are equivalent if the graphs G_ψ and G_φ are isomorphic.*

Now, let us consider whether a transition state candidate of an atom-atom mapping uniquely defines the full mapping.

Definition 5 (Completion of Partial Mapping). *Given a partial AAM $\psi'\colon A \subset V_1 \to B \subset V_2$, a completion of ψ' is an AAM such that $\psi|_A = \psi'$ and outside A, ψ preserves all properties of G_1 and G_2.*

Note that such a completion need not exist for a given partial AAM.

Proposition 2 (Partial Mapping). *If ψ and φ are two completions of a partial AAM ψ', ψ and φ are equivalent.*

Proof. Consider the two AAMs ψ and φ and their associated graphs G_ψ and G_φ. By assumption, they are both completions of the same partial AAM ψ'; therefore the two induced sub-graphs $G_\psi[\mathrm{dom}(\psi')]$ and $G_\varphi[\mathrm{dom}(\psi')]$ are identical. Consider the subgraphs $G' := G_\psi \setminus E(G_\psi[\mathrm{dom}(\psi')])$ and $G'' := G_\varphi \setminus E(G_\varphi[\mathrm{dom}(\psi')])$ without edges in $G_\psi[\mathrm{dom}(\psi')]$. G' and G'' both are identical to $G_1 \setminus E(G_\psi[\mathrm{dom}(\psi')])$ if only considering the labels from l_1 in each of G_ψ and G_φ. As both ψ and φ preserve all properties of G_1 and G_2 outside $\mathrm{dom}(\psi')$, the labels from l_2 are always identical to the labels from l_1 outside $\mathrm{dom}(\psi')$.

Thus G_ψ and G_φ are isomorphic and by definition ψ and φ are equivalent. $\square$

Proposition 2 can be applied in different ways. In `AltCyc` it shows we only have to complete each candidate transition state once in order to enumerate all mappings. In `ILP2` it can be used to exclude solutions based on mapping variables defining the transition states instead of all mapping variables.

4 Results

The RHEA [24] database (v. 50), which provides access to a large set of expert-curated biochemical reactions, has been used to test our suggested AAM algorithms, and to underline the necessity of graph transformation methods for network completion. We exclude all reactions with unspecified repeating units and wildcards, resulting in a set of 19753 reactions involving a set, M, of 3786 non-isomorphic molecular graphs. We performed a statistical analysis of RHEA, that shows how often molecules are used in the reaction listed in the database, and how many non-isomorphic isomers are stored in RHEA. Interestingly, terpene chemistry [12] clearly dominates the high frequency isomers (see Appendix A). Due to space limitations, we focus on a brief runtime analysis and network completion results. As `AltCyc` constructs solutions in a stepwise fashion, a chemical mechanism explaining the bond changes as subsequent transformations is naturally inferred. An example for a mechanistic inference of Stork's cyclisation of farnesyl acetic acid to ambreinolide [33] is given in Fig. 3.

Runtime. We compared `AltCyc`, `ILP2`, and a naïve ILP-implementation with $O(n^4)$ constraints, `ILP4`, with regard to their ability of enumerating all non-equivalent AAMs within a fixed runtime (see Appendix B). We found that `ILP2` drastically outperforms the naïve ILP-implementation and also is systematically more efficient than `AltCyc`. The latter has a (small) advantage for instances with small transitions states. For both methods we see an exponential decline in ratio of quickly solved instances as size of instances grow, this corresponds well with the expected exponential runtime.

Network Completion. Databases of metabolic networks are by no means complete because the enzymes catalyzing many of the reactions in particular in the so-called secondary metabolism have remained unknown. Furthermore, for almost one third of the known metabolic activities, no protein sequences are known that could encode the corresponding enzyme. *Network completion* is an important task to fill gaps i.e. missing reaction steps, in genome-scale metabolic networks. Reaction perception, i.e. finding AAMs, is the only technique capable of finding possible candidates for the missing reactions, where homology based methods fail, due to lack of data.

Inferring all candidate *2-to-2* reactions addresses this issue by determining for all disjoint pairs A, B of multisets A (one or two educt molecules, potentially isomorphic) and multisets B (one or two product molecules, also potentially isomorphic), whether there is a chemically plausible reaction transforming A to B. By Proposition 1, *any* reaction satisfying mass and charge balance has a cyclic transition state.

Let $R_{2,2}$ denote the set of all sets $\{A, B\}$ such that A and B are disjoint multi-subsets of the set of molecules M, both of size at most 2 with A and B containing the same vertex labels, charges, etc. The set of test instances $R_{2,2}$ of *2-to-2* reactions that satisfy mass balance can be extracted from a database with molecule set M in time $O(|M|^2 \log |M| + |R_{2,2}|)$ using Algorithm 5 (see Appendix C). We obtain a set of $|R_{2,2}| = 114,429,849$ balanced reaction candidates with at most two molecules on either side of the reaction.

It is not feasible to test 100 million candidates for chemical feasibility in an exact manner. Using the length of the transition state as a filter, however, will remove implausible candidates as well as multi-step mechanisms. The length of the transition state can be bounded in both `AltCyc` as well as `ILP2`. We used

`AltCyc` because of its performance advantage with small transition states. In a random sample of 10.000 instances drawn from $R_{2,2}$ we found 34, 59, and 167 reactions with transition states of length 4, 6, and 8, respectively. Extrapolating from this sample we have to expect approximately three million candidate reactions with AAMs that will need to be examined in more detail. Clearly this number is too large for a biochemical network. Further pruning of the candidate list will thus require additional information, e.g., on the energetics of the reactions and on these reaction mechanisms plausibly catalyzed by enzymes. However, it underlines the need for graph transformation techniques for computing realistic candidate sets.

5   Conclusion

Graph transformation systems have great potential as a model of chemistry in particular in the context of large reaction networks. Their practical usefulness, however, stands and falls with the ability to produce collections of transformation rules that closely reflect chemical reality. We have shown here that the extraction of AAMs from educt/product pairs is a necessary first step because the restriction of the graph transformation to the reaction center, which is uniquely determined by the AAM, provides a minimal description of the corresponding reaction rule. We have shown formally that it is not sufficient to solve a general graph editing problem. Instead, the cyclic nature of the transition states must be taken into account as additional constraints. With `AltCyc` and `ILP2` we have introduced two complementary approaches to solve this chemically constrained maximum subgraph problem. The constructive `AltCyc` approach performs better on short cycle instances. If more complex transition states need to be considered or if flexibility in the cost function is required `ILP2` becomes the method of choice.

Advances in high-throughput sequencing technologies drives the reconstruction of organism-specific large-scale metabolic networks from genomic sequence information. Reaction perception, as illustrated here on the Rhea database, is currently the only computational technique to suggest missing reactions in the reconstructed networks once the methods of comparative genomics to infer enzyme activities are exhausted. We have demonstrated here that efficient computation of AAMs serves as first effective step. Much remains to be done, however. Most importantly, the AAM determines only a minimal reaction rule confined to the reaction center. The feasibility of chemical reactions, however, also depends on additional context in the vicinity of the reaction center. While graph grammar systems readily accommodate non-trivial context [3,5], we have yet to develop methods to infer the necessary contexts from the huge body of chemical reaction knowledge. Once this is solved, such more elaborate rules will form a highly efficient filter for the candidate AAMs. In this context the stepwise construction of the transition state in `AltCyc` holds further promise: context information could be used efficiently already in the AAM construction step to prune its search tree, simultaneously increase the chemical realism of the solutions and its computational efficiency.

Acknowledgments. This work was supported in part by the Volkswagen Stiftung proj. no. I/82719, and the COST-Action CM1304 "Systems Chemistry" and by the Danish Council for Independent Research, Natural Sciences.

A   Statistical Analysis of Rhea

Of the $M = 3786$ non-isomorphic molecular graphs in RHEA, 2204 are identified uniquely by their sum formula. While 2030 of the molecules appear only in a minimum of 4 reactions, some compounds take part in a very large fraction of all reactions in RHEA, e.g., H^+ participates in 11,1147 reactions, some of which are different descriptions of similar reactions where only the direction of the reaction differs, 5055 of these are truly distinct, adenosine di-, and tri-phosphate (and it derivatives), water, and dioxide each participate in more than 2000 reactions (depicted as red dots in Fig. 4 (right)). The maximum number of isomers (i.e., compounds that have the same sum formula but a non-isomorphic graph representation) is 63. The corresponding sum formula is C_15H_24. Interestingly, most of the large sets of isomers in RHEA are terpenes, condensates of identical five carbon atom building blocks. The terpenes form a combinatorial class of polycyclic ring-systems via complex sequences of cyclisation and isomerization reactions. Figure 4 (left) summarizes the results (terpenes marked with red).

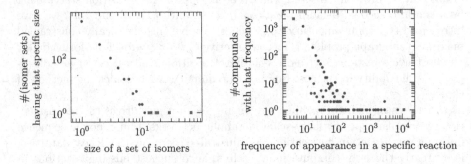

Fig. 4. Distribution of isomers and frequency of participation in reactions in Rhea. Left plot shows a few sets of isomers are very large, while most compounds in Rhea are unique up to sum formula of those compounds. Right plot shows the frequency with which a compound participates in reactions. (Color figure online)

B   Analysis of Runtime

As we are mainly interested in single step reactions, we restricted our algorithms to only look for connected, vertex-disjoint transition states during the comparison. Figure 5 shows the fraction of instances where AltCyc, ILP2 and a naïve ILP-implementation with $O(n^4)$ constraints, ILP4, are able to enumerate all non-equivalent atom-atom mappings for different instance size categories as well as absolute number of instances solved divided by solution size.

Only very few instances that are not completely solved within the first 60 s are solved within reasonable time (one hour). So there seems to be a sharp divide between easy and hard instances. From the plot in Fig. 5 (left) of the fraction of instances solved fast we observe an exponential decline in ratio of solved instances. This corresponds well with the expected exponential runtime of the algorithms.

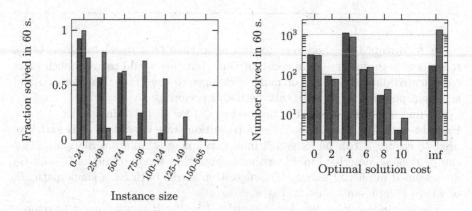

Fig. 5. Fraction and number of instances where all optimal atom-atom maps are found in 60 s (user time) by instance size and optimal solution cost for `AltCyc` (magenta), `ILP2` (cyan) and `ILP4` (gray). (Color figure online)

As we restricted the solution set certain instances are proven infeasible by `ILP2`, while `AltCyc` will continue searching for solutions until the parameter k, the number of weight changes, gets arbitrarily high. We chose to deem instances where `AltCyc` found no solutions for $k \leq 10$ infeasible and terminate the search. These two classes of solutions are marked in the rightmost column in Fig. 5. Note that the performance of `AltCyc` on the infeasible class of instances depends heavily on the somewhat arbitrary choice of maximum k.

Both ILP models are implemented using CPLEX, an efficient state of the art MIP-solver. `AltCyc` and `ILP2` has been tested on a total of 4295 Rhea instances, while `ILP4` has only been tested on a subset of these of size 250.

C   Algorithmic Details

For completeness we include pseudo-code for the sub-procedures used in the paper.

Pseudo-code for `WeightAlongPath`: In `AltCyc*` (see Algorithm 2) we need to find all previous changes to an edge $\{i, j\}$ currently under examination, $w_P(\{i, j\})$.

Algorithm 3. WeightAlongPath($\{i, j\}, P$)

$w_P \leftarrow 0$
$\sigma \leftarrow 1$
for i' **from** 0 **to** $|P| - 2$ **do**
$\quad j' \leftarrow i' + 1$
$\quad$ **if** $\{i', j'\} = \{i, j\}$ **then**
$\quad\quad w_P \leftarrow w_P + \sigma$
$\quad\quad \sigma \leftarrow -1 \cdot \sigma$

In Algorithm 3 we show how to do this in time $O(|P|)$, where $|P| \in O(k)$. It is possible to find $w_P(e)$ in constant time, but this would require much more complicated data structures or making changes to the graphs we work on and as k is in practice very small, this method is preferred.

To find $w_P(e)$ for a list of paths, add $w_P(e)$ for all paths in the list.

Pseudo-code for Complete**:** When a transition state candidate ψ' is found we need to ensure it can be extended into a complete atom-atom mapping. This can be done as described in Algorithm 4. Note that the two graphs G_1 and G_2 are assumed implicitly known. The algorithm works both for a single path, P, or where P represents a list of paths.

The only non-trivial detail in Algorithm 4 is that it is not correct to remove all edges in the induced subgraph on the domain of ψ', the weight change needs to be sufficient, and there may be unchanged cords to consider.

Algorithm 4. Complete(ψ', P)

---

for $e \in P$ **do**{Here P is considered a set of edges}
$\quad w_P \leftarrow$ WeightAlongPath$:(e, P)$
$\quad$ **if** $w_P = w_2(\psi(e)) - w_1(e)$ **then**
$\quad\quad$ Remove e from G_1 and $\psi(e)$ from G_2
$\quad$ **else**
$\quad\quad$ **fail**
for $(i, p) \in V_1 \times V_2$ **where** $\psi'(i) = p$ **do**
$\quad$ Relabel i and p to have identical, otherwise unique labels
return an isomorphism from G_1 to G_2

Finding 2-to-2 Candidates in $O(n^2 \log n)$ Comparisons. In order to generate all $O(n^4)$ candidate reactions with no more than two molecules in the educts or products we use Algorithm 5. A set of molecules, M, is given, as well as a method to obtain the distribution of atoms and charges of the molecules h, in practice some implementation of sparse vectors. We assume we keep pointers to the original molecules that resulted in each distribution, and we get these with the function mol.

The algorithm is dominated by one of two things, either the sorting of the length n^2 array H (where $n = |M|$), or the time to output candidates $k \in O(n^4)$, the resulting runtime is then $O(n^2 \log n + k)$.

Algorithm 5. 2to2(M)

$\mathcal{H} \leftarrow h(M) \cup \{\mathbf{0}\}$
generate $H = \{h_1 + h_2 \mid (h_1, h_2) \in \mathcal{H} \times \mathcal{H} \wedge h_1 \leq h_2\}$ as an array
Sort(H)
for $i \leftarrow 1$ **to** $\text{len}(H) - 1$ **do**
    $j \leftarrow i + 1$
    while $j \leq \text{len}(H) \wedge H[i] = H[j]$ **do**
        output $(mol(H[i]), mol(H[j]))$
        $j \leftarrow j + 1$

References

1. Akutsu, T.: Efficient extraction of mapping rules of atoms from enzymatic reaction data. J. Comp. Biol. **11**, 449–462 (2004)
2. Akutsu, T., Tamura, T.: A polynomial-time algorithm for computing the maximum common connected edge subgraph of outerplanar graphs of bounded degree. Algorithms **6**(1), 119 (2013)
3. Andersen, J.L., Flamm, C., Merkle, D., Stadler, P.F.: 50 shades of rule composition. In: Fages, F., Piazza, C. (eds.) FMMB 2014. LNCS, vol. 8738, pp. 117–135. Springer, Heidelberg (2014)
4. Bahiense, L., Mani, G., Piva, B., de Souza, C.C.: The maximum common edge subgraph problem: a polyhedral investigation. Discrete Appl. Math. **160**(18), 2523–2541 (2012). v Latin American Algorithms, Graphs, and Optimization Symposium Gramado, Brazil, 2009
5. Benkö, G., Flamm, C., Stadler, P.F.: A graph-based toy model of chemistry. J. Chem. Inf. Comput. Sci. **43**, 1085–1093 (2003). presented at MCC 2002, Dubrovnik CRO, June 2002; SFI # 02–09–045
6. Biggs, M.B., Papin, J.A.: Metabolic network-guided binning of metagenomic sequence fragments. Bioinformatics (2015)
7. Breitling, R., Vitkup, D., Barrett, M.P.: New surveyor tools for charting microbial metabolic maps. Nat. Rev. Microbiol. **6**, 156–161 (2008)
8. Burkard, R., ela, E., Pardalos, P., Pitsoulis, L.: The quadratic assignment problem. In: Du, D.Z., Pardalos, P. (eds.) Handbook of Combinatorial Optimization, pp. 1713–1809. Springer, US (1999)
9. Chen, W.L., Chen, D.Z., Taylor, K.T.: Automatic reaction mapping and reaction center detection. WIREs Comput. Mol. Sci. **3**, 560–593 (2013)
10. Cordella, L.P., Pasquale, F., Sansone, C., Vento, M.: A (sub)graph isomorphism algorithm for matching large graphs. IEEE Trans. Pattern Anal. Mach. Intell. **26**(10), 1367–1372 (2004)
11. Crabtree, J., Mehta, D., Kouri, T.: An open-source java platform for automated reaction mapping. J. Chem. Inf. Model **50**, 1751–1756 (2010)
12. Degenhardt, J., Köllner, T.G., Gershenzon, J.: Monoterpene and sesquiterpene synthases and the origin of terpene skeletal diversity in plants. Phytochem **70**, 1621–1637 (2009)
13. Ehrlich, H.C., Rarey, M.: Maximum common subgraph isomorphism algorithms and their applications in molecular science: a review. WIREs Comput. Mol. Sci. **1**, 68–79 (2011). doi:10.1002/wcms.5
14. Feist, A.M., Herrgøard, M.J., Thiele, I., Reed, J.L., Palsson, B.Ø.: Reconstruction of biochemical networks in microorganisms. Nat. Rev. Microbiol. **7**, 129–143 (2009)

15. First, E.L., Gounaris, C.E., Floudas, C.A.: Stereochemically consistent reaction mapping and identification of multiple reaction mechanisms through integer linear optimization. J. Chem. Inf. Model **52**, 84–92 (2012)
16. Fujita, S.: Description of organic reactions based on imaginary transition structures. 1. Introduction of new concepts. J. Chem. Inf. Comput. Sci. **26**, 205–212 (1986)
17. Gao, X., Xiao, B., Tao, D., Li, X.: A survey of graph edit distance. Pattern Anal. Appl. **13**(1), 113–129 (2010)
18. Hendrickson, J.B.: Comprehensive system for classification and nomenclature of organic reactions. J. Chem. Inf. Comput. Sci. **37**, 852–860 (1997)
19. Herges, R.: Organizing principle of complex reactions and theory of coarctate transition states. Angew. Chem. Int. Ed. **33**, 255–276 (1994)
20. Jeltsch, E., Kreowski, H.J.: Grammatical inference based on hyperedge replacement. In: Ehrig, H., Kreowski, H.-J., Rozenberg, G. (eds.) Graph Grammars 1990. LNCS, vol. 532, pp. 461–474. Springer, Heidelberg (1991)
21. Justice, D., Hero, A.: A binary linear programming formulation of the graph edit distance. IEEE Trans. Pattern Anal. Mach. Intell. **28**(8), 1200–1214 (2006)
22. Latendresse, M., Malerich, J.P., Travers, M., Karp, P.D.: Accurate atom-mapping computation for biochemical reactions. J. Chem. Inf. Model **52**, 2970–2982 (2012)
23. Mann, M., Nahar, F., Schnorr, N., Backofen, R., Stadler, P.F., Flamm, C.: Atom mapping with constraint programming. Alg. Mol. Biol. **9**, 23 (2014)
24. Morgat, A., Axelsen, K.B., Lombardot, T., Alcntara, R., Aimo, L., Zerara, M., Niknejad, A., Belda, E., Hyka-Nouspikel, N., Coudert, E., Redaschi, N., Bougueleret, L., Steinbeck, C., Xenarios, I., Bridge, A.: Updates in rhea a manually curated resource of biochemical reactions. Nucleic Acids Res. **43**(D1), 459–464 (2015)
25. Prigent, S., Collet, G., Dittami, S.M., Delage, L., Ethis de Corny, F., Dameron, O., Eveillard, D., Thiele, S., Cambefort, J., Boyen, C., Siegel, A., Tonon, T.: The genome-scale metabolic network of Ectocarpus siliculosus (EctoGEM): a resource to study brown algal physiology and beyond. Plant J. **80**(2), 367–381 (2014)
26. Raymond, J.W., Willett, P.: Maximum common subgraph isomorphism algorithms for the matching of chemical structures. J. Comput. Aided Mol. Des. **16**(7), 521–533 (2002)
27. Riesen, K., Bunke, H.: Approximate graph edit distance computation by means of bipartite graph matching. Image Vis. Comput. **27**(7), 950–959 (2009). 7th IAPR-TC15 Workshop on Graph-based Representations (GbR 2007)
28. Schaub, T., Thiele, S.: Metabolic network expansion with answer set programming. In: Hill, P.M., Warren, D.S. (eds.) ICLP 2009. LNCS, vol. 5649, pp. 312–326. Springer, Heidelberg (2009)
29. Veblen, O.: An application of modular equations in analysis situs. Ann. Math. **14**, 86–94 (1912)
30. Warr, W.A.: A short review of chemical reaction database systems, computer-aided synthesis design, reaction prediction and synthetic feasibility. Mol. Inform. **33**, 469–476 (2014)
31. Wittig, U., Rey, M., Kania, R., Bittkowski, M., Shi, L., Golebiewski, M., Weidemann, A., Müller, W., Rojas, I.: Challenges for an enzymatic reaction kinetics database. FEBS J. **281**, 572–582 (2014)
32. Yadav, M.K., Kelley, B.P., Silverman, S.M.: The potential of a chemical graph transformation system. In: Ehrig, H., Engels, G., Parisi-Presicce, F., Rozenberg, G. (eds.) ICGT 2004. LNCS, vol. 3256, pp. 83–95. Springer, Heidelberg (2004)
33. Yoder, R.A., Johnston, J.N.: A case study in biomimetic total synthesis: polyolefin carbocyclizations to terpenes and steroids. Chem. Rev. **105**, 4730–4756 (2005)

Using Graph Transformation for Puzzle Game Level Generation and Validation

David Priemer, Tobias George, Marcel Hahn,
Lennert Raesch, and Albert Zündorf[✉]

Kassel University Germany, Kassel, Germany
david@jumpsuit-entertainment.com,
{george,hahn,raesch,zuendorf}@uni-kassel.de

Abstract. As part of a student's project, computer science student David Priemer and former graphics design student Daniel Goffin started to develop a nice puzzle based computer game called Perlinoid. In his Bachelor thesis, David developed a level generator for Perlinoid. The challenge was to generate interesting puzzles with a reasonably small number of elements and reasonably complex series of steps required for the solution. Being educated in graph transformations, David modeled the possible puzzles as graphs and possible user steps as graph transformations. First a random level is generated, for which the level generator computes a reachabilty graph by applying all graph transformations in all possible combinations on all possible matches. Next, the reachability graph is analyzed for distances between start graph and end graph, number of possible choices, number of different solution paths, etc. and a number of metrics are applied. Finally, the result gets evaluated. In case the level is considered insufficient, it will be discarded and the process starts from the beginning. This is done within the Perlinoid application on demand at user side. To minimize the footprint of the level generator, a new dedicated graph transformation engine has been built and incorporated into the Perlinoid game. This paper reports about the concepts and experiences with this approach.

Keywords: Graph transformations · Reachability graphs · Fujaba · SDM-Lib

1 Introduction

Generating pseudo-random content has always been an important part of the video game industry but becomes even more important with ever growing worlds and multiplayer games. Automating the process of content generation does not only resonate well with the players who constantly long for more challenges, but takes a lot of work from developers. Some of the biggest challenges include validation, classification and quality assurance of such generated content as well as keeping the generator flexible enough to adjust for changes and additions to the game mechanics that occur during development.

© Springer International Publishing Switzerland 2016
R. Echahed and M. Minas (Eds.): ICGT 2016, LNCS 9761, pp. 223–235, 2016.
DOI: 10.1007/978-3-319-40530-8_14

Perlinoid is a puzzle game by computer science student David Priemer and former graphics design student Daniel Goffin. Perlinoid is scheduled for release in 2016. The finished game currently consists of 40 levels, amounting for two hours of gameplay. Such a relatively short play time can be received with negative criticism by both players and press. This paper describes the attempt to add automatically generated content to the game Perlinoid by means of graph transformation. On top of the 40 basic levels there is a quick play mode where on the fly generated levels are offered to the player.

The content generating system is based on a new graph transformation engine that is easily adjustable to incorporate new game rules as well as to modify existing rules. It is also used for verification and analysis of the generated content with the help of reachability graphs a la Groove [Ren03].

The paper is structured as follows. Section 2 introduces the basic rules of Perlinoid. Section 3 describes the steps and contained rewrite rules, which have to be applied for generating a level. Section 4 briefly presents the graph transformation engine before Sect. 5 concludes the paper and gives an outlook on future work.

2   Game Rules

Figure 1 shows a riddle of the game Perlinoid, the gameplay elements are named according to the included notation.

In order to solve a level, the player has to navigate each character to one of the exits. An exit, however, can only be used by a single character. This leads to the difficulty of assigning each of the characters to the right exit.

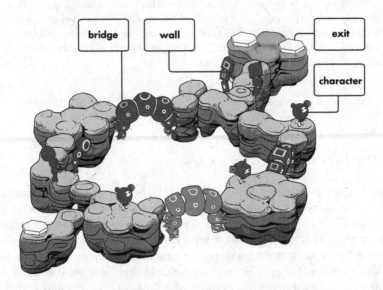

Fig. 1. A complete level of Perlinoid (Color figure online)

The levels consist of platforms, which are connected by obstacles. Characters can cross obstacles to get from one platform to another, if certain conditions are met.

Characters and obstacles are colored with one of the eight colors red, blue, yellow, purple, orange, green, black and white. Red, blue and yellow are described as 'primary colors', purple, orange and green are referred to as 'secondary colors'. Each color can be described as a set of primary-colors, which is referred to as 'sub-colors'.

Secondary colors consist of two primary colors. For example, the sub-colors of purple are red and blue. Primary colors only contain themselves in their sub-colors, the sub-colors of white are empty and black contains all primary-colors in its sub-colors.

There are two types of obstacles: bridges and walls. A bridge can be crossed by a character if all sub-colors of the bridge are contained in the sub-colors of the character.

For example, a red bridge as shown in Fig. 2 can be crossed by a red character. Purple characters can cross red bridges as well, since purple also contains red as a sub-color. The red bridge, however, will reject a blue character for its lack of red in its sub-colors. Figure 2 shows a small level in two different states. The red and the purple character can move from the left to the right platform and back again, while the blue character is not able to leave the platform by itself.

Fig. 2. Cross-Bridge-Rule (Color figure online)

Walls, as depicted in Fig. 4, act exactly the opposite of bridges. Characters can only cross a wall if none of their sub-colors are contained in the sub-colors of the wall. As seen in Fig. 3 this allows the blue character to cross the red wall, while the red and purple characters will be rejected by it.

Fig. 3. Cross-Wall-Rule (Color figure online)

Fig. 4. Combine-and-Split-Rule (Color figure online)

Two characters can be combined if they are colored in different primary colors. The resulting character will have a secondary color with the two primary colors as sub-colors. As seen in Fig. 4, the red and blue characters are combined and replaced by a purple character. This action can be reversed at any time, which is referred to as 'splitting'. A purple character, for example, can be split into a red and blue character.

3   Level Generation

The level generator is able to create a new level from multiple configurable parameters. With these parameters it is possible to influence the number of platforms, the maximal amount of obstacles and the difficulty of the level.

The generation process works in four phases. First, in the layout phase, a number of platforms are placed on the game field and connected by abstract obstacles. This generates the basic topology of our riddle. The second phase, the content generation phase, replaces abstract obstacles by specific obstacles, i.e. bridges or walls. In the third phase the characters and exits are placed. The last phase analyzes the generated riddles and removes unattractive and easy ones. All steps of the generation are done by a graph transformation engine.

The rewrite rules used in this paper are in a *Fujaba* story diagram notation [FNTZ98]. This notation describes rewrite rules as object diagrams. In general, a rewrite rule consists of a left-hand side, the pattern graph, application context conditions and a right-hand side, the replacement graph. A rewrite rule in *Fujaba* notation combines all of these elements in a single extended object diagram. Matching nodes and links, colored black (e.g. the *e2:Element* in Fig. 5), and delete nodes and links, annotated with ≪delete≫ and colored red (e.g. the *e1:Element* and the link between *e1* and *e2* in Fig. 5) make up the left-hand side. The application context conditions are a set of boolean conditions and variable assignments, depicted next to the objects. All boolean conditions have to be true to execute the graph transformation. The matching nodes and links, colored black, and green creation nodes (annotated with ≪create≫) and link (e.g. the element *e3* and the links between *e2* and *e3* in Fig. 5) describe the right hand side.

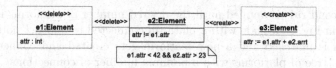

Fig. 5. Example rewrite rule (Color figure online)

3.1   Level Layout Generation

Due to restrictions on screen size, user perspective and art work, the basic level layout is based on three rows of hexagonally connectable platforms with three or four platforms per row, cf. Fig. 6. In the first step this hexagon grid is modeled as the basic topology graph. Note that the left and right center nodes are linked to the maximum amount of neighbors.

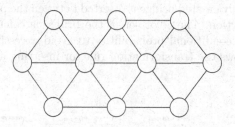

Fig. 6. Hexagonal layout host graph

Three graph transformation rules are applied on the basic topology graph in order to place the desired amount of platforms and platform connections into the basic topology. By applying the first rule, an initial platform is placed on a topology node as shown in Fig. 7a. One of the topology nodes, e.g. the left center node is matched and a new platform gets created and connected to the topology node.

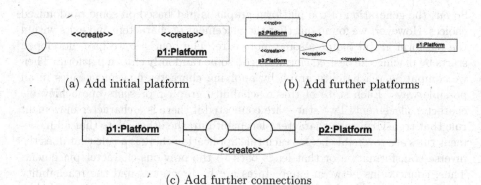

(a) Add initial platform   (b) Add further platforms

(c) Add further connections

Fig. 7. Rulegraphs for 'Add platforms on game field'

The second rule is applied until the desired number of platforms is reached. As shown in Fig. 7b a node with a platform connected to another node without a platform is matched and a new platform gets created and connected. This results in a tree of platforms with a minimal number of connections.

To add more complexity by obtaining more connections, rule three is applied a random number of times creating up to a full mesh of connections. As shown in Fig. 7c two platforms on adjacent nodes get matched and connected.

3.2   Level Content Generation

When the underlying platform graph has been generated, precolored bridges and/or walls get placed in the platform graph which is the new host graph for this generation step. This is achieved by applying one of two transformation rules in the platform graph. The transformation rule for inserting a bridge is shown in Fig. 8a where platform *p1* and platform *p2* are matched as two connected platforms and a newly created bridge is inserted between the platforms, replacing the previous connection. The color set of the bridge is calculated by calling a method which uses conditional probabilities to avoid excessive use of a certain color. Figure 8b shows the transformation rule for inserting a wall, which works correspondingly.

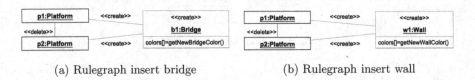

(a) Rulegraph insert bridge   (b) Rulegraph insert wall

Fig. 8. Rulegraph for inserting bridges and walls

3.3   Level Reachability Graph

So far, the generation of the platform graphs is just based on some randomized choices. However, we found out that the placement of characters and exits within the platform graph requires some more care. The heuristic we have developed starts by placing the desired characters on some randomly chosen platform. Then we compute a reachability graph by applying character movement rules in all possible ways. Each state of this reachability graph represents one achievable character placement. Two states are connected if there is a character movement rule that transforms one character placement into the other. Note that all movement rules are reversible i.e. for each rule application there is a rule that does the inverse transformation or that leads back to the previous character placement. Thus, connections between game states are bi-directional and the reachability graph contains no dead ends and from each state every other state is reachable. Figure 12 shows two example structures of such reachability graphs.

Later on, we search this reachability graph for a pair of game states with a large distance to each other. A pair of such states is then chosen as start and as end state, i.e. the characters are placed according to the start state and exits are added according to the end state. The idea is that the player has to go through the whole path of game states from the start to the end state in order to solve the riddle.

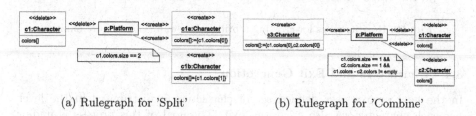

(a) Rulegraph for 'Split'   (b) Rulegraph for 'Combine'

Fig. 9. Rewrite rules for splitting and combining characters (Color figure online)

Figure 9a shows the graph transformation rule for splitting a character according to the game rule in Fig. 4. The matching condition of the rule says, there has to be a character $c1$ on a platform p. If a match is found, the application condition is validated. The condition $c1.colors.size == 2$ says, that the matching character $c1$ is required to have a color set of exactly two primary colors, e.g. blue and red for a purple character. If the application condition holds true, the graph transformation is executed. Two new character objects $c1a$ and $c1b$ will be created and linked to platform p. The first new character $c1a$ gets the first sub-color of the previously matched character $c1$, e.g. red. The second new character $c1b$ gets the second sub-color, e.g. blue. The transformation execution is finished after deleting the old character $c1$.

Figure 9b shows the graph transformation rule for combining two characters $c1$ and $c2$, each colored a different primary color, matching Fig. 4, this time considering the arrow from left to right, labeled "Combine". The first and second application condition $c1.colors.size == 1$ and $c2.colors.size == 1$ ensure that the matching characters are colored in a primary color. The third part of the condition $c1.colors - c2.colors != empty$ is a set operation. It says that all colors of $c1$ excluding all colors of $c2$ should not result in the empty set. This ensures that the characters are colored in different colors. For example, excluding the colors from a red character, from the colors of another red character would result in the empty set. Therefore they cannot be combined.

Figure 10a shows the graph transformation for the 'Cross-Bridge' rule shown in Fig. 2. The application condition $b1.colors - c1.colors == empty$ ensures that the character's colors contain all colors of the bridge.

Figure 10b shows the game rule for crossing a wall. It says that excluding the colors of $w1$ from the colors of $c1$, needs to result in the colors of $c1$.

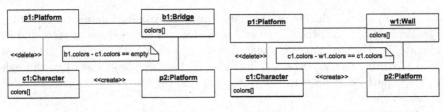

(a) Rulegraph for 'Cross Bridge'   (b) Rulegraph for 'Cross Wall'

Fig. 10. Rewrite rules for crossing a bridge or a wall (Color figure online)

3.4   Characters and Exit Generation

In the next step a user defined set of characters is distributed in the level. For each character we also create an exit. The goal of this process is finding a character and exit distribution that makes best use of the given structure of platforms and obstacles. To achieve this, at first all characters are placed on one platform. After this, the reachability graph for the level is calculated. Inside of this graph, two nodes with the largest possible distance need to be found. The path between those two nodes can already be used as the solution path for the level.

In most cases, however, this procedure will lead to a shorter solution depth than expected. Figure 11 shows a generated level with a solution depth of zero, despite the chosen pair of nodes having a distance of 4. The problem occurs, because of the exits not being bound to specific colors. To avoid generating levels with solution-shortcuts, a token for each state of the solution graph is

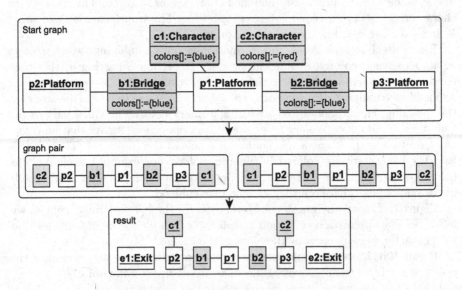

Fig. 11. Generated level with solution depth zero (Color figure online)

calculated, which stores information about the number of characters on each platform. The nodes of the pair shown in Fig. 11 for example would both be labeled with the token '1-0-1', each number refering to the amount of characters on one of the platforms.

To avoid solution-shortcuts, it is required to choose a solution-path, where the token, which labels the solution-state, is not assigned to any other states in the solution path.

3.5   Level Filter

The amount of levels that can possibly be constructed by platforms, obstacles, characters and exits is huge, even if the boundaries are very limited. Not all of the constructable levels are solvable, and most of them are not fun to play at all. In order to provide players with enjoyable levels, the riddles have to fulfill certain criteria. They should be challenging, while keeping the amount of used gameplay elements as small as possible. Therefore, generated levels need to be evaluated, in order to discard deficient riddles.

To evaluate the difficulty of a given level, information derived from the reachability graph can be analyzed.

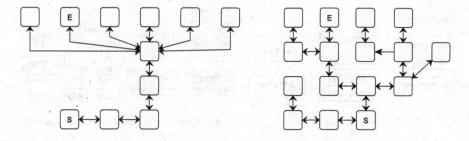

Fig. 12. Reachability graphs

The two reachability graphs in Fig. 12 represent the states of two constructable levels. Nodes marked with an 'S' contain the initial state, an 'E' marks the solution. An important indication for the difficulty of a level is the distance between the start and the end state, which will be referred to as 'solution depth'. For example, the graph on the left in Fig. 12 has a solution depth of 5, the right graph has a depth of 4. However, the level represented by the left graph is not considered the more difficult level. This is because most of its states are trivial, since they are only connected to two or less other nodes. In such states, the player does not have to make an actual decision, as he only can apply one single rule to manipulate the current state of the game. Therefore, the average number of neighbors for each node is another important aspect of evaluating the difficulty of a level. Last but not least, the evaluation is influenced by the total number of nodes in the reachability graph. A higher amount of possible states in a riddle requires to remember and analyze more information, which the developers consider to be the most challenging part of the game in general.

In generated levels it is possible, that some of the used platforms or obstacles are irrelevant for solving the riddle. Figure 13a shows a simple riddle, that can be solved in one step, by navigating the character $c1$ from platform $p1$ to $p3$. This demonstrates, that the elements $b1$ and $p2$ can be ignored by the player, which is considered a deficiency of the riddle. The level in Fig. 13b shows a more complex example for unused elements. The shortest solution for it requires the player to use all of its elements. However, the riddle alternatively can be solved by ignoring the purple bridge $b3$ and simply moving all characters over the red bridges. Therefore, finding unused elements for the riddles requires analyzing all possible solutions, not only the fastest one.

Levels also need to be discarded in case of so-called 'action-chain-repetitions'. An 'action-chain' shall be defined as a specific application of a sequence of rules. If the reachability graph of a level contains multiple similar action-chains in one of its solution-paths, this is called 'action-chain-repetition'. An example for this can be found in Fig. 14, where the red character c3 has to combine with a blue character and walk back over the bridge b1, to split up into red and blue again. This sequence has to be repeated two times in order to solve the level.

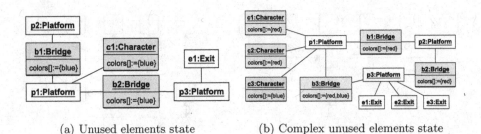

(a) Unused elements state   (b) Complex unused elements state

Fig. 13. Two examples for unused elements (Color figure online)

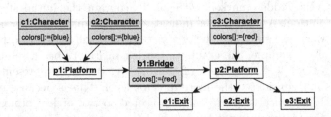

Fig. 14. Example for an action-chain-repetition (Color figure online)

4   Graph Transformation Engine

Perlinoid was developed with the game engine Unity3D and written in C#. In order to tightly integrate the generation of content into the game, the graph transformation engine had to be written in C# as well.

Unfortunately, Fujaba [FNTZ98] and SDMLib [SDM14] as of now only support Java. Existing C# graph engines like GrGen [GK07] and GraphSynth [Gra06] had licensing issues for usage in a commercially published product like Perlinoid or were not suitable. However, after analyzing the alternatives the developers of Perlinoid decided that developing a new graph rewriting system would be a smaller amount of work, considering the special requirements for the targeted system. The obstacle generation for example requires the use of graph-rewrite rules that support creating a node with a random set of attributes, as well as applying rules by a certain probability. Furthermore, when creating the reachability graphs information about the applied rules have to be stored. This information is used in the filters, for example when action-chain-repetitions are identified.

Developing a new graph rewriting system was facilitated by tailoring it for the Perlinoid game. We did not need to build e.g. code generation for the implementation of user defined graph models but it sufficed to deal with the Perlinoid specific graph model. Similarly, our graph rewriting system does not need to support general attribute conditions but Perlinoid requires only special conditions on sets of colors. By building a special purpose graph rewriting system, it was easy to package it with the rest of the Perlinoid game and to deliver it to the end user such that the end user may generate new levels on demand within the game.

Building a new graph rewriting system was also facilitated by following the simple architecture proposed in [Gra]. [Gra] is a regular master course given by Albert Zündorf. In the first lecture it covers the implementation of a generic graph model. The second lecture covers the implementation of a model for graph rewrite rules. Then there are two lectures for implementing graph pattern matching. Each week there is an assignment to implement the covered functionality in your own graph engine implementation. This needs about 6 h per lecture. Thus, in about 24 h of development work you have a basic graph engine. Reachability graphs need another two lectures as you need to implement graph certificate computation for hashing the graphs and a graph isomorphism check. Thus, within about 40 h you are done. Note, this does not include a graphical user interface, but graphs and rules are created by normal setup code. Fortunately Perlinoid only needs basic graph rewrite rules. Complex negative application conditions or amalgamation concepts require more effort.

5   Conclusions and Future Work

Overall, modeling the Perlinoid game with graph transformation rules was very simple and intuitive. In addition, the usage of graph transformation rules enabled the computation of reachability graphs for the analysis of the complexity of generated riddles. The reachability graphs are an ideal basis for the analysis of solution possibilities of a given game level. Different kinds of graph algorithms and metrics like shortest paths, distances, cyclomatic numbers, etc. may be applied. This idea is easily re-used in future puzzle based games. One just models game states as graphs and possible user actions as graph transformation rules and then a reachability graph may be computed and analyzed for interesting paths and features. One may also add some heuristics for the exploration of the reachability graph e.g. to employ an A* algorithm for discovering only high rated parts of the reachability graph. In general, graph transformations provide an ideal basis for such an analysis. Validation of generated content is done by solving the riddles. The levels are then classified by calculating the solution depth and more on the reachability graph. Quality assurance is done by the filters, that are running at the very end of the generation process.

The level generator will be shipped together with the game. The implementation uses Unity3D and the first version will run on PCs and usual gaming consoles (supported by Unity3D by default). As players will not wait very long, level generation needs to be done in some seconds. This is feasible by restricting us to limited level sizes.

The final game incorporates more colors, more mechanics and more rules. Some of these have already been added to the graph-transformation engine, while others still need to be added to the game and to be shipped with the level generator. A bigger challenge remains, namely adjusting the generator for not immediately obvious features, such as creating memorable levels. This is something that a human level designer does when manually creating levels, sometimes adding useless parts from a game mechanics perspective. Even more challenging is the creation of combinations that are suggestive in a way that has the player try the wrong things first, as these levels tend to be more rewarding once solved. It remains to be proven that these combinations can be either generated by graph transformation or at least identified in the filter based level selection process. There are more features like these, e.g. creating levels with only one color or creating levels with a pleasing combination of obstacles.

Acknowledgments. Perlinoid is (c) 2016 by David Priemer and Daniel Goffin. All ingame assets were created by Daniel Goffin.

References

[FNTZ98] Fischer, T., Niere, J., Torunski, L., Zündorf, A.: Story diagrams: a new graph rewrite language based on the unified modeling language and java. In: Ehrig, H., Engels, G., Kreowski, H.-J., Rozenberg, G. (eds.) TAGT 1998. LNCS, vol. 1764, pp. 296–309. Springer, Heidelberg (2000)

[GK07] Geiß, R., Kroll, M.: GrGen.NET: a fast, expressive, and general purpose graph rewrite tool. In: Schürr, A., Nagl, M., Zündorf, A. (eds.) Applications of Graph Transformations with Industrial Relevance. LNCS, vol. 5088, pp. 568–569. Springer, Heidelberg (2008)

[Gra]   Graph Engineering Course at Kassel University
[Gra06] GraphSynth. Open-source research software for creating generative grammars (2006). http://designengrlab.github.io/GraphSynth/. 17 Feb 2016
[Ren03] Rensink, A.: The GROOVE Simulator: A Tool for State Space Generation. In: Pfaltz, J.L., Nagl, M., Böhlen, B. (eds.) Applications of Graph Transformations with Industrial Relevance. LNCS, vol. 3062, pp. 479–485. Springer, Heidelberg (2003)
[SDM14] SDMLib.org. Story Driven Modeling Library at GitHub (2014). https://github.com/fujaba/SDMLib, 17 Feb 2016

Graph Transformation Meets Reversible Circuits: Model Transformation and Optimization

Hans-Jörg Kreowski, Sabine Kuske, Aaron Lye$^{(\boxtimes)}$, and Caro von Totth

Department of Computer Science, University of Bremen,
P.O.Box 33 04 40, 28334 Bremen, Germany
{kreo,kuske,lye,caro}@informatik.uni-bremen.de

Abstract. Reversible circuits provide the subject of a new promising direction of circuit design. Reversible circuits are cascades of reversible gates specifying bijective functions on Boolean vectors. As one encounters quite a variety of reversible gates in the literature, there are many classes of reversible circuits. Two main problems are considered: (1) How can circuits of one class be transformed into the ones of another class? (2) How can circuits within one class be optimized with respect to certain measures? While reversible circuits are studied on the functional level and on the level of propositional calculus, there is also a visual representation used frequently for illustrative purposes in an informal way. In this paper, the visual description of reversible circuits is formalized by means of graph transformation. In particular, it is shown that the problems of model transformation and optimization can be investigated within the graph-transformational framework. This continues the authors' earlier work on the generation, evaluation and synthesis of reversible circuits as graphs.

1 Introduction

Reversible circuits including quantum circuits provide the subject of a new promising direction of circuit design. Reversible circuits are cascades of reversible gates specifying bijective functions on Boolean vectors. As one encounters quite a variety of reversible gates in the literature like Toffoli gates with positive and negative control lines [1], Fredkin gates [2] and others, there are many classes of reversible circuits. These underlying gate classes are also called gate libraries. Two main problems are considered concerning model transformation, optimization and verification. (1) How can circuits of one class be transformed into the ones of another class? For example, Toffoli circuits with positive and negative control lines can be transformed into Toffoli circuits with positive control lines only. And the latter can be transformed into Toffoli circuits with two or less control lines. (2) How can circuits within one class be optimized with respect to certain measures? A typical measure is the number of gates in a circuit. If the gate costs matter, one tries to find equivalent circuits of shorter length. And there are other more sophisticated measures and criteria in which respects

© Springer International Publishing Switzerland 2016
R. Echahed and M. Minas (Eds.): ICGT 2016, LNCS 9761, pp. 236–251, 2016.
DOI: 10.1007/978-3-319-40530-8_15

optimization is considered. A very interesting side effect of model transformation and optimization is that the underlying procedures preserve functional equivalence. Hence, circuits that can be transformed into each other are proved to be equivalent at the same time.

While reversible circuits are studied on the functional level and on the level of propositional calculus, there is also a visual representation used frequently for illustrative purposes in an informal way. In this paper, the visual description of reversible circuits is formalized by means of graph transformation. In particular, it is shown that the problems of model transformation and optimization can be investigated within the graph transformational framework. This continues the authors' earlier work on the generation, evaluation and synthesis of reversible circuits as graphs [3]. Besides the topic of model transformation that is not addressed in [3], there are further differences. We consider in this paper arbitrary reversible gates rather than Toffoli gates only. Moreover, an algebraic operation is introduced for the generation of reversible circuit graphs as an alternative to a rule-based definition. As far as we know, model transformation is not yet studied systematically in the area of reversible circuits. But in recent years, several rewrite rules have been proposed in the literature (see, e.g., [4–7]) establishing a spectrum of single transformations of reversible circuits. Confer Sect. 5 for examples.

Considering reversible circuits as graphs and processing on circuits as graph transformation has several advantages: (1) The set-theoretical description is advanced to a more descriptive and visible formulation that supports the intuition. (2) The rewrite rules that one encounters in the literature as already mentioned above provide a spectrum of examples of circuit transformations on the graph level so that they can be studied in our framework in a uniform way. (3) Our approach allows to apply the wealth of graph transformation methods and results to the area of reversible circuits. In particular, the parallelization theorems, the critical-pair analysis and the results on termination and confluence are expected to lead to new insights. (4) Encoded graph transformation rules can be executed by graph transformation tools. These tools are optimized to finding matches and replacing subgraphs.

The paper is organized as follows. After the graph-transformational preliminaries in Sects. 2 and 3 presents reversible functions, gates, and circuits. Section 4 introduces graph representations of reversible circuits. Section 5 shows exemplarily how reversible circuits can be transformed using graph transformation units. In Sect. 6, circuit graph transformation units are discussed in detail and some fundamental properties are shown. Section 7 contains a short conclusion.

2   Graph-Transformational Preliminaries

In this section, the basic notions and notations of graph transformation units are recalled as far as needed (see, e.g., [8]).

A *graph* over a set Σ is a system $G = (V, E, s, t, l)$ where V is a finite set of *nodes*, E is a finite set of *edges*, $s, t: E \rightarrow V$ are mappings assigning a *source*

$s(e)$ and a *target* $t(e)$ to every edge in E, and $l: E \rightarrow \Sigma$ is a mapping assigning a *label* to every edge in E. An edge e with $l(e) = x$ is an *x-edge*; if $s(e) = t(e)$, it is also called an *x-loop* or a *loop*. The components V, E, s, t, and l of G are also denoted by V_G, E_G, s_G, t_G, and l_G, respectively. The set of all graphs over Σ is denoted by $\mathcal{G}_\Sigma$. We reserve a specific label $*$ which is omitted in drawings of graphs.

For graphs $G, H \in \mathcal{G}_\Sigma$, G is a *subgraph* of H, denoted by $G \subseteq H$, if $V_G \subseteq V_H$, $E_G \subseteq E_H$, $s_G(e) = s_H(e)$, $t_G(e) = t_H(e)$, and $l_G(e) = l_H(e)$, for each $e \in E_G$. A *graph morphism* $g: G \rightarrow H$ is a pair of mappings $g_V: V_G \rightarrow V_H$ and $g_E: E_G \rightarrow E_H$ such that $g_V(s_G(e)) = s_H(g_E(e))$, $g_V(t_G(e)) = t_H(g_E(e))$, and $l_H(g_E(e)) = l_G(e)$ for all $e \in E_G$. If the mappings g_V and g_E are bijective, then G and H are *isomorphic*, denoted by $G \cong H$. For a graph morphism $g: G \rightarrow H$, the image $g(G) \subseteq H$ of G in H is called a *match* of G in H.

A *rule* $r = (L \supseteq K \subseteq R)$ consists of three graphs $L, K, R \in \mathcal{G}_\Sigma$ such that K is a subgraph of L and R. The components L, K, and R of r are called *left-hand side*, *gluing graph*, and *right-hand side*, respectively.

The application of $r = (L \supseteq K \subseteq R)$ to a graph $G = (V, E, s, t, l)$ yields a directly derived graph H and consists of the following three steps.

1. A match $g(L)$ of L in G is chosen subject to the following conditions.
   - *dangling condition*: $v \in g_V(V_L)$ with $s_G(e) = v$ or $t_G(e) = v$ for some $e \in E_G - g_E(E_L)$ implies $v \in g_V(V_K)$.
   - *identification condition*: $g_V(v) = g_V(v')$ for $v, v' \in V_L$ implies $v = v'$ or $v, v' \in V_K$ as well as $g_E(e) = g_E(e')$ for $e, e' \in E_L$ implies $e = e'$ or $e, e' \in E_K$.
2. Now the nodes of $g_V(V_L - V_K)$ and the edges of $g_E(E_L - E_K)$ are removed yielding the *intermediate graph* $Z \subseteq G$.
3. Let $d: K \rightarrow Z$ be the restriction of g to K and Z. Then H is constructed as the disjoint union of Z and $R - K$ where all edges $e \in E_Z + (E_R - E_K)$ keep their labels and their sources and targets except for $s_R(e) = v \in V_K$ or $t_R(e) = v \in V_K$ which is replaced by $d_V(v)$.

The application of a rule r to a graph G is denoted by $G \underset{r}{\Longrightarrow} H$ and called a *direct derivation*. The sequential composition of direct derivations

$$d = G_0 \underset{r_1}{\Longrightarrow} G_1 \underset{r_2}{\Longrightarrow} \cdots \underset{r_n}{\Longrightarrow} G_n \quad (n \in \mathbb{N})$$

is called a *derivation* from G_0 to G_n. If $r_1, \ldots, r_n \in P$ (for some set P of rules), d can be denoted as $G_0 \underset{P}{\overset{*}{\Longrightarrow}} G_n$.

A *graph class expression* may be any syntactic entity X that specifies a class $SEM(X) \subseteq \mathcal{G}_\Sigma$. A typical example is $reduced(P)$ for some set of rules P specifying the graphs that are reduced with respect to P, i.e. $SEM(reduced(P)) = \{G \mid$ there is no $G \underset{r}{\Rightarrow} H$ for $r \in P\}$. Another example of a graph class expression specific for reversible circuits is introduced in Sect. 5.

A *control condition* may be any syntactic entity that restricts the non-determinism of the derivation process. A typical example is *as long as possible*.

It requires that the rules are applied as long as possible. The fact that a derivation $G \stackrel{*}{\Rightarrow}_{P} H$ is permitted by a control condition C is denoted by $G \stackrel{*}{\Rightarrow}_{P,C} H$.

A *graph transformation unit* is a system $gtu = (I, P, C, T)$, where I and T are graph class expressions which specify initial and terminal graphs respectively. P is a finite set of rules, and C is a control condition. The operational semantics of gtu is the set of all derivations from initial to terminal graphs that apply rules in P and satisfy the control condition, i.e.

$$DER(gtu) = \{G \stackrel{*}{\Rightarrow}_{P,C} H \mid (G, H) \in SEM(I) \times SEM(T)\}.$$

Restricting the operational semantics to the initial and terminal graphs yields the input-output semantics, i.e. $SEM(gtu) = \{(G, H) \mid G \stackrel{*}{\Rightarrow}_{P,C} H \in DER(gtu)\}$.

In explicit examples, the components I, P, C, and T of a graph transformation unit are preceded by the keywords *initial*, *rules*, *control*, and *terminal*, respectively.

3 Reversible Functions, Gates and Circuits

In this section, the basic notions and notations of reversible functions and their specification by reversible gates and circuits are recalled (see, e.g., [1]).

Let $\mathbb{B} = \{0, 1\}$ be the set of truth values with the negations $\bar{0} = 1$ and $\bar{1} = 0$, and let ID be a set of identifiers serving as a reservoir of Boolean variables. Let $\mathbb{B}^X$ be the set of all mappings $b: X \to \mathbb{B}$ for some $X \subseteq ID$ where the elements of $\mathbb{B}^X$ are called *Boolean assignments*. Then a bijective Boolean (multi-output) function $f: \mathbb{B}^X \to \mathbb{B}^X$ is called *reversible*.

If $X = [n] = \{1, \ldots, n\}$ for some $n \in \mathbb{N}$ with $[0] = \emptyset$, then one may write $\mathbb{B}^n$ instead of $\mathbb{B}^{[n]}$ and the elements $b \in \mathbb{B}^n$ can be denoted as Boolean vectors $(b_1, \ldots, b_n) = (b(1), \ldots, b(n))$.

Each reversible function f can be represented by its truth table $(b, f(b))_{b \in \mathbb{B}^X}$. As this table has 2^n entries if X has n elements, one is interested in smaller representations which are given by reversible gates and reversible circuits in many cases.

For example, the function $C^k NOT: \mathbb{B}^{k+1} \to \mathbb{B}^{k+1}$ for some $k \in \mathbb{N}$ is defined by $C^k NOT(b_1, \ldots, b_{k+1}) = (b_1, \ldots, b_k, b_1 \cdots b_k \oplus b_{k+1})$ for all $(b_1, \ldots, b_{k+1}) \in \mathbb{B}^{k+1}$ where xy is the logical *AND* of x and y and $x \oplus y$ the logical *XOR*. In other words, $C^k NOT$ negates b_{k+1} provided that $b_1 = \cdots = b_k = 1$. All other values are kept invariant. Therefore, $C^k NOT$ is fully specified by its name and the parameter k where the components $1, \ldots, k$ are called *control lines* and the component $k+1$ is called *target line*. Note that $C^0 NOT$ is defined as the negation NOT, and $C^1 NOT$ and $C^2 NOT$ are often denoted by $CNOT$ and $CCNOT$.

Similarly, the function $C^k \overline{C}^l NOT: \mathbb{B}^{k+l+1} \to \mathbb{B}^{k+l+1}$ for some k, $l \in \mathbb{N}$ is defined by $C^k \overline{C}^l NOT(b_1, \ldots, b_{k+l+1}) = (b_1 \ldots, b_{k+l}, b_1 \cdots b_k \overline{b}_{k+1} \cdots \overline{b}_{k+l} \oplus b_{k+l+1})$ for all $(b_1, \ldots, b_{k+l+1}) \in \mathbb{B}^{k+l+1}$. This means that b_{k+l+1} is negated if $b_1 = \cdots = b_k = 1$ and $b_{k+1} = \cdots = b_{k+l} = 0$ and all other values are kept invariant. The components $1, \ldots, k$ are called *positive control lines* and $k+1, \ldots, k+l$ *negative control lines*.

Another example is the conditional swap $C^m SWAP \colon \mathbb{B}^{m+2} \to \mathbb{B}^{m+2}$ for some $m \in \mathbb{N}$ which swaps the values of the lines $m+1$ and $m+2$ if and only if the control lines $1, \ldots, m$ carry the value 1, i.e. $C^m SWAP(1, \ldots, 1, b_{m+1}, b_{m+2}) = (1, \ldots, 1, b_{m+2}, b_{m+1})$ and $C^m SWAP(b) = b$ for all other $b \in \mathbb{B}^{m+2}$.

Such simple functions with simple representations can be used to build more complicated functions by renaming the variables and employing the obvious fact that the sequential composition $g \circ f$ of reversible functions f and g over the same set of variables is again a reversible function.

Let $f \colon \mathbb{B}^U \to \mathbb{B}^U$ be a reversible function and $X \subseteq ID$. Then an injective mapping $r \colon U \to X$ is called *renaming* which induces a *reversible extension* $f_r \colon \mathbb{B}^X \to \mathbb{B}^X$ of f with $b \in X$ by $f_r(b)(r(y)) = f(b \circ r)(y)$ for all $y \in U$ and $f_r(b)(x) = b(x)$ for $x \in X - r(U)$.

The pair (f, r) (and any representation of it) is called a *reversible gate* over X. Usually, one tries to keep gates small by using small sets U and simple functions on $\mathbb{B}^U$ like the examples above.

Let $(f_i \colon \mathbb{B}^{U_i} \to \mathbb{B}^{U_i}, r_i \colon U_i \to X)$ for $i = 1, \ldots, n$ be reversible gates over X, then the sequence $rc = (f_1, r_1) \cdots (f_n, r_n)$ is a *reversible circuit* over X which specifies the reversible function $f_{rc} = (f_n)_{r_n} \circ \cdots \circ (f_1)_{r_1}$. This means that each choice of a class RF of reversible functions and a set X of Boolean variables induces a class $\Gamma(RF, X) = \{(f \colon \mathbb{B}^U \to \mathbb{B}^U, r \colon U \to X) \mid f \in RF, r \text{ injective}\}$ of reversible gates and a class $\Gamma(RF, X)^+$ of reversible circuits over X. Two such circuits are called *(functionally) equivalent* if they represent the same function.

Consider, for example, the set of reversible functions $TOF_2 = \{C^k NOT \mid k \leq 2\}$. Then $\Gamma(TOF_2, X)$ and $\Gamma(TOF_2, X)^+$ are the classical classes of Toffoli gates and Toffoli circuits respectively (cf. [1]). Similarly, $TOF_k = \{C^i NOT \mid i \leq k\}$ for some $k \in \mathbb{N}$ and $TOF_{k,l} = \{C^i \overline{C}^j NOT \mid i \leq k, j \leq l\}$ for some $k, l \in \mathbb{N}$ yield the Toffoli gates and circuits with up to k control lines and up to k positive as well as up to l negative control lines respectively.

It should be noted that a Toffoli gate of the form $(C^k NOT, r)$ is usually represented by its target line $t = r(k+1) \in X$ and by its set of control lines $C = \{r(i) \mid i = 1, \ldots, k\} \subseteq X$ and is denoted by $TG(t, C)$. Similarly, $(C^k \overline{C}^l NOT, r)$ may be represented by $TG(t, C, \overline{C})$ with $t = r(k+l+1), C = \{r(i) \mid i = 1, \ldots, k\}$ and $\overline{C} = \{r(i) \mid i = k+1, \ldots, k+l\}$. The Toffoli circuits over X based on TOF_k for $k \geq 2$ and $TOF_{k,l}$ for $k+l \geq 2$ are universal in the sense that every reversible function on $\mathbb{B}^X$ can be specified by such a circuit [1].

Another typical example is the Fredkin gate of the form $(C^m SWAP, r)$ for some $m \in \mathbb{N}$ and a renaming $r \colon [m+2] \to X$ which is often represented by $FG(t_1, t_2, C)$ with $t_1 = r(m+1), t_2 = r(m+2)$ and $C = \{r(i) \mid i = 1, \ldots, m\}$. Let $FRE_m = \{C^i SWAP \mid i \leq m\}$ for some $m \in \mathbb{N}$. Then $\Gamma(FRE_m \cup TOF_{k,l}, X)^+$ is the set of reversible circuits consisting of Fredkin gates with up to m control lines and Toffoli gates with up to k positive and up to l negative control lines.

4 Graph Representation of Reversible Circuits

In the literature on reversible circuits, one often finds visual representations of gates and circuits for illustrative purposes. A typical example is depicted in Fig. 1.

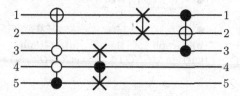

Fig. 1. The circuit $TG(1, \{5\}, \{3, 4\}) FG(3, 5, \{4\}) FG(1, 2, \emptyset)\ TG(2, \{1, 3\})$

The variables $1, \ldots, 5$ are drawn as horizontal lines, target lines are indicated by $\oplus$, positive control lines by $\bullet$, negative control lines by $\circ$ and the swapped lines are denoted by $\times$. The vertical lines connect the components that belong to a gate. These representations can be formalized as graphs which provide the basic structures for the rule-based approach to circuit transformation and optimization as considered in the next sections.

The circuit graphs are defined by sequential composition of gate graphs. For this purpose, the gate graphs carry *in-* and *out-*loops. Moreover, a gate graph extends the graph representation of a reversible function in the following way.

Let $f: \mathbb{B}^U \to \mathbb{B}^U$ be a reversible function. Then the graph $gr(f)$ of f is depicted in Fig. 2(a) for $U = [k]$ and in 2(b) for the special case of $CCNOT$. For an arbitrary U, it has a node with an f-loop and incoming edges for each variable $x \in U$ labeled by x as well as outgoing edges for each variable labeled by this variable. The sources of the incoming edges and the targets of the outgoing edges are separate nodes. So one has a node and a loop corresponding to f and two copies of the variables as nodes and two such copies as edges, i.e. $gr(f) = (U \times \{in, out\} \cup \{f\}, U \times \{in, out\} \cup \{f\}, s_f, t_f, l_f)$ with $s_f((y, in)) = (y, in)$, $t_f((y, in)) = f = s_f((y, out))$, $t_f((y, out)) = (y, out)$, $l_f((y, in)) = l_f((y, out)) = y$ for all $y \in U$ and $s_f(f) = t_f(f) = l_f(f) = f$. In drawings, the node with the f-loop is represented as a rectangle with an f inside.

Let $(f, r: U \to X)$ be a reversible gate. Then $gr(f)$ can be extended to a graph $gr(f, r)$ by adding nodes for each variable $X - r(U)$ and by adding *in-*loops and *out-*loops for each variable to corresponding nodes, i.e. $gr(f, r) = gr(f) + (X - r(U), X \times \{in, out\}, s_{f,r}, t_{f,r}, l_{f,r})$ with $s_{f,r}((x, in)) = s_{f,r}((x, out)) = x = t_{f,r}((x, in)) = t_{f,r}((x, out))$ and $l_{f,r}((x, in)) = in_x$, $l_{f,r}((x, out)) = out_x$ for $x \in X - r(U)$ as well as $s_{f,r}((r(y), in)) = (y, in) = t_{f,r}((r(y), in))$, $s_{f,r}((r(y), out)) = (y, out) = t_{f,r}((r(y), out))$ and $l_{f,r}((r(y), in)) = in_{r(y)}$, $l_{f,r}((r(y), out)) = out_{r(y)}$ for all $y \in U$.

In Fig. 2(c), $gr(CCNOT, r_4)$ with the renaming $r_4(1) = 1, r_4(2) = 3$ and $r_4(3) = 2$ for $X = [5]$ is depicted.

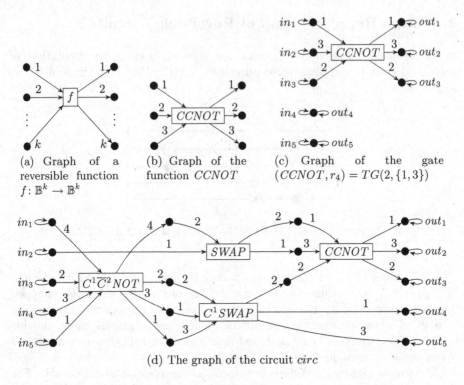

(a) Graph of a reversible function $f \colon \mathbb{B}^k \to \mathbb{B}^k$

(b) Graph of the function $CCNOT$

(c) Graph of the gate $(CCNOT, r_4) = TG(2, \{1,3\})$

(d) The graph of the circuit $circ$

Fig. 2. Graph representations of functions, gates and circuits

Graphs with exactly one in_x-loop and one out_x-loop for each $x \in X$ are called *in-out*-graphs. They allow a simple sequential composition. Let G and H be two such graphs, then the sequential composition $G \circ H$ is an *in-out*-graph obtained by merging the node with the out_x-loop in G with the corresponding node with the in_x-loop in H for each $x \in X$ and by removing these loops. This composition is obviously associative. Moreover, each set of *in-out*-graphs $\mathcal{G}$ induces a closure $\mathcal{G}^{\circ}$ under sequential composition (cf. [9, Chap. 2]), i.e. $\mathcal{G}^{\circ} = \bigcup_{i \geq 1} \mathcal{G}^i$ with $\mathcal{G}^1 = \mathcal{G}$ and $\mathcal{G}^{i+1} = \mathcal{G}^i \circ \mathcal{G}^1$.

Therefore, reversible circuits can be represented by graphs, called *circuit graphs*, that are sequential compositions of the graphs representing the gates of the circuits. Let $rc = rg_1 \cdots rg_n$ be a reversible circuit with the reversible gates rg_i for $i = 1, \ldots, n$. Then rc is represented by the *in-out*-graph $gr(rc) = gr(rg_1) \circ \cdots \circ gr(rg_n)$. Obviously, a circuit graph $gr(rc)$ specifies a reversible functions $f_{gr(rc)} = f_{rc}$. The class of circuit graphs over X is denoted by $\mathcal{CG}_X$.

In Fig. 2(d), the graph

$$gr(circ) = gr(C^1\overline{C}^2 NOT, r_1) \circ gr(C^1 SWAP, r_2) \circ gr(SWAP, r_3) \circ gr(CCNOT, r_4)$$

is shown where $circ = (C^1\overline{C}^2 NOT, r_1)(C^1 SWAP, r_2)(SWAP, r_3)(CCNOT, r_4)$ using the renamings $r_1(1) = 5$, $r_1(2) = 3$, $r_1(3) = 4$, $r_1(4) = 1$, $r_2(1) = 4$,

$r_2(2) = 3$, $r_2(3) = 5$, $r_3(1) = 2$, $r_3(2) = 1$, and r_4 as above. It should be noted that this graph is the formal counterpart to the informal drawing in Fig. 1 as $TG(1, \{5\}, \{3, 4\}) = (C^1\overline{C}^2NOT, r_1)$, $FG(3, 5, \{4\}) = (C^1SWAP, r_2)$, $FG(1, 2, \emptyset) = (SWAP, r_3)$ and $TG(2, \{1, 3\}) = (CCNOT, r_4)$.

It is not difficult to see that the graph representation for reversible circuits preserves the sequential composition. For example, as Toffoli circuits are sequential compositions of Toffoli gates, the graph representations of Toffoli circuits are obtained as the sequential composition of the graphs representing Toffoli gates. In general, the following observation holds.

Observation 1. 1. Let rc_1 and rc_2 be two reversible circuits over X. Then
$gr(rc_1 rc_2) = gr(rc_1) \circ gr(rc_2)$.
2. Let RG be a set of reversible gates. Then $gr(RG^+) = gr(RG)^\circ$ where $gr(RC) = \{gr(rc) \mid rc \in RC\}$ for each set RC of reversible circuits.

While the closure of a set of reversible gates under sequential composition is a free semigroup, the corresponding closure of gate graphs is not free in general. Consider, for example, the gates (C^1SWAP, r_2) and $(SWAP, r_3)$. Both their sequential orders induce the same part of the graph in Fig. 2(d) because they use disjoint sets of lines. Hence, in general, the following observation holds.

Observation 2. Let $(f, r\colon U \to X)$ and $(f', r'\colon U' \to X)$ be two reversible gates with $r(U) \cap r'(U') = \emptyset$. Then $gr((f, r)(f', r')) \cong gr((f', r')(f, r))$.

5   Typical Examples of Reversible-Circuit Transformation

In the literature, one encounters quite a spectrum of transformations on reversible circuits that serve various purposes [1, 4–7]. In this section, we recall two typical examples. Toffoli circuits with positive and negative control lines are transformed into Toffoli circuits without negative control lines. And Fredkin gates are replaced by Toffoli circuits of length 3.

5.1   Getting Rid of Negative Control Lines and Swapping

Given a Toffoli circuit with a negative control line in one of its gates, then this control line can be turned into a positive one without changing the semantics, if one negates the respective line before and after the control takes place. More formally, a Toffoli gate $tg = TG(t, C, \overline{C})$ for $t \in X$ and $C \cup \overline{C} \subseteq X$ and the Toffoli circuit $tc(\overline{c}) = TG(\overline{c}, \emptyset) \, TG(t, C \cup \{\overline{c}\}, \overline{C} - \{\overline{c}\}) \, TG(\overline{c}, \emptyset)$ for some $\overline{c} \in \overline{C}$ describe the same reversible function. Given a Toffoli circuit, the replacement of such gates by the corresponding circuits may be iterated as long as one finds negative control lines. One ends up with a Toffoli circuit that is equivalent to the initial one, but has no longer any negative control lines. The transformation process terminates because the number of negative control lines decreases in each step by 1. As these negative control lines become positive ones, the circuits in $\Gamma(\bigcup_{k+l \leq m} TOF_{k,l}, X)^+$ for each $X \subseteq ID$ and $m \in \mathbb{N}$ are transformed into equivalent circuits in $\Gamma(TOF_m, X)^+$. The transformation yields Toffoli circuits without negative control lines as normal forms of arbitrary Toffoli circuits. It is also an optimization with respect to the number of negative control lines ending up in local minima with 0 negative control lines.

Another example is the replacement of Fredkin gates by Toffoli circuits. More formally, a Fredkin gate $FG(t_1, t_2, C)$ is equivalent to the Toffoli circuit $TG(t_1, \{t_2\}) TG(t_2, C \cup \{t_1\}) TG(t_1, \{t_2\})$. Therefore, every reversible circuit with Fredkin gates can be transformed into one without Fredkin gates. If all other gates are Toffoli gates, then one ends up with Toffoli circuits where the maximum number of control lines is increased by 1 provided that there is a Fredkin gate with the maximum number of control lines. Circuits in $\Gamma(FRE_k \cup TOF_l, X)^+$ are transformed into equivalent circuits in $\Gamma(TOF_{max(k+1,l)}, X)^+$. The transformation can be seen as a local optimization with respect to the number of Fredkin gates. The transformation can be used to get rid of swaps in particular.

5.2   A Graph-Transformational View

Both examples of reversible-circuit transformations can be adequately modeled as graph-transformation units using rules that can be applied to the graph representations of reversible circuits.

The first unit removes negative control lines and may be called *remove-ncl* therefore. The class of initial graphs is $gr(\Gamma(\bigcup_{k+l \leq m} TOF_{k,l}, X)^+)$ for some $X \subseteq ID$ and $m \in \mathbb{N}$ containing all graphs of Toffoli circuits with up to m control lines. It is properly specified by $TOF_{posneg}(m, X)$ as graph class expression, i.e. $SEM(TOF_{posneg}(m, X)) = gr(\Gamma(\bigcup_{k+l \leq m} TOF_{k,l}, X)^+)$. The class of terminal graphs is $gr(\Gamma(TOF_m, X)^+)$ specified by $TOF(m, X)$.

To reflect the replacement of $tg = TG(t, C, \overline{C})$ by $tc(\overline{c}) = TG(\overline{c}, \emptyset) TG(t, C \cup \{\overline{c}\}, \overline{C} - \{\overline{c}\}) TG(\overline{c}, \emptyset)$ by a graph transformation rule, one may use the graph representation of the gate and the circuit as left-hand side and right-hand side respectively. One can ignore the lines outside $C \cup \overline{C} \cup \{t\}$ as they are kept unchanged automatically if they do not belong to the matching of a rule. One can also ignore the renaming as this is provided by a matching. Moreover, it is not meaningful that the right-hand side adds *in-* and *out-*loops. Finally, the *in-* and *out-*nodes can be used as gluing part of the rule. Altogether, the reasoning leads to the following rule

$$r(k, l) = (gr(C^k \overline{C}^l NOT)^- \supseteq inout(k + l + 1)^- \subseteq rhs(k, l)) \text{ with}$$

$$rhs(k, l) = (gr((NOT, r_0)) \circ gr((C^{k+1} \overline{C}^{l-1} NOT, id)) \circ gr((NOT, r_0)))^- \text{ and}$$

$$r_0 : [1] \rightarrow [k + l + 1] \text{ given by } r_0(1) = k + 1$$

where the left-hand side is the graph of the function $C^k \overline{C}^l NOT$ for some $k, l \in \mathbb{N}$ with $l > 0$; the gluing graph is the discrete subgraph of the left-hand side consisting of *in-* and *out-*nodes (i, in) and (i, out) for all $i \in [k+l+1]$, and the right-hand side is the graph of the circuit $(NOT, r_0)(C^{k+1} \overline{C}^{l-1} NOT, id)(NOT, r_0)$ without *in-* and *out-*loops where the removal of these loops from an *in-out-*graph G is

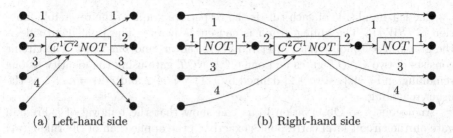

(a) Left-hand side   (b) Right-hand side

Fig. 3. The rule $r(1, 2)$

denoted by G^-. The set of lines of this circuit is $[k+l+1]$. The middle gate keeps the lines identical so that its reversible function is $C^{k+1}\overline{C}^{l-1}NOT$. The first and the third gate negate the line $k + 1$. Without loss of generality, the node from which the in_i-loop is removed can be seen as (i, in) and the node from which the out_i-loop is removed can be seen as (i, out) for all $i \in [k + l + 1]$ so that the gluing graph is a subgraph of the right-hand side. Figure 3 shows the respective rule exemplarily. Altogether, we get the unit

> **remove-ncl**(m, X)
>   initial: $TOF_{posneg}(m, X)$
>   rules: $r(k, l)$ for $k + l \leq m$ and $l > 0$
>   terminal: $TOF(m, X)$

It is easy to show that the replacement of a Toffoli gate $tg = TG(t, C, \overline{C})$ within a circuit $tc = tc'tgtc''$ by the circuit $tc(\overline{c}) = TG(\overline{c}, \emptyset) TG(t, C \cup \{\overline{c}\}, \overline{C} - \{\overline{c}\}) TG(\overline{c}, \emptyset)$ yielding the circuit $\overline{tc}$ corresponds to the application of the rule $r(k, l)$ to $gr(tc)$ in the case $k = \#C$ and $l = \#\overline{C}$. Using this notation, the following observation holds.

Observation 3. 1. The rule $r(k, l)$ is applicable to $gr(tc)$ in such a way that the derived graph is isomorphic to $gr(\overline{tc})$.
2. Let $gr(tc) \underset{r(k,l)}{\Longrightarrow} H$ be an application of the rule $r(k, l)$ to the graph $gr(tc)$. Then $tc = tc'tgtc''$ for some Toffoli gate tg and Toffoli circuits tc' and tc'' and $H \cong gr(tc'tc(\overline{c})tc'')$.

The removal of Fredkin gates replacing them by Toffoli circuits as recalled in Sect. 5.1 can be modeled by a graph transformation unit employing the same ideas.

> **remove-fg**(k, l, X)
>   initial: $FRE(k, X) \cup TOF(l, X)$
>   rules: $r'(m) = (gr(C^m SWAP) \supseteq inout(m + 2)^- \subseteq$
> $gr((CNOT, \overline{r})(C^{m+1}NOT, id)(CNOT, \overline{r}))^-)$
>   for $m \leq k$
>   terminal: $TOF(max(k + 1, l), X)$

The left-hand side of each rule is the respective graph of the reversible function $C^m SWAP$. The gluing graph is chosen as before. The right-hand side is the graph of a Toffoli circuit after removing the *in*- and *out*-loops. The circuit consists of two $CNOT$ gates and one $C^{m+1}NOT$ gate using the identity as one renaming and $\bar{r}\colon [2] \to [m+2]$ defined by $\bar{r}(1) = m+2$ and $\bar{r}(2) = m+1$ as the other renaming.

Analogously to Observation 3 one can show that the removal of a Fredkin gate on the circuit level corresponds exactly to the application of the rule $r'(m)$ on the graph level.

In both examples, the terminal graphs are exactly those that are reduced with respect to the respective rules. Therefore, one could replace the given specification of terminal graphs by the graph class expression *reduced* or use the control condition *as long as possible* instead.

Finally, one may notice that it is also meaningful to reverse the rule $r(k,l)$ and $r'(m)$ by exchanging left- and right-hand side. The reversed rules introduce negative control lines and Fredkin gates respectively while the number of gates decreases. This illustrates that the left-hand side of circuit graph transformation rules are not necessarily graphs of reversible functions but can also be circuit graphs (without *in*- and *out*-loops).

6   Circuit Graph Transformation Units

In the literature on reversible circuits, one encounters quite a spectrum of further examples of circuit transformations that could be mirrored on the level of graph transformation. This motivates to propose a general framework for the transformation of circuit graphs. We tailor graph transformation units to serve this purpose.

Definition 1. 1. Let $X \subseteq ID$. Then each set RF of reversible functions specifies the set of graph representations of all circuits given as non-empty sequences of the gates induced by RF and X. i.e. $SEM(RF) = gr(\Gamma(RF, X)^+)$.
2. Let rc and rc' be reversible circuits over $U \subseteq ID$ such that none of their graph representations contains an isolated node. Then the *rule induced by the pair* (rc, rc') is given by

$$r(rc, rc') = gr(rc)^- \supseteq inout(U)^- \subseteq gr(rc')^-.$$

3. Let $X \subseteq ID$. A *circuit graph transformation unit* over X is a graph transformation unit $cgtu(X) = (IRF, R, C, TRF)$ where IRF and TRF are sets of reversible functions, R is a set of rules each induced by a pair of reversible circuits, and C is a control condition.

A gate graph contains an isolated node if the corresponding variable is not used by the function of the gate. Hence, a circuit graph contains an isolated node if there is a variable that is not used by any of the gate functions. We forbid such nodes because $inout(U)$ is not a subgraph of both $gr(rc)$ and $gr(rc')$ if there is a variable such that one of them has an isolated node and the other not. If the variable node is isolated in both cases, then it has no effect.

The relational semantics of a circuit graph transformation unit is a binary relation between the initial circuit graphs and the terminal circuit graphs and as such a model transformation of initial models into terminal ones. Without further assumption, the transformation may be partial, i.e. there are initial circuit graphs that cannot be derived into terminal circuit graphs, or there may be infinite derivation sequences so that one can run into a bad track. Clearly, the transformation may also be nondeterministic relating various terminal models to an initial model. But nondeterminism is not necessarily a handicap with respect to reversible circuits because many transformation processes in this context are nondeterministic or choose a particular result out of many possible results. Nevertheless, in most cases one would like to assure further properties like (1) consistency meaning that each circuit graph derives only circuit graphs, (2) semantics preservation in the sense that the semantic function of each initial circuit graph is equal to the semantic function of each derived terminal graph, (3) termination, (4) total definedness, i.e. each initial circuit graph derives some terminal one, or (5) optimality in the sense that resulting circuit graphs are optimal - or at least locally optimal - with respect to some valuation function. In the latter case, a circuit graph transformation unit can be seen as an optimization. In the remainder of this section, these properties are formally defined and sufficient conditions are provided that assure them. With respect to optimization, we restrict the consideration to minimization in this paper.

Definition 2. Let $cgtu(X) = (IRF, R, C, TRF)$ be a circuit graph transformation unit. Then it is

1. *consistent* if, for each initial circuit graph G and each derivation $G \underset{R,C}{\overset{*}{\Longrightarrow}} H$, H is a circuit graph,
2. *semantics-preserving* if, for each derivation $G \underset{R,C}{\overset{*}{\Longrightarrow}} H \in DER(cgtu(X))$, $f_G = f_H$,
3. *terminating*, if there is no infinite derivation starting with an initial circuit graph using the rules in R,
4. *totally defined* if there is a derivation $G \underset{R,C}{\overset{*}{\Longrightarrow}} H \in DER(cgtu(X))$ for each initial circuit graph,
5. *locally minimizing* with respect to a valuation function $val: \mathcal{CG}_X \to \mathbb{N}$ if, for each derivation $G \underset{R,C}{\overset{*}{\Longrightarrow}} H \in DER(cgtu(X))$, there is no derivation $H \underset{R}{\overset{*}{\Longrightarrow}} I$ such that $G \underset{R,C}{\overset{*}{\Longrightarrow}} H \underset{R}{\overset{*}{\Longrightarrow}} I \in DER(cgtu(X))$ and $val(I) < val(H)$.

The following example shows that the induced rules do not preserve circuit graphs necessarily.

Example 1. Consider the rule $gr(NOT) + gr(NOT) \supseteq K \subseteq gr(f)$ for some $f: \mathbb{B}^2 \to \mathbb{B}^2$ where K consists of the *in-* and *out-*nodes properly renamed. This rule can be applied to $gr(NOT) \circ gr(NOT) \circ gr(NOT)$ matching the first and third negations. Then the derived graph is

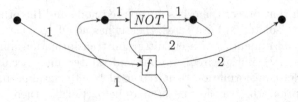

It has a cycle so that it is not a circuit graph. This motivates the consistency condition.

Lemma 1. Let $cgtu(X) = (IRF, R, C, TRF)$ be a circuit graph transformation unit. Then the following holds:

1. Let $G \underset{r(rc,rc')}{\Longrightarrow} H$ be a direct derivation for a circuit graph G and a rule induced by reversible circuits rc and rc'. Then H is a circuit graph if the following *consistency condition* holds: For each *in*-node v and *out*-node v' that are connected by a path in $gr(rc')$, but not in $gr(rc)$, there is no path from $g(v')$ to $g(v)$ in G for the nodes $g(v)$ and $g(v')$ in the match of $gr(rc)$ in G that are the images of v and v' respectively.
2. Let $G \underset{r(rc,rc')}{\Longrightarrow} H$ be a direct derivation for some circuit graphs G and H. Let $f_{rc} = f_{rc'}$. Then $f_G = f_H$.

Proof (sketch). 1. By assumption, there is a graph morphism $g\colon gr(rc) \to G$. Without loss of generality, one can assume, that the rule is not a parallel rule of two induced rules. Otherwise the proof can be done separately for the component rules. Due to their constructions, the graphs $gr(rc)$ and $gr(rc')$ share the *in*-nodes (x, in) and the *out*-nodes (x, out) for $x \in U$ (where U is the set of variables of rc and rc'). Then g_V must map the nodes (x, in) and (x, out) to nodes on the same line $y \in X$ defining a renaming $r\colon U \to X$ in this way. Using the consistency condition, one can decompose G into $G' \circ gr(rc, r) \circ G''$ where the components are obtained as follows: (1) Let $rc = (f_1, r_1) \ldots (f_k, r_k)$. Then $gr(rc, r) = gr((f_1, r \circ r_1)) \circ \cdots \circ gr((f_k, r \circ r_k))$. (2) G' is an initial section of G that contains at least all function nodes that precede $g(gr(rc))$ in G. (3) G'' is an terminal section of G containing the remainder parts. As the given direct derivation can be restricted to $gr(rc, r)$ yielding $gr(rc', r)$, one gets $H = G' \circ gr(rc', r) \circ G''$ proving that H is a circuit graph.

2. Due to the decomposition above, we get $f_G = f_{G''} \circ f_{gr(rc,r)} \circ f_{G'} = f_{G''} \circ (f_{rc})_r \circ f_{G'} = f_{G''} \circ (f_{rc'})_r \circ f_{G'} = f_{G''} \circ f_{gr(rc',r)} \circ f_{G'} = f_H$.

Note that the consistency condition holds always if one considers only rules where the right-hand sides do not create new paths between *in*- and *out*-nodes as it is the case with all rules in the previous section. Hence one may restrict the consideration to such rules to avoid any trouble with consistency. But this would exclude interesting cases. For example, we have studied sequentialization, parallelization and shift operations on multi-target Toffoli circuits in [10]. As these operations are defined by pairs of reversible circuits, they induce circuit transformation units. But the rules corresponding to parallelization and shift create new paths sometimes. Actually, the example above is a parallelization of two NOTs.

Theorem 1. Let $cgtu(X) = (IRF, R, C, TRF)$ be a circuit graph transformation unit where C requires in particular to use only direct derivations that fulfill the consistency condition. Then the following holds.

1. $cgtu(X)$ is consistent.
2. If each rule of R is induced by equivalent circuits rc and rc', i.e. $f_{rc} = f_{rc'}$, then cgtu(X) is semantics-preserving.

Proof. According to the assumption, only direct derivations are considered that fulfill the consistency condition. Let $G \overset{*}{\underset{R,C}{\Longrightarrow}} H$ be a derivation and G a circuit graph. Then H is a circuit graph due to Lemma 1 if the derivation has length 1. By a simple induction on the length of derivations, H is always a circuit graph so that $cgtu(X)$ is consistent. If, in addition, the assumption of the second part holds, then due to Lemma 1 each derivation step and, by induction, each derivation preserve the semantic function so that $f_G = f_H$. This applies to the derivations in the derivation semantics of $cgtu(X)$ proving its correctness.

As in other cases, one can speak about optimization of reversible circuits whenever a quantitative valuation is assumed. In the literature, one encounters various examples like the number of gates, the nearest-neighbor costs [11,12], the number of control lines and others. As all such valuations can be carried over to circuit graphs, one can study optimization problems in the framework of circuit graph transformation units. As mentioned before, we concentrate on minimization. The basic idea (which is often used in many contexts) is to find a valuation that decreases under derivation. Such valuations yield also termination and total definedness.

Theorem 2. Let $cgtu(X) = (IRF, R, C, TRF)$ be a circuit graph transformation unit where the control condition includes the consistency condition for all rule applications. Let $val: \mathcal{CG}_X \to \mathbb{N}$ be a valuation function subject to the following *decreasing property*: If $G \underset{r}{\Longrightarrow} H$ for some $r \in R$ and $G, H \in \mathcal{CG}_X$, then $val(G) > val(H)$. Then the following holds.

1. $cgtu(X)$ is terminating.
2. $cgtu(X)$ is totally defined and locally minimizing provided that reduced circuit graphs are terminal.

Proof. 1. As the values along a derivation decrease monotonicly, every derivation must end up in a reduced circuit graph.
2. Deriving an initial circuit graph as long as possible yields a reduced circuit graph that is also terminal by assumption. Hence, the unit is totally defined. At the same time, the reduced circuit graphs are locally minimal because they cannot lead to a better value by further derivation steps.

Finally, we want to mention, that the examples in Sect. 5 satisfy the conditions in the theorems.

7   Conclusion

In this paper, we have continued the development of a graph-transformational framework for the modeling, processing and analysis of reversible circuits with the emphasis on model transformation and optimization. In particular, we have introduced a notion of circuit graph transformation units to transform initial into terminal circuit graphs by means of rules that are induced by pairs of circuits. This concept covers many examples in the literature, and the very first results on consistency, semantics preservation, termination and optimization look promising enough to continue the study including the following topics:

1. The circuit graph transformation rules may be generalized and made more flexible in such a way that the underlying set of variables is not kept invariant, but may allow to remove or add variables.
2. The sufficient conditions to assure termination, optimization and equivalence should be strengthened to cover more cases.
3. One very important, but hard problem for reversible circuits is to check their equivalence. Hence special cases with efficient equivalence tests are of interest. Here Theorem 1 can help. If a circuit graph transformation unit is shown to be semantics-preserving, then two circuits are equivalent if they can be transformed into each other. This observation should be worked out in the future.
4. The nodes representing reversible functions together with their incoming and outgoing edges can be considered as hyperedges. Therefore, circuit graph transformation units are hyperedge replacement systems whenever the left-hand sides of rules consist of single graphs of functions (as in the two main examples in Sect. 5). In such cases, the well-developed theory of hyperedge replacement applies (see, e.g., [13]) – a fact to be studied in more detail.
5. More case studies can shed more light on the significance of the approach. A good candidate may be a graph-transformational counterpart of a model transformation of so-called multi-target Toffoli circuits into canonical ones that we investigated recently on the set-theoretical level of description [10]. The transformation by shift rules optimizes circuits with respect to the waiting degree. The rules can be carried over as induced rules, but the waiting degree depends on the sequential order of gates so that a new optimization criterion is needed.
6. The graph-transformational modeling of reversible circuits offers the perspective to employ graph transformation tools. This would allow to get more insight into large circuits where the intuition fails.

Acknowledgment. We are greatful to the anonymous reviewers for their valuable comments that led to various improvements.

References

1. Toffoli, T.: Reversible computing. In: de Bakker, J., van Leeuwen, J. (eds.) Automata, Languages and Programming. LNCS, pp. 632–644. Springer, Heidelberg (1980)
2. Fredkin, E., Toffoli, T.: Conservative logic. Int. J. Theor. Phys. **21**(3/4), 219–253 (1982)
3. Kreowski, H.-J., Kuske, S., Lye, A., Luderer, M.: Graph transformation meets reversible circuits: generation, evaluation, and synthesis. In: Giese, H., König, B. (eds.) ICGT 2014. LNCS, vol. 8571, pp. 237–252. Springer, Heidelberg (2014)
4. Maslov, D., Dueck, G.W., Michael Miller, D.: Toffoli network synthesis with templates. IEEE Trans. Comput. Aided Des. Integr. Circ. Syst. **24**(6), 807–817 (2005)
5. Arabzadeh, M., Saeedi, M., Zamani, M.S.: Rule-based optimization of reversible circuits. In: Proceedings of 15th Asia South Pacific Design Automation Conference, ASP-DAC 2010, pp. 849–854. IEEE (2010)
6. Soeken, M., Thomsen, M.K.: White dots *do* matter: rewriting reversible logic circuits. In: Dueck, G.W., Miller, D.M. (eds.) RC 2013. LNCS, vol. 7948, pp. 196–208. Springer, Heidelberg (2013)
7. Thomsen, M.K., Kaarsgaard, R., Soeken, M.: Ricercar: a language for describing and rewriting reversible circuits with ancillae and its permutation semantics. In: Krivine, J., Stefani, J.B. (eds.) Reversible Computation. LNCS, vol. 9138, pp. 200–215. Springer, Switzerland (2015)
8. Kreowski, H.-J., Klempien-Hinrichs, R., Kuske, S.: Some essentials of graph transformation. In: Esik, Z., Martín-Vide, C., Mitrana, V. (eds.) Recent Advances in Formal Languages and Applications. SCI, vol. 25, pp. 229–254. Springer, Heidelberg (2006)
9. Courcelle, B., Engelfriet, J.: Graph Structure and Monadic Second-Order Logic - A Language-Theoretic Approach. Cambridge University Press, Cambridge (2012)
10. Kreowski, H.-J., Kuske, S., Lye, A.: Canonical multi-target toffoli circuits. In: Dediu, A.H., et al. (eds.) LATA 2016. LNCS, vol. 9618, pp. 603–616. Springer, Heidelberg (2016). doi:10.1007/978-3-319-30000-9_46
11. Saeedi, M., Wille, R., Drechsler, R.: Synthesis of quantum circuits for linear nearest neighbor architectures. Quantum Inf. Process. **10**(3), 355–377 (2011)
12. Lye, A., Wille, R., Drechsler, R.: Determining the minimal number of swap gates for multi-dimensional nearest neighbor quantum circuits. In: Proceedings of 20th Asia and South Pacific Design Automation Conference, ASP-DAC 2015, pp. 178–183. IEEE (2015)
13. Drewes, F., Habel, A., Kreowski, H.-J.: Hyperedge replacement graph grammars. In: Rozenberg, G. (ed.) Handbook of Graph Grammars and Computing by Graph Transformation: Foundations, vol. 1, pp. 95–162. World Scientific, Singapore (1997)

Author Index

Printed in the United States
By Bookmasters